Springer Tracts in Modern Physics 108

W0232079

Springer Tracts in Modern Physics

---

* denotes a volume which contains a Classified Index starting from Volume 36

Particles
and Detectors

Festschrift for Jack Steinberger

Editors: K. Kleinknecht and T. D. Lee

With 91 Figures

Springer-Verlag
Berlin Heidelberg GmbH

Professor Dr. Konrad Kleinknecht
Johannes Gutenberg-Universität, Postfach 3980
D-6500 Mainz, Fed. Rep. of Germany

Professor Dr. Tsung Dao Lee
Columbia University, New York, NY 10027, USA

Manuscripts for publication should be addressed to:
Gerhard Höhler
Institut für Theoretische Kernphysik der Universität Karlsruhe
Postfach 6380, D-7500 Karlsruhe 1, Fed. Rep. of Germany

*Proofs and all correspondence concerning papers in the process of publication
should be addressed to:*
Ernst A. Niekisch
Haubourdinstrasse 6, D-5170 Jülich 1, Fed. Rep. of Germany

ISBN 978-3-662-15193-8   ISBN 978-3-540-39768-7 (eBook)
DOI 10.1007/978-3-540-39768-7

Library of Congress Cataloging in Publication Data. Particles and detectors. (Springer tracts in modern physics; V. 108) 1. Particles (Nuclear physics) 2. Nuclear counters. 3. Steinberger, J. I. Steinberger, J. II. Kleinknecht, K. (Konrad), 1940–. III. Lee, T. D. (Tsung D.) IV. Series. QC1.S797 vol. 108 [QC793.28]   539 s   [539.7]   86-3817

© Springer-Verlag Berlin Heidelberg 1986
Originally published by Springer-Verlag Berlin Heidelberg New York Tokyo in 1986
Softcover reprint of the hardcover 1st edition 1986

2153/3150-543210

Jack Steinberger

Preface

On May 25, 1986, Jack Steinberger will be 65 years old. This volume is intended to be part of the celebration of that joyful occasion, and, above all, to pay homage to Jack. For almost four decades he has made so many important contributions to high energy physics that the development of that field has indeed been inseparable from his scientific career. In addition to his own contribution, Jack has shaped generations of younger physicists. His students and others he has influenced have literally permeated the field. The effect of Jack Steinberger can be felt in all the major high energy laboratories throughout the world.

Those of us who have had the pleasure and the privilege of being Jack's friends appreciate not only his science, but his personal and professional integrity as well. We hope these twenty papers written for him capture some of this feeling. In a small way, the diversity of their subject matter reflects the extent of Jack's own interests.

This volume is dedicated to Jack as a sincere expression of our admiration and affection.

Dortmund and Mainz
Konrad Kleinknecht

New York
T.D. Lee

Contents

Biographical Note

Jack Steinberger was born in Bad Kissingen in Franconia in 1921. His father, Ludwig, was cantor and religious teacher for the little jewish community, his mother gave English and French lessons to the tourists of the spa. The childhood he shared with his two brothers was simple, Germany was living through the postwar depression. Things took a dramatic turn when Jack was entering his teens. In 1933 the Nazis came to power, and the systematic persecution of the Jews followed quickly. When, in 1934, the American jewish charities offered to find homes for 300 German refugee children, Steinberger's father applied for Jack and his older brother. Christmas 1934, they left for New York. In Chicago, Barnett Faroll, the owner of a grain brokerage house, took Steinberger into his house, parented his high school education, and made it possible also for his parents and younger brother to come to the United States in 1938. The reunited family settled down in Chicago. Jack Steinberger was able to continue his education for two years at the Armour Institute of Technology (now Illinois Institute of Technology), where he studied chemical engineering.

After very difficult years, working day time in a pharmaceutical laboratory and studying in the evening with the help of a scholarship from the University of Chicago, Steinberger finished an undergraduate degree in chemistry at the University of Chicago in 1942. In December 1942, after the Japanese attack on Pearl Harbour, Steinberger joined the Army and was sent to the MIT radiation laboratory, after a few months of introduction to electronics and electromagnetic wave theory in a special course given for army personnel at the University of Chicago. The radiation laboratory was engaged in the development of radar bomb sights; Steinberger was assigned to the antenna group. Among the outstanding physicists in the laboratory were Ed Purcell and Julian Schwinger. The two years there offered Steinberger the opportunity to take some basic courses in physics.

After Germany surrendered in 1945, Steinberger spent some months on active duty in the army, but he was released after the Japanese surrender to continue his studies at the University of Chicago. The professors there were Enrico Fermi, W. Zachariasen and Edward Teller. Fellow students included Yang, Lee, Goldberger, Rosenbluth, Garwin, Chamberlain, Wolfenstein and Chew.

For his thesis Fermi asked Steinberger to look into a problem raised in an experiment by Rossi and Sands on stopping cosmic ray muons. They did not find the expected number of decays. After correcting for geometrical losses, there was still a missing factor of two and Steinberger suggested to Sands that this might be due to the fact that the decay electron had less energy than expected in the two body decay, and that one might test this experimentally. When this suggestion was not followed, Fermi suggested that Steinberger himself do the experiment. The cosmic ray experiment required less than a year from conception to its conclusion at the end of the summer of 1948. It showed that the muon decay is three body, probably into an electron and two neutrinos, and helped lay the experimental foundation for the concept of a universal weak interaction.

There followed an interlude at the Institute of Advanced Study in Princeton at the invitation of Oppenheimer. Steinberger calculated the decay rate of a neutral pion to two gamma rays via intermediate nucleons. This calculation paved the way for later developments in theory, notably the triangle anomaly.

In 1949, Gian Carlo Wick invited Steinberger to be his assistant at the University of California in Berkeley. There the experimental possibilities in the Radiation Laboratory, created by E.O. Lawrence, were so great, that, instead of doing theory, as Wick might have expected, Steinberger reverted to experimentation. During the year there he had the opportunity of working on the just-completed electron synchrotron of Edward McMillan. It enabled him to do the first experiments on the photo production of pions (with A.S. Bishop), to establish the existence of neutral pions (with W.K.H. Panofsky and J. Steller), as well as to measure the pion mean life (with D. Chamberlain, R.F. Mozley and C. Wiegand).

After only one year in Berkeley, Steinberger left, partly because he declined to sign the anticommunist loyalty oath, and moved on to Columbia University in the summer of 1950. At its Nevis Laboratory, Columbia had just completed a 380 MeV cyclotron, which, for the first time, offered the possibility of experimenting with beams of pi mesons. In the next years he exploited these beams to determine the spins and parities of charged and neutral pions, to measure the $\pi^- - \pi^0$ mass difference and to study the scattering of charged pions. This work leaned heavily on the collaboration of Profs. D. Bodansky and A.M. Sachs, as well as several Ph.D. students: R. Durbin, H. Loar, P. Lindelfeld, W. Chinowsky and S. Lokanathan.

These experiments all utilised small scintillator counters. In the early 50s the bubble chamber technique was discovered by Donald Glaser, and in 1954 Steinberger and three graduate students, J. Leitner, N.P. Samios and M. Schwartz, began to study this technique which had not as yet been exploited to do physics. Their first effort was a 10-cm diameter propane chamber. The first bubble chamber result to be published was an experiment at the newly built Brookhaven Cosmotron, using a 15-cm propane chamber without magnetic field. It yielded a number of results on the properties of the new unstable (strange) particles at a level unattainable with older techniques, and so dramatically demonstrated the power of the new technique which was to dominate particle physics for the next dozen years. Only a few months later, this group published three events of the type $\Sigma^0 \to \Lambda^0 + \gamma$, which demonstrated the existence of the Σ^0 hyperon and gave a measure of its mass. This experiment used a new propane chamber, eight times larger in volume, and with magnetic field. This chamber also introduced the use of more than two stereo cameras, a development which is crucial for the rapid, computerized analysis of events, and has been incorporated into all subsequent bubble chambers.

In the decade which followed, the same collaborators, together with Profs. Plano, Baltay, Franzini, Colley and Prodell, and a number of new students, constructed three more bubble chambers: a 12″ H_2 chamber as well as 30″ propane and H_2 chambers, developed the analysis techniques, and performed a series of experiments to clarify the properties of the new particles. Some important experiments of this period are:

- the demonstration of parity violation in Λ decay, 1957
- the demonstration of the β decay of the pion, 1958
- the determination of the π^0 parity on the basis of angular correlation in the double internal conversion of the gamma rays, 1962
- the determination of the ω and ϕ decay widths (lifetimes), 1962
- the determination of the $\Sigma^0 - \Lambda^0$ relative parity, 1963
- the demonstration of the validity of the $\Delta S - \Delta Q$ rules in K^0 and in hyperon decays, 1964.

This long chain of bubble chamber experiments, done partially with the collaboration of two Italian groups, the Bologna group of G. Puppi, and the Pisa group of M. Conversi, was interrupted in 1961, in order to perform, at the suggestion of Mel Schwartz, and with Leon Lederman, Gordon Danby and others, the first experiment using a high energy neutrino beam. The detector was a 10 ton arrangement of optical spark chambers, which had just been invented. The intensity was just enough to make it possible to get a few dozen events, and the detector had sufficient discrimination between electrons and muons to allow the demonstration that the neutrinos produced muons but not electrons, and that therfore the neutrinos emitted in pion decay are distinct from the neutrino emitted in β decay.

In 1964, CP violation was discovered by Christenson, Cronin, Fitch and Turlay. Soon after, Jack Steinberger went on sabbatical leave to CERN, and proposed, together with Rubbia and others, to look for the interference between K_S^0 and K_L^0 amplitudes in the time dependence of K^0 decay. Such interference was expected in the CP-violation explanation of the results of Christenson et al., but not in other explanations which had also been proposed. The experiment was successful, and marked the beginning of a set of experiments to learn more about CP violation, which was to last a decade. The next result was the observation of the small, CP-violating, charge asymmetry in K_L^0 leptonic decay, in 1966. Measurement of the time dependence of this charge asymmetry, following a regenerator, permitted a determination of the regeneration phase, which together with the earlier interference experiments, yielded, for the first time, the CP-violating phase $\phi_{\eta+-}$ and, in consequence, together with the observed magnitudes of the CP-violating amplitudes in the two pi and the leptonic decays, certain checks on the superweak model. The same experiments also gave a sensitive check of the $\Delta S = \Delta Q$ rule, an ingredient of the present standard model.

In 1968, Steinberger joined CERN. Charpak had just invented proportional wire chambers, and this development offered a much more powerful way to study the K_0 decay. Two identical detectors were constructed, one at CERN together with H. Filthuth, K. Kleinknecht and others, and one at Columbia together with J. Christenson, D. Nygren, W. Carithers and students. The BNL beam was long, and therefore contained no K_S but only K_L; the CERN beam was short, and therefore contained a mixture of K_S and K_L. It was contaminated by a large Λ^0 flux, and so was also a hyperon beam and permitted the first measurement of Λ^0 total cross sections as well as the Coulomb excitation of Λ^0 to Σ^0, a difficult and interesting experiment carried out chiefly by P. Steffen and F. Dydak. The more important result to come from the Columbia experiment was the observation of the rare decay $K_L \to \mu^+ + \mu^-$ with a branching ratio compatible with theoretical predictions based on unitarity. Previously, a Berkeley experiment had searched in vain for this decay and had claimed an upper limit in violation of unitarity. Since unitarity is fundamental to field theory, this result had a certain importance.

The CERN-Heidelberg experiment, which extended to 1975, produced a series of precise measurements on the interference of K_S and K_L in the two pion and leptonic decay modes, to obtain highly precise results of the CP-violating parameters in K^0 decay as well as checks on the $\Delta S = \Delta Q$ rule. The results were all in excellent agreement with the superweak model.

In 1972, the K^0 collaboration of CERN, Dortmund and Heidelberg was joined by a group from Saclay under R. Turlay to study the possibilities for a neutrino experiment at the CERN SPS then under construction. The CDHS detector, a modular array of magnetized iron disks, scintillation counters and drift chambers, 3.75 m in diameter, 20 m long, and weighing 1200 tons, was designed, constructed and exposed to different neutrino beams at the SPS during the period 1976–1984. It provided a large body of data on the charged current and neutral current inclusive reactions in iron, which permitted first of all the clearing away of a number of incorrect results, e.g. the "high y anomaly", produced at Fermilab, permitted the first precise and correct determination of the Weinberg angle, demonstrated the existence of right-handed neutral currents, and provided measurements of the structure functions which gave quantitative support to the quark constituent model of the nucleon, and to QCD through the measurement of scaling violations in the Q^2 evolution of the structure functions. The study of multimuon events gave results in quantitative agreement with the GIM model of the Cabibbo current through its prediction of charm production.

In the CDHS experiment there were about thirty physicists. Since 1983 Jack Steinberger has played a leading role in the design and construction of a detector for the $100 + 100$ GeV e^+e^- collider LEP, to be ready at the beginning of 1989. In this collaboration there are 200 physicists. In the meantime Steinberger also helped to design an experiment to compare CP violation in the charged and neutral two pion decay of the K_L^0, in the hope that this will shed some light on the origins of CP violation. This experiment is now on the floor, but his commitment to the LEP detector ALEPH is keeping Steinberger from effective participation in the CP experiment. The ALEPH experiment will remain his prime interest in the years to come.

Limits on Like-Sign Dilepton Production in Neutrino Interactions

C. Baltay

Columbia University, New York, NY 10027, USA

This article is dedicated to Jack Steinberger. His common interest in these same problems in neutrino interactions has been a constant stimulation throughout this experiment. His high standards and integrity set an example that many of us tried to follow.

1. INTRODUCTION

The production of a pair of leptons with the same electric charge in neutrino interactions is of some interest because it could be the signature of new phenomena not predicted by the Standard Model. The observed production of opposite sign dileptons ($\mu^- e^+$ and $\mu^- \mu^+$) on the other hand is explained in terms of the Standard Model as the semileptonic decay of charmed particles produced in charged current neutrino interactions. Both the production rate and the kinematic parameters of these events are well described by the GIM mechanism [1]; in particular, the abundance of strange particles associated with the opposite sign dilepton events as observed in the bubble chamber experiments is a confirmation of the Cabibbo favoured coupling of charm to strangeness.

The production of like-sign dileptons ($\mu^- e^-$ or $\mu^- \mu^-$) could be explained by the Standard Model as coming from associated charm production or from the decays of bottom quarks. The rates expected from these sources have been estimated [2] by many authors. Several experiments have reported the production of like-sign dileptons in neutrino interactions (see Tables I and II and References 8-16), some of them with rates more than an order of magnitude higher than the estimates [2] based on the Standard Model. There was thus no satisfactory explanation of this high rate of like-sign dilepton production without invoking some new physics.

In this paper we present our final results [3] for the search for $\mu^- e^-$ pairs produced in ν_μ interactions,

$$\nu_\mu + Ne \rightarrow \mu^- + e^- + \ldots \quad .$$

Most of the previous experiments have looked at $\mu^- \mu^-$ pairs produced in ν_μ interactions. The dominant background in these experiments, charged current events with a π^- (or K^-) $\rightarrow \mu^- + \bar{\nu}_\mu$ decay, are at a comparable level to the observed signals and thus require a difficult background calculation. The search for $\mu^- e^-$ pairs does not suffer from this background since the corresponding decays, π^- (or K^-) $\rightarrow e^- + \bar{\nu}_e$ are suppressed. The dominant backgrounds in this case, due to Compton electrons or asymmetric Dalitz pairs in ν_μ interactions and fake muons from π or K decays in the less frequent ν_e interactions ($\nu_e/\nu_\mu \approx 0.01$), are considerably smaller, so a more sensitive search is possible.

2.   THE EXPERIMENTAL ARRANGEMENT

The results presented in this paper come from a recent run of 15-ft bubble chamber equipped with an External Muon Identifier (EMI) at Fermilab exposed to a wideband neutrino beam. The neutrino beam was a single magnetic horn focussed wideband beam, with an average of 1.1×10^{13} protons per pulse on the target. The ν_μ event spectrum has an average energy of around 50 GeV, but there is a useful number of events extending up to 200 GeV. The chamber was in a 30 kG magnetic field, allowing precise momentum measurement and charge identification of charged tracks. It was filled with a heavy neon-hydrogen mixture (62 atomic % neon), with a density of 0.8 grams/cm^3, a radiation length of 40 cm, and a hadronic interaction length of 125 cm. The short radiation length provides excellent electron identification, and the hadronic inter-action length (small compared to the 4 meter chamber diameter) gives reasonable muon identification since most hadrons will interact in the chamber while muons leave without interacting. The EMI improves the muon-hadron separation by an addi-tional factor of 50. The EMI consists of two planes of multiwire proportional cham-bers downstream of the bubble chamber [4]. The first plane, which covers 21 m^2, was wrapped cylindrically around the vertical axis of the chamber 3.6 m from the axis. The second plane (18 m^2) was mounted perpendicular to the beam axis 3 m downstream of the first plane. Muons leaving the chamber transversed 3 to 5 hadronic interac-tion lengths of copper and zinc before crossing the first EMI plane, and an addi-tional 5.5 interaction lengths of concrete and lead before crossing the second plane. The readout of the EMI chambers, when reconstructed using the programme [4] EMIKE gave an x and y position in each plane with a 3 mm accuracy and a prompt time in each plane with a precision of 60 ns.

3.   ANALYSIS OF THE DATA

The exposure consisted of 264.000 stereo-triplet photographs, of which 245.000 had good EMI data. All of this film was scanned for events with an e^+ or e^- coming di-rectly from the interaction vertex with or without muons, and with any number of hadrons in the event. In addition, events with apparent Dalitz pairs were recorded and later used in estimating backgrounds. In these scans $e^\pm$ tracks were defined to be minimum ionizing tracks with at least two of the following signatures: spiraling at the end, radiation followed by conversion into e^+e^- pairs, sudden change of cur-vature, large δ ray or trident, or annihilation (e^+ only). Negative tracks which left the chamber without interacting were taken to be μ^- candidates. The charged current normalization was determined from an unbiased measurement of all events on 2% of the frames.

The raw numbers of events found in the scan were $\approx 10^5$ events with a μ^- (extrapolated to all frames), $\approx 10^3$ events with an e^- (presumably mostly ν_e interactions), $\approx$ 250 events with a μ^- and an e^+, and $\approx$ 150 events with an e^- and a leaving negative track (L^-). All events with electron candidates were measured and processed by the geometrical reconstruction programme, TVGP. The $e^- L^-$ events, the candidates for the $e^- \mu^-$ search, were reexamined by a physicist. At this stage various background categories were removed and some selection criteria were applied. First a restricted fiducial volume [5] was imposed on all of the samples to insure good visibility and track measurability. A cut of $P_{L^-} > 5$ GeV/c was made on the candidate muon momentum to ensure reasonable geometrical acceptance in the EMI and a cut of $P_{e^-} > 1$ GeV/c was made on the electron momentum to remove low energy Dalitz pairs and Compton electrons. In addition the leaving negative tracks (muon candidates) had to extrapolate toward the EMI planes. Remaining Dalitz pairs and close-in γ conversions were removed. Events where the e^- could be a δ ray (knock-on electron) on a fast muon or hadron track were removed by eliminating 12 events with e^- tracks within 4^0 of a fast charged track. This selection, in addition to removing all of the δ rays, also removed a small percentage of the potential $\mu^- e^-$ sample, and a correction of $(7 \pm 3)\%$ based on the angular distributions of the $\mu^- e^+$ sample, was made. Five events had an e^- with an additional $e^+ e^-$ pair. These events were subjected to trident selection criteria developed to distinguish tridents (single electrons with an $e^+ e^-$ pair on them) from a Dalitz pair or close-in photon conversion with a δ ray close to the vertex [6]. All five events were classified as γ's and were removed from the sample. This selection may also remove potential $\mu^- e^-$ candidates [6], and a correction of $(3.0 \pm 0.5)\%$ was made. Two events, in which the apparent electron candidates were identified as $K^- \to \pi^0 e^- \bar{\nu}_e$ decays, were removed. In both cases the electron candidate traveled greater than three radiation lengths with no observable energy loss and the converted photons from the π^0 decay pointed to the decay point rather than being tangential to the primary track as one would expect for true electrons. One events was a trilepton candidate. In addition to the μ^- ($P_\mu = 11$ GeV/c) and the identified e^- ($P_{e^-} = 13$ GeV/c) there is a spiraling positive track (P = 0.8 GeV/c) with significant energy loss. The effective mass of the $e^- e^+$ pair is consistent with the ϕ mass ($m_{e^+ e^-} = 1$ GeV/c^2). Other experimental results on like-sign dilepton production exclude possible contributions from dilepton sources since they are most likely due to Drell-Yan or vector meson decay backgrounds. Consequently this event is removed from the sample. A total of 65 $\mu^- e^-$ candidates remained after all these cuts were imposed.

We have reported [7] a result on like-sign dilepton production from an earlier run of this experiment which had essentially the same statistics as our new run discussed in this paper. In that earlier run, the EMI data were not useful. With tighter cuts of $P_{e^-} \geq 2$ GeV/c and $P_{\mu^-} \geq 10$ GeV/c, we had 20 events, with an estimated background of 9 events. This signal of 11 $\pm$ 6 events we did not consider significant. If we analyze our new data in the same way without using the EMI, we have

28 events with $P_{e^-} \geq 2$ GeV/c and $P_{\mu^-} \geq 10$ GeV/c, with an estimated background of 16 events. The remaining signal of 12 ± 7 events is again not significant and is quite consistent with the result from the earlier run.

At this point in the analysis, we make use of the EMI data to separate events with true muons from those with negative hadrons that left the bubble chamber volume without interacting (L⁻ tracks). Using the measured momenta and positions in the chamber, the trajectories of these L⁻ tracks were extrapolated, using the programme XTRAP [4], to the two EMI planes taking multiple scattering effects into account in calculating the errors on the expected x and y coordinates in the two EMI planes. A track was called a good muon if these extrapolated x and y positions agreed within errors with the hits found in the two EMI planes (with a χ^2 probability larger than 10^{-3}) and the prompt times of the hits in the two EMI planes agreed with each other to within ten least counts (360 ns). We find that only 12 of our sample of 65 candidates have a true muon. The distribution of the e⁻ momenta for these 12 events is shown in Fig.1. It is worth noting that of these 12 events, 7 have P_{e^-} between 1 and 2 GeV/c, 5 have P_{e^-} between 2 and 4 GeV/c, and none have an e⁻ over 4 GeV/c.

In view of this result, it was important to check the efficiency of the EMI. This was done in two steps. The geometrical acceptance, i.e. the probability that

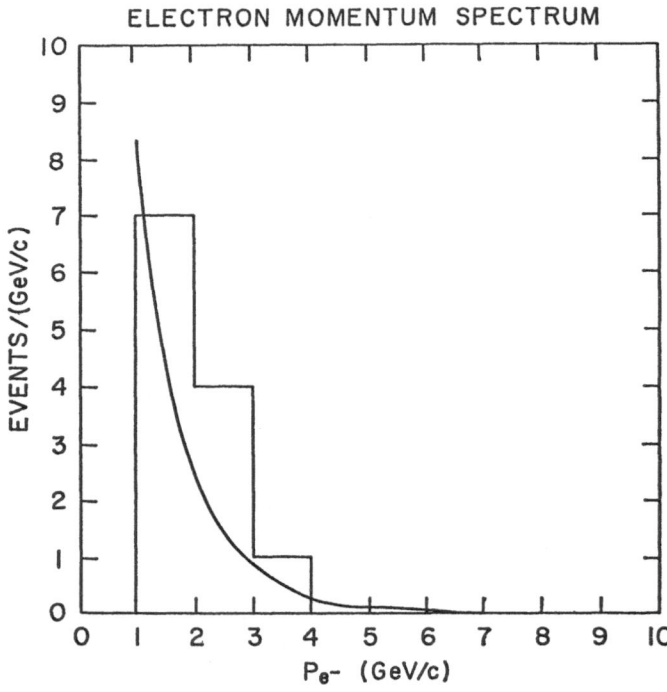

ELECTRON MOMENTUM SPECTRUM

Fig.1: Histrogram displays the momentum spectrum of the electrons, for those events containing a muon identified in the EMI. The curve displays the expected momentum spectrum for electrons arising from conventional sources

a μ^- produced by a ν_μ interaction in the chamber will travel in a direction covered by the EMI planes, was measured using a sample of 1230 ν_μ + Ne $\rightarrow$ μ^- + ... interactions. For muons with $P_{\mu^-} \geq 5$ GeV/c the acceptance was found to be $(89 \pm 1)\%$. The electronic efficiency, i.e. the probability that a muon traversing both planes of the EMI will give the correct x, y, and prompt time information and pass the 'good muon' probability and prompt time criteria, was measured to be $(94 \pm 1)\%$ using a sample of 881 'straight-through' muons. In fact, the EMI efficiency cancels in our final limits on like-sign dilepton production since the muons in the normalization sample were required to satisfy the identical EMI criteria as the muons in the dilepton samples.

4. BACKGROUNDS

We now turn to the estimation of the backgrounds. We first discuss the background from ν_e interactions with a negative hadron faking a muon. The fake muon rate in the EMI, i.e. the probability that a negative hadron leaving the chamber is called a good EMI muon, was measured using a sample of 110 events that had an EMI identified fast μ^-, with an additional leaving negative track with $P_{L^-} \geq 5$ GeV/c. These second leaving tracks had a high probability of being hadrons. It was found that $(2.0 \pm 1.7)\%$ of them fake a good EMI muon. This total fake muon probability includes hadrons penetrating the EMI without interacting, π^-, $K^- \rightarrow \mu^- + \bar{\nu}_\mu$ decays, accidental associations in the EMI, etc. To estimate the total background from this source, we start with our sample of $\approx 10^3$ ν_e interactions. Of these, 65 have an e^- with $P_e \geq 1$ GeV/c and a leaving negative track with $P_{L^-} \geq 5$ GeV/c, as discussed above. Applying the above fake muon probability to this sample yields an expected background of 1.1 ± 0.9 events.

To estimate the backgrounds due to ν_μ interactions with asymmetric Dalitz pairs or close-in external $\gamma \rightarrow e^+ e^-$ pairs and close-in Compton electrons ($\gamma + e \rightarrow \gamma + e^-$), we measured a sample of 1218 ν_μ events with Dalitz pairs and external γ conversions where the external $e^+ e^-$ pairs appear to originate at the ν_μ interaction vertex. The rate for these events was found to be $(6.9 \pm 0.6)\%$ per charged current ν_μ event. We estimate that of this $(1.8 \pm 0.4)\%$ is due to Dalitz pairs and $(5.1 \pm 0.7)\%$ is due to close-in external pairs. From the measured energy distribution of these $e^+ e^-$ pairs the background from both asymmetric Dalitz pairs and external photon conversions was calculated to be $(4.3 \pm 0.5) \times 10^{-5}$ per charged current event where the electron momenta are greater than 1 GeV/c and the positron momentum is less than 5 MeV/c (the limit of visibility in the chamber). A similar calculation for the Compton background gives $(17.8 \pm 2.8) \times 10^{-5}$ electrons with momentum greater than 1 GeV/c per charged current event. Using the measured number of ν_μ interactions, we obtain the sum of asymmetric pairs plus Compton electron background to be 8.4 ± 1.4 events with $P_{e^-} \geq 1$ GeV/c. Adding this to the 1.1 ± 0.9 fake muon background gives a total expected background of 9.5 ± 1.7 events, to be compared to the 12 $\mu^- e^-$

11

events observed. The shape of the background is similar to the P_e distribution of the observed events, as shown in Fig.1.

5.   RESULTS AND CALCULATION OF RATES

We thus conclude that there is no significant signal for like-sign dilepton production with P_{e^-} between 1 and 4 GeV/c, and emphasize that there are no dilepton
events observed (regardless of background considerations) with $P_{e^-} \geq 4$ GeV/c. To
compare with previous experiments which typically use lepton momentum cuts of 4 or
5 GeV/c, we will give upper limits for the like-sign dilepton rates for $P_{e^-} \geq 4$
GeV/c and $P_{\mu^-} \geq 5$ GeV/c, relative to both the total charged current rate and the
opposite sign dilepton rate. For these cuts the backgrounds discussed above reduce
to (0.9 ± 0.8) fake muon events and (0.7 ± 0.1) events from asymmetric pairs and
Comptons, or a total expexted background of 1.6 ± 0.8 events.

The observed number of events have to be corrected for several inefficiencies
and losses: i) e^- identification efficiency of (94.0 ± 1.5)%, ii) the criteria to
resolve the trident ambiguity retains (97 ± 1)% of the good electrons, iii) the cut
to remove δ rays retains (93 ± 3)% of the good electrons. There are some corrections
that are needed relative to the total charged current normalization, but cancel out
relative to the opposite sign dilepton normalization: i) the scanning efficiency
for finding events containing an $e^{\pm}$ was found to be (77 ± 2)% from a double scan of
a part of the film, ii) the efficiency of finding $e^{\pm}$ due to confusion in some fraction of the events, was estimated to be (96.0 ± 1.5)%. Thus the net efficiency was
(62 ± 3)% with respect to the total charged current rate and (85 ± 3)% with respect
to the opposite sign dilepton rate.

The normalization samples were subjected to the same fiducial volume cuts, and
were required to have a good EMI muon with $P_{\mu^-} \geq 5$ GeV/c using identical criteria
to those used for the like-sign dilepton sample so that the EMI efficiencies can-
celled out. This resulted in 48.800 ± 3.300 total ν_μ + Ne → μ^- + ... events. In
the $\mu^- e^+$ sample, the e^+ tracks were selected with the same criteria as the e^-
tracks. After all corrections there were (51 ± 8) ν_μ + Ne → μ^- + e^+ + ... events
with $P_{e^+} \geq 4$ GeV/c and $P_{\mu^-} \geq 5$ GeV/c.

Since we observe no $\mu^- e^-$ events with $P_{e^-} > 4$ GeV/c and $P_{\mu^-} > 5$ GeV/c the 90%
confidence level upper limit is 2.3 events.

The resulting 90% confidence level upper limits on the rates are:

$$\frac{\nu_\mu + Ne \to \mu^- + e^- + \ldots}{\nu_\mu + Ne \to \mu^- + \ldots} \leq 0.76 \times 10^{-4}$$

$$\frac{\nu_\mu + Ne \to \mu^- + e^- + \ldots}{\nu_\mu + Ne \to \mu^- + e^+ + \ldots} \leq 5.3 \times 10^{-2} \quad .$$

These limits are valid for the wideband neutrino energy spectrum of this experiment.

We can separate the data into two energy bins, $E_\nu \leq 100$ GeV, and $100 \leq E_\nu \leq 200$ GeV. The 90% confidence level limits for these energy bins are 0.88×10^{-4} and 5.4×10^{-4} relative to the total charged current rate, and 7.2×10^{-2} and 2.3×10^{-1} relative to the opposite sign dilepton rate, respectively.

6. COMPARISON WITH OTHER EXPERIMENTS AND CONCLUSIONS

These results are summarized and compared to other experiments in Tables I, II. The experiments listed in Table I used wideband beams with an energy spectrum very similar to ours, so the overall rates can be compared. Our upper limit is significantly below the rate for the CHARM experiment which is the only result in this table with more than two standard deviation significance.

The experiments listed in Table II have different energy spectra so the energy dependence is relevant. The only significant result in this table is from the HPWFOR [15] experiment. Their measured rate is $(3.4 \pm 0.9) \times 10^{-4}$ compared to our limit of $\leq 0.76 \times 10^{-4}$. It is not possible to directly compare these rates as a function of energy in a meaningful way since the HPWFOR energy dependence is given for a data set with much less statistical significance than that used for the overall rate.

TABLE I   Comparison with Other Experiments

a) Experiments in Wideband Neutrino Beams with similar E_ν spectra

EXPERIMENT	THIS EXPT	CDHS[8]	CHARM[9]	BCS[10]
P_μ cut(GeV/c)	5	6.5	4	4.5
P_ℓ cut(GeV/c)	4	6.5	4	0.8
$\frac{\mu^- \ell^-}{\mu^-} \times 10^4$ all E_ν	≤ 0.76	$0.34\pm.18$	–	5.3 ± 2.9
$\frac{\mu^- \ell^-}{\mu^- \ell^+} \times 10^2$ all E_ν	≤ 5.3	4.1 ± 2.2	$14\pm4\pm3$	22 ± 12

b) Experiments in Wideband Antineutrino Beams

EXPERIMENT	FSMM[11]	BEBC[12]
P_μ cut(GeV/c)	4	4
P_ℓ cut(GeV/c)	0.4	0.8
$\frac{\mu^+ \ell^+}{\mu^+} \times 10^4$ all E_ν	48^{+58}_{-32}	< 20
$\frac{\mu^+ \ell^+}{\mu^+ \ell^-} \times 10^2$ all E_ν	–	< 6

13

TABLE II   Experiments with Narrow Band (NB) or Quad Triplet (QT) Beam.

EXPERIMENT BEAM	THIS EXP WB	CDHS[13] NB	NCFRR[14] QT	HPWFOR[15]* QT	BFHWW[16] QT	CFRR[17] NB
P_μ cut (GeV/c)	5	4.7	9	10	4	9
P_ℓ cut (GeV/c)	4	4.7	9	10	0.8	9
$\dfrac{\mu^-\ell^-}{\mu^-}$ x10^4						
all E_ν	≤ 0.76	3±2	2.0±1.1	3.4±.9	–	1.4±0.8
$E_\nu < 100$	≤ 0.88	–	0.25±0.64	0.45±0.45	–	0.1±0.5
$100 < E_\nu < 200$	≤ 5.4	–	2.2±2.4	6.0±3.0	–	3.6±2.1
$200 < E_\nu < 300$	–	–	9.5±6.5	10.0±7.0	–	2.5±3.1
$\dfrac{\mu^-\ell^-}{\mu^-\ell^+}$ x10^2						
all E_ν	≤ 5.3	5±3	–	–	<7	
$E_\nu < 100$	≤ 7.2	–	–	–	<15	
$100 < E_\nu < 200$	≤ 23.0	–	–	–	<12	

*In this experiment the division between the two two lower energy bins is at 80 GeV. The overall rate is for events originating in the iron target while the energy dependence is taken from Fig. 4(a) of Ref. 15 which is for events originating in the liquid scintillator and iron calorimeters.

14

There is no evidence in this high statistics bubble chamber experiment for same-sign dilepton production by neutrinos. This is to be contrasted with most previous experimental results which have an excess of candidates compared to their expected background. Since the wideband beam used in this experiment peaks at $\approx$ 30 GeV, the experiment primarily yields a significant limit at low energy ($\lesssim$ 100 GeV).

7.　　ACKNOWLEDGEMENTS

We thank the scanning and measuring staffs of our institutions, and the operating crews of the 15-ft chamber and the neutrino beam at Fermilab. We give special thanks to the groups from the University of Hawaii, Lawrence Berkeley Lab., and Fermilab who built and maintained the EMI, which played such an important role in this analysis. This research was supported by the Department of Energy and the National Science Foundation.

REFERENCES

1　S.L.Glashow, J.Iliopoulos, L.Maiani, Phys. Rev. D2, 1285 (1970)

2　H.Goldberg, Phys. Rev. Lett. 39, 1598 (1977)
　B.L.Young et al., Phys. Lett. 74B, 111 (1978)
　V.Barger et al., Phys. Rev. D18, 2308 (1978)

3　This experiment was carried out by a collaboration from Columbia University, Brookhaven National Lab., and Rutgers University consisting of N.J.Baker, C.Baltay, M.Bregman, E.B.Brucker, P.L.Connolly, M.Hibbs, P.F.Jacques, S.A.Kahn, M.Kalelkar, E.L.Koller, J.T.Liu, M.J.Murtagh, J.Okamitsu, R.B.Palmer, R.J.Plano, N.P.Samios, A.C.Schaffer, K.Shastri, P.E.Stamer and M.Tanaka. In particular, it was the Ph.D. Thesis of A.C.Schaffer (Columbia University Nevis Report 250, 1985), where a detailed description of the work can be found.

4　See R.Cence et al., NIM 138, 245 (1976)

5　We used our standard fiducial volume defined as $|Z| < 125$ cm, $\sqrt{x^2+y^2+z^2} < 170$ cm, and $\sqrt{(x+70)^2+y^2+z^2} < 180$ cm, where x is in the beam direction, z is vertical and the origin is at the center of the chamber.

6　See A.M.Cnops et al., Phys. Rev. Lett. 41, 357 (1978)

7　M.Murtagh, Proceedings of the Fermilab Lepton-Photon Symposium, p.277 (1979)

8　J.G.H.deGroot et al., Phys. Lett. 86B, 103 (1979)

9　M.Jonker et al., Phys. Lett. 107B, 241 (1981)

10　A.Haatuft et al., Nucl. Phys. B222, 365 (1983)

11　V.V.Ammosov et al., Phys. Lett. 106B, 151 (1981)

12　P.Marage et al., Z. Phys. 21C, 307 (1984)

13　M.Holder et al., Phys. Lett. 70B, 396 (1977)

14　N.Nishikawa et al., Phys. Rev. Lett. 46, 1555 (1981), Errata Phys. Rev. Lett. 54, 1336 (1985)

15　T.Trinko et al., Phys. Rev. D23, 1889 (1981)

16　H.C.Ballagh et al., Phys. Rev. D24, 7 (1981)

17　K.Lang et al., Proc. XXth Rencontre de Moriond (1985)

Electroproduction at Very Small Values of the Scaling Variable

J.D. Bjorken

Fermi National Accelerator Laboratory, P.O. Box 500,
Batavia, IL 60510, USA

I. Introduction

The description of electroproduction dynamics at very small values of the scaling variable x (I am thinking of $x < 10^{-2}$) poses special challenges for theory. The issues are quite complementary to those at large x, where analyses of structure functions in terms of moments provide a direct link to the small-distance structure of current correlation functions, and nearly incontrovertible links to QCD predictions. At wee x, the process involves large longitudinal distances along the light cone[1] and hence issues of the geometry of the collision process, including A-dependence and morphology of the final-state hadron phase-space distribution.

Furthermore, perturbative QCD calculations show[2] that higher order processes are asymptotically very important; there are a plethora of contributions of order $(\alpha_s \log^2)^n$. Thus the Q^2 and x dependences should exhibit strong scaling violations (on logarithmic scales).

At present, the limitation of $x \leq .02$ implies $Q^2 \leq$ 1-10 GeV^2. The future of such studies with CERN or Fermilab neutrino beams is therefore somewhat limited, although the Fermilab muon beam should be quite useful in extending our understanding of this kinematic regime. But this situation in principle changes dramatically at HERA, where the leverage in Q^2 goes out to $\leq$ 1000 GeV^2. The HERA events at $x \sim .02$ have the kinematic structure of a 30 GeV electron scattering at large angles from a parton in the 10 GeV momentum range. These appear in principle very accessible. But, for the time being, the Q^2 leverage is not so great and we will set aside the high-order QCD effects and concentrate on the moderate Q^2 behavior of the phenomena. In what follows we first outline the kinematics and two contrasting dynamical mechanisms. One is scattering from the "naive" ocean parton distribution, and the other is quark-pair production via "photon-gluon fusion". Thereafter we discuss implications for A-dependence studies and properties of the hadron final states. We also speculate on implications for hadroproduction processes.

II. Kinematics and Mechanisms

The two mechanisms to be discussed are shown in Fig. 1.

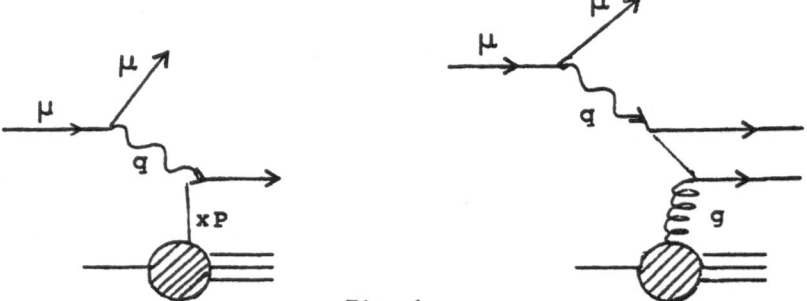

Fig. 1

Fig. 1   a) Naive parton model   b) Photon-gluon fusion

The cross-sections for these may be easily computed and we record them here:

(a) Naive scattering from the "ocean" (u, d, s, $\bar{u}$, $\bar{d}$, $\bar{s}$)

$$\lim_{E_\mu \to \infty} \nu \frac{d\sigma}{dQ^2 d\nu} = \frac{4\pi\alpha^2}{Q^4} \left(\frac{4}{9} + \frac{1}{9} + \frac{1}{9} \right) \frac{F_q(x)}{3} \tag{2.1}$$

where $x^{-1}F_q(x)$ is the quark distribution summed over quark types $(u,d,s,\bar{u},\bar{d},\bar{s})$. We ignore charm because of the strong threshold dependence in the relevant range of Q^2. The scaling variable

$$x = \frac{Q^2}{2M\nu} \tag{2.2}$$

18

is interpreted as usual and is defined by the lepton kinematics. This comment is relevant for the photon-gluon fusion mechanism we now discuss.

(b) Photon-gluon Fusion

We obtain the cross-section from the analysis of charm electroproduction of Barger et al as quoted by Gollin, et al.[3] Setting $m_c = 0$ wherever possible, one obtains

$$\lim_{E_\mu \to \infty} \nu \frac{d\sigma}{dQ^2 d\nu} = \frac{4\pi\alpha^2}{Q^4} \cdot \frac{2}{9} \frac{\alpha_s}{\pi} F_g (x) \, \phi(x) \qquad \text{with} \qquad (2.3)$$

$$\phi(x) = \int_x^1 \frac{d\xi}{x} \left(\frac{x}{\xi}\right)^2 \frac{F_g(\xi)}{F_g(x)} \left[2\left(\ln \frac{m^2}{m_Q^2}\right) \left\{1 - 2\left(\frac{x}{\xi}\right) + 2\left(\frac{x}{\xi}\right)^2\right\} - \left\{1 - 8\left(\frac{x}{\xi}\right) + 8\left(\frac{x}{\xi}\right)^2\right\} \right] . \quad (2.4)$$

Here $F_g (x)$ is the gluon distribution, summed over colors. We have taken this factor out in order to facilitate comparison of the normal "naive" ocean mechanism with the photon-gluon mechanism, since F_g and F_q (at small x) should not be too different, and since the only other differences in normalization are the factors α_s/π and $\phi(x)$ appearing in the photon-gluon-fusion expression.

The heart of the matter is the factor ϕ, which asymptotically contains logarithms galore. Hence $\frac{\alpha_s}{\pi} \phi$ may be ≥ 1 and the two mechanisms competitive. The parameter

$$\xi = x \left(1 + \frac{m^2}{Q^2}\right) \qquad (2.5)$$

appearing in the integral is the momentum fraction of the fused gluon; the parameter m is the mass of the $q\bar{q}$ system to which photon and gluon fuse. The logarithm within the integral comes from the integration over the angular distribution of the produced $q\bar{q}$ pair at fixed mass m; more about that later. The remaining polynomials in x/ξ are inconsequential.

19

All this appears - and thus far is - quite straightforward. Nevertheless, when one contemplates A-dependence effects, subtleties arise, to which we now turn.

III. Space-Time Properties of the Amplitudes

Essential to both mechanisms is the fact that at small x these electroproduction amplitudes involve large longitudinal distances. This implies the relevance of a (generalized) "vector-dominance" mechanism. That is, we think of the evolution of the system in two stages:

i) Well upstream of the target nucleon or nucleus, the virtual photon dissociates into a $q\bar{q}$ pair, which then is free to possibly evolve further before arrival at the target.

It is easiest to use an old-fashioned perturbation-theory estimate of the energy difference ΔE between virtual photon and virtual $q\bar{q}$ system of mass m to estimate the propagation distance of this system.

$$\Delta t \sim \frac{1}{\Delta E} = \frac{2\nu}{Q^2+m^2} = \frac{1}{xM} \cdot \frac{1}{\left(1+\frac{m^2}{Q^2}\right)}$$

$$\approx \left(\frac{10^{-2}}{x}\right) \cdot \frac{1}{\left(1+\frac{m^2}{Q^2}\right)} \cdot (20f.) \tag{3.1}$$

ii) Upon arrival at the target this virtual system interacts and is liberated, thus forming the final system of produced hadrons.

The origin of the distinction between production mechanisms lies in the structure of the virtual intermediate state, not at birth (there it is always a "bare" $q\bar{q}$ pair), but rather at arrival at the nucleon or nuclear target. Here we may distinguish three possible descriptions:

1) "Naive" Vector Dominance:

This option, which we shall rapidly dismiss, imagines the intermediate virtual system as a typical hadron, e.g. ρ, ρ', ρ'', ..., which is absorbed with a typical hadronic cross-section on the target. As pointed out by Gribov[4], such a model is inconsistent with scaling by a whole power of Q^2. Elementary calculations give, for absorption cross-section on a large target.

$$ \sigma_T + \sigma_L = [1 - Z_3(Q^2)] \pi R^2 \ . \tag{3.2} $$

Here R is the target radius, and $(1-Z_3)$ is the probability (which "runs" with Q^2) that the photon is a hadron at arrival. This is directly connected to the hadronic vacuum polarization contribution to the photon propagator; hence to the dimensionless colliding-beam cross-section parameter R:

$$ 1 - Z_3 = \frac{\alpha}{3\pi} \int_0^{\tilde{s}} \frac{dm^2 \ m^2 \ R(m^2)}{(Q^2 + m^2)^2} \ . \tag{3.3} $$

For large Q^2, the factor $1-Z_3$ contains no intrinsic scale. Therefore we have, from dimensional analysis alone, the result that this "naive" vector-dominant picture predicts $\sigma_T + \sigma_L \gtrsim \alpha R^2$ while scaling predicts $\sigma_T + \sigma_L \sim Q^{-2}$. Thus this picture is experimentally and conceptually wrong. We now turn to the mechanisms by which the models in question evade this result.

2) Photon-gluon Fusion

The photon-gluon fusion picture is perhaps the easiest to describe. It is kinematically similar to the QED Bethe-Heitler pair-production process, the only distinction being the virtuality of the incident photon. (This makes the typical mass of the produced pair order $\sqrt{Q^2}$, not $2m_e$.) For photon gluon fusion the role of the Weizsaecker-Williams Coulomb photon is of course replaced by the gluon cloud of the target.

A typical final state will leave the q and $\bar{q}$ (or e^+e) with comparable longitudinal momenta. Therefore the transverse momenta of the quarks will be $\leq m/2$ which in turn is typically of order $(1/2)\sqrt{Q^2}$, i.e. <u>large</u> in the scaling limit. This in turn implies that, at arrival at the target, the transverse separation Δx_T of the q and $\bar{q}$ is small, or order $1/m$, or $1/\sqrt{Q^2}$. This has been verified by calculating, via "old-fashioned" light-cone perturbation theory, the wave function of the pair at arrival. It is also consistent with a simple classical geometrical estimate:

$$\Delta x_T \sim \left(\frac{p_t}{\nu}\right) x_L \sim \left(\frac{p_t}{\nu}\right) \frac{2\nu}{(Q^2+m^2)} \sim \frac{2m}{Q^2+m^2} \leq \frac{1}{Q} \quad . \tag{3.4}$$

This is in turn consistent with the quantum uncertainty relations.

Because, at arrival at the target, the q and $\bar{q}$ have hardly separated, they constitute a small color-dipole. Hence the interaction with the target is suppressed, relative to a typical strong interaction, by the square of the dipole moment. This gives a full power of Q^2, and restores the scaling behavior which was lost in the "naive" vector dominant approach. Furthermore, it should be a good first approximation to treat the intermediate $q\bar{q}$ system as free particles. Nonperturbative effects, e.g. formation of a string between q and $\bar{q}$, should be unimportant — although the <u>hard</u>-gluon radiative corrections of perturbative QCD, as usual, should be appended. At very large Q^2 and ν they do become important.

In analogy with QED and the Bethe-Heitler process, we expect the inelastic interaction of this color-dipole with nucleons in the nucleus to be incoherent and additive, due to the coulomb-like interaction of the dipole with the distinct gluon-clouds of the individual nucleons in the nucleus. Thus the basic dependence of this process on nuclear size is A^1, not $A^{2/3}$.

We also note that the typical final state consists, asymptotically, of two balanced high-p_T quark-jets and <u>no</u> beam-jet. This is hardly what is anticipated from ordinary parton-model considerations, where no final-state large-p_T secondaries are, to first approximation, expected.

3) "Naive" Parton Model

The parton-model description is usually not carried out in the laboratory frame but in, say, a center-of-mass frame of the system of target and incident lepton. In such a frame the momentum of exchanged photon is transferred to a single "ocean" quark. For a nuclear target this "ocean" quark is found in a cloud of longitudinal extent $\Delta x \lesssim 1f$, which is <u>large</u> compared to the thickness of the Lorentz-contracted pancake containing the nuclear matter. It is then expected that the number of ocean-partons per unit of transverse area (and per unit rapidity) saturates. Hence the electroproduction cross-section for lepton-nuclear scattering from wee ocean-partons would be expected to scale as $A^{2/3}$.

How does all this look in the laboratory frame? In that frame, the previous mechanism of photon-gluon fusion is in general inoperative, because the "naive" parton model includes only the nonperturbative strong interactions of partons of comparable momentum (or rapidity) and ignores long-range correlations in rapidity, such as the gluon-exchange interaction of the fast quarks in the color-dipole with the slow quarks in the target.

However, in the laboratory frame, the general vector-dominance picture still applies; the virtual photon first dissociates into the $q\bar{q}$ pair which then may evolve further.* What is different? It is simply that in this case the partition of virtual-photon longitudinal momentum to q and $\bar{q}$ is highly <u>asymmetric</u> – sufficiently asymmetric that the quarks no longer posses high p_T. When this is the case, on arrival the transverse separation Δx_T of q and $\bar{q}$ is no longer small, and can be of order $\langle p_T \rangle^{-1} \sim 1f$. Under

---

* It is quite appropriate that the initial ocean quark of the center-of-mass description is found, in the laboratory frame, in the negative-energy Dirac sea.

these circumstances, there can be non-perturbative dynamical evolution during the propagation of the virtual state - e.g. string formation, creation of a cloud of wee partons, etc. We repeat, for $q\bar{q}$ initial configurations sufficiently asymmetric in longitudinal momentum, non-perturbative evolution can occur, and the system on arrival at the target may be "hadron-like" and be absorbed by the target with a typical nuclear mean free path.

The angular distribution of the virtual $q\bar{q}$ pair of given mass m, in its rest frame, is essentially isotropic. This means that the requirement of sufficient alignment along the beam-direction for the transverse momentum of q and $\bar{q}$ to be "typically" small, say $\leq$ $\langle p_T \rangle$ ~ 300 MeV, is simply that the center of mass angle θ^* satisfy

$$\theta^* \leq \frac{\langle p_T \rangle}{m} \qquad \text{or} \qquad (3.5)$$

$$1 - \cos \theta^* \sim \frac{\langle p_T^2 \rangle}{m^2} \qquad . \qquad (3.6)$$

Isotropy implies a distribution uniform in $\cos \theta^*$. Hence the alignment probability is

$$\sim \frac{\langle p_T^2 \rangle}{m^2} \sim \frac{\langle p_T^2 \rangle}{Q^2} \qquad . \qquad (3.7)$$

It is this feature that restores the scaling behavior for the "naive" parton model[4].

The longitudinal momentum fraction z of the slow member of the quark pair is, under these conditions,

$$z = \frac{1 - \cos \theta^*}{2} \leq \frac{\langle p_T^2 \rangle}{m^2} \qquad . \qquad (3.8)$$

We may again check via the classical calculation that the transverse separation Δx_T of the pair on arrival is large. It is

24

$$\Delta x_T \sim \frac{p_T}{p_L}(\Delta x_L) \sim \frac{\langle p_T \rangle}{z \nu} \cdot \frac{2}{Q^2 + m^2} \sim \frac{2}{\langle p_T \rangle} \cdot \frac{m^2}{(Q^2 + m^2)} \qquad (3.9)$$

which typically is large. This estimate again is compatible with the quantum uncertainty relations. With this amount of $q\bar{q}$ separation there is enough time available to dress the original $q\bar{q}$ system with wee partons – indeed with partons of momentum $\leqq \frac{M}{x}$. (This is because the time required to dress the system with partons of momentum p is proportional to that momentum).

Thus the structure which arrives at the target is complex, and partially dressed. Its constituents which have momenta $\leqq$ M/x are hadron-like. The remaining high-momentum portion is carried by the single leading quark which, <u>after</u> the collision with the target, may be expected to evolve and dress as does the final system in e^+e^- annihilation. Hence we recover the parton-model view of the structure of final state hadrons in longitudinal phase-space shown in Fig. 2.

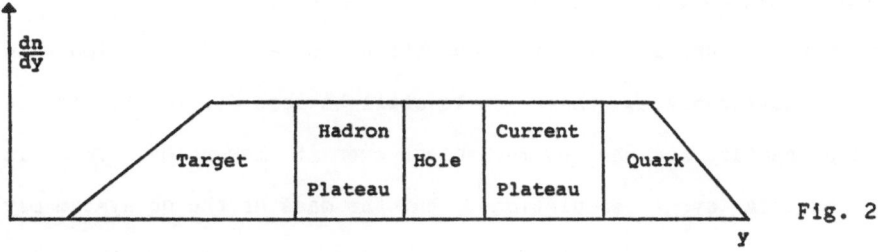

Fig. 2

We reiterate that the nuclear dependence of this process is expected to be $A^{2/3}$, and that the final state is predominantly low-p_T. This follows from the laboratory frame description as well as from the more conventional "infinite-momentum" frame of the parton model.

IV.   Recapitulation of Experimental Consequences

A.   Electroproduction

Let us now summarize the inferences we have made. First of all, at moderate Q^2 and wee $x \leq .02$, we expect the "naive" parton-model contributions to be the dominant contributor to electroproduction from a nucleon, with the photon-gluon contribution a relatively small correction (~ 10-20%). However, the A-dependence of the former is ~ $A^{2/3}$, while that of the latter is ~ A^1. Hence the two mechanisms may be quite comparable in very heavy nuclei.

While it may be difficult to untangle these contributions from measurement of structure functions alone, it should in principle be possible to do better by examination of the final state. In the idealized limit of small x and quite high Q^2 ($\gtrsim 100$ GeV^2) one should, for the naive parton-model mechanism, see only the typical beam-jet fragmentation, while for the photon-gluon mechanism one should see two leading, balanced high-p_T jets (p_T ~ $(1/2)\sqrt{Q^2}$ and no beam jet).

Unfortunately, until HERA is operating this idealization is unreachable. Nevertheless, there could still be some distinction in the final state properties for the two mechanisms even at lower Q^2. This is best examined via event simulations. But the mass of the $q\bar{q}$ system may have to be larger than 5-6 GeV in order to discern a jet orientation transverse to beam for photon-gluon fusion and parallel to beam for naive parton mechanism - just as it is in e^+e^- annihilation.

An additional distinction may occur in the multiplicity of low-momentum secondaries. The small color dipole present in the photon-gluon-fusion mechanism is not likely to suffer multiple collisions in traversing nuclear matter. Thus the final state may be relatively "diffractive," with less nuclear excitation and production of slow secondaries. Indeed, one may conjecture that the only produced hadrons in

26

the typical photon-gluon fusion collisions - even in nuclei- are those associated with the fragmentation of the $q\bar{q}$ system. In other words the prescription is for produced $q\bar{q}$ system of mass m and cms production angle θ^* as follows: take the final hadron state for e^+e^- hadrons at cms energy m and jet angle θ^* and boost it until it has the momentum ν of the incident virtual photon. These are then conjectured to be <u>all</u> the particles produced in electroproduction via this mechanism.

In the "naive" parton mechanism, on the other hand, the $q\bar{q}$ system on arrival at a nuclear target does contain a low-momentum parton component and can be expected to suffer multiple nuclear collisions like an ordinary hadron. Thus the multiplicity of lower energy $(E \le M/x)$ hadrons should be characteristic of what is observed for a comparable π-nucleus collision.

B. Dijet Photoproduction

An upcoming Fermilab experiment[5] (E-683) which will use real photons is also very relevant to these considerations. According to vector-dominance phenomenology, roughly half the time a real photon which dissociates into hadrons essentially may be regarded, on arrival at a target, as a low-mass vector meson ρ, ω, ϕ. However there is a finite but small probability for the dissociated system to arrive as a massive $q\bar{q}$ system with symmetric momentum partition. If this occurs, the final system should again be of the photon-gluon-fusion character: two balanced high-p_T jets and no beam jet. The Fermilab experiment intends to observe this final state and test the QCD estimates of the production cross-section. The relevance of the remarks in this note is mainly in the final-state morphology. Will the production be a "diffractive" phenomenon with an A^1 dependence, as described above?

C. Heavy Quark Photoproduction and Electroproduction

Do these same considerations apply to the electroproduction of heavy quarks Q such as charm and bottom? The arguments clearly generalize as

27

long as the mass of the produced $Q\bar{Q}$ system is large compared to the threshold value $2m_Q$; none of the previous kinematic estimates are significantly affected. When the mass is near threshold, one needs to estimate the size of the $Q\bar{Q}$ system on arrival as well as how much residual nuclear interaction occurs. Here it would seem that the theoretical issue can be largely finessed: the size can hardly be larger than that of typical $Q\bar{Q}$ onium systems. But it is known that ψ suffers very little absorption in nuclear matter; hence for low mass systems the nuclear dependence of photoproduction and/or electroproduction should be A^1. The only possibility for non-perturbative evolution seems to be for those $Q\bar{Q}$ configurations in which the relative p_T of Q and $\bar{Q}$ is small, ≤ 300 MeV. However these are probably power - law suppressed, just from phase-space arguments alone. We conclude that the process should have a linear dependence on atomic number, even in the "diffractive" limit.

D. Hadroproduction of Heavy Quarks

It is tempting to try to apply these ideas to heavy - quark hadroproduction, a subject which is today still somewhat confused. The two issues of relevance to the contents of this note are the dynamics of forward production and the question of A dependence. And, in brief, the question comes down to whether there appears anywhere a candidate for a nonperturbative production mechanism.[6]

After several false starts, I can offer only one candidate, illustrated in Fig. 3. As shown there an initial quark or gluon radiates a gluon which virtually converts to a $Q\bar{Q}$ heavy quark system which is in a color octet (Note this mechanism has no analogue in electroproduction). If the p_T of the $Q\bar{Q}$ system, and hence of the companion q or g is small, their relative impact parameter can become large, and a string and/or wee - parton cloud again will have the opportunity to form upstream of the target.

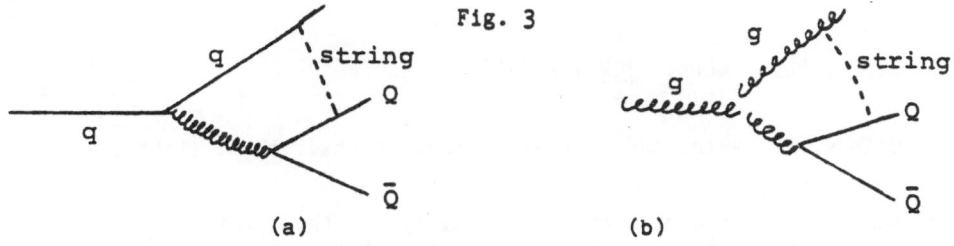

Fig. 3

This can lead to an $A^{2/3}$ contribution to the cross-section. Furthermore, since the momentum partition to the $Q\bar{Q}$ system is asymmetric, at least half the time the $Q\bar{Q}$ system will be leading, i.e. carry the full momentum of the initial q or g. Thus there may be an argument that production of a leading $Q\bar{Q}$ system is in part nonperturbative, and possesses an $A^{2/3}$ dependence. But the subject is tricky and I am not too sure of my ground here.

V. Conclusion

It is a real pleasure to dedicate these remarks to my colleague and friend, Jack Steinberger, whose work with neutrino production of hadrons has so much advanced our knowledge of the interior structure of the nucleon.

To be sure, the content of this paper is not very much directed to neutrino physics. There the situation at large x is in good shape. And, while there is still room for progress in the direction of higher energy and smaller x, it is probably the case that ultimately this subject is best attacked by muon and electron scattering experiments at the highest possible energies. I trust that this work will proceed with the same high standards and thoroughness as we have come to expect from Jack and his CDHS colleagues.

I thank Al Mueller for helpful comments and criticism.

29

References

1. B. Ioffe, Phys. Lett. <u>30B</u> ,123(1969).

2. L. Gribov, E. Levin, and M. Ryskin, Phys. Repts. <u>100</u> , 1(1984).

3. V. Barger, W. Keung, and R. Phillips, Phys. Lett. <u>91B</u> , 253(1980); G. Gollin et. al. , Phys. Rev. <u>D24</u> , 559(1981).

4. Many of these ideas are discussed in J. Bjorken, <u>Proceedings of the 1971 Symposium on Electron and Photon Physics at High Energies</u>, Cornell Univ. 1971, ed. N. Mistry.

5. L. Cormell et. al. , Fermilab Proposal E-683.

6. A recent discussion of perturbative mechanisms is given by J. Collins, D. Soper, and G. Sterman Univ. of Oregon preprint OITS-292(1985).

Space Resolution of a Large Time Projection Chamber

W. Blum[1], *J. May*[2], *and L. Rolandi*[3]

[1]Max-Planck-Institut für Physik und Astrophysik,
  D-8000 München 40, Fed. Rep. of Germany
[2]CERN, CH-1211 Genève 23, Switzerland
[3]Dipartimento di Fisica and Sezione INFN, I-Trieste, Italy

1. Introduction

When in 1980 Jack Steinberger initiated the formation of a big collaboration now known as 'ALEPH', dedicated to do experiments at the LEP Collider, we soon began to think about the best central detector for the purpose. It had to be almost as large as the magnet we could afford, because a high premium was set on the measurement accuracy of high particle momenta. Should the detector of the charged particles be of the well-proven type of the drift chamber where now a volume of 45 m³ would be filled with wires stretched over 5 m? Or should we consider the time projection chamber (TPC) pioneered in Berkeley [1], where the particle tracks are drifted over several metres through an empty volume of gas before they are finally measured in the end plate?

Once it was decided to seriously study a TPC because of its measuring principle by which each section of a particle track has its coordinates independently measured in all three dimensions, it soon became apparent that we would enjoy the advantages of doing pattern recognition in three dimensions and of a considerable simplicity of the whole construction. The parallelism of the electric and magnetic fields would permit using the highest possible magnetic field. To optimize the momentum resolution even more, and to keep the interaction of charged particles and photons at the minimum level, we had in mind to operate the chamber at atmospheric pressure. Could we understand and limit the distortions of the tracks during their long drift? What spatial and momentum resolution could we then achieve? These were some of the problems we faced. We began immediately with the construction of several prototypes.

2. Principle of Coordinate Measurement

Figure 1 shows a schematic view of the TPC as developed in Berkeley and adapted to our dimensions. The cylindrical chamber is immersed in a uniform axial magnetic field parallel to an electric drift field.

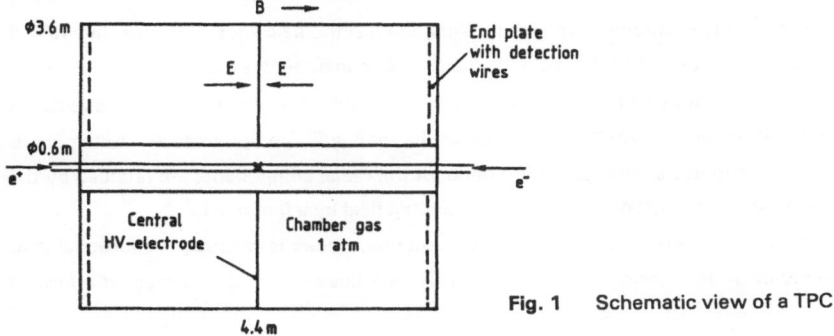

Fig. 1   Schematic view of a TPC

After the passage of a charged particle through the sensitive gas volume, the ionization electrons drift axially towards the end plates and are detected by proportional wires stretched across the end planes. It is a fundamental idea of the TPC that the magnetic field will not only bend the particles but also reduce the transverse diffusion of the drifting electrons.

For each track segment the two coordinates in the end plane are given by the position of the cathode 'pads' which transmit the electric pulses from the avalanches at the sense wires into the amplifiers connected to each pad. The z coordinate is found from the drift time and the known drift velocity.

The direction in which the magnetic field bends the particles is at right angles to the cylinder-radius vector, and the particle momentum is determined by a precise measurement of the sagitta of the particle helix projected onto the end plane. The pads in the cathode are arranged such that the induced signal spreads over a few pads. The precision on $r\phi$ is obtained by interpolating between them.

The $r\phi$ measuring accuracy depends on several factors. Some of them are connected to the transport of the electrons in the drift volume. Some others influence directly the avalanche localization along the wire. We treat the errors, according to their physical origin, in three categories:

i) Displacement of the drift trajectories due to field inhomogeneities.

ii) Statistical variation of the position of the centre of the electron swarm due to the finite number of electrons and their clustering.

iii) Mechanical and electrical imperfections in the localization of this centre.

3. Displacement of the Drift Trajectories
In the case of inhomogeneous $\vec{E}$ and $\vec{B}$ fields, the drift velocity vector

$$\vec{v} = [\mu E/(1 + \omega^2\tau^2)] \ [\hat{E} + \omega\tau \ [\hat{E} \times \hat{B}] + \omega^2\tau^2(\hat{E}\cdot\hat{B})\hat{B}] \tag{1}$$

depends on the two gas constants μ and $\omega\tau$; μ is the electron mobility and $\omega\tau$ the dimensionless product of the electron cyclotron frequency ω = (e/m)B and the mean free time τ between collisions of the electrons with the gas molecules. From this equation, where E is the magnitude of the electric field, and $\hat{E}$ and $\hat{B}$ are unit vectors, one deduces that field inhomogeneities will create distortions of the drift trajectories. Knowing the fields, these distortions could be corrected with the help of (1). Whilst the magnetic field can be measured and is stable, this may not be the case for the electric field. It could be distorted by uncontrollable and local charge accumulations [2]. The influence of the electric field is strongest for $\omega\tau \ll 1$, since the electrons travel mainly along the electric field lines, whereas they follow mainly the magnetic field lines for $\omega\tau \gg 1$. For $\omega\tau = 1$, the ($\hat{E} \times \hat{B}$) component is maximal.

The distortion in the azimuthal direction, which influences the measurement of the sagitta, is induced by the azimuthal components of the fields and also by their radial components due to the E × B term of (1). There are advantages at large $\omega\tau$: the effects of the radial components are reduced by the factor $1/\omega\tau$ and those of the azimuthal component of the electric field by a factor $1/(\omega^2\tau^2)$.

In order to obtain large $\omega\tau$ the gas pressure should be kept low, since τ is inversely proportional to it. Argon–methane mixtures at atmospheric pressure can have $\omega\tau$ values up to 10 in a magnetic field of 1.5 T [1].

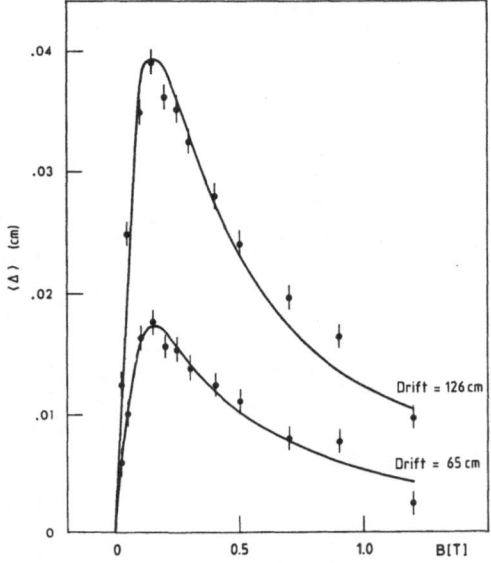

Fig. 2 Antisymmetric displacement of measured coordinate as a function of magnetic field

Fig. 3 Symmetric displacement of measured coordinate as a function of magnetic field

A big effort has been put into the design of the magnet to achieve the best field uniformity. This is obtained by an extra coil at each end of the solenoid and some shims. Even with these precautions, given the expected value of $\omega\tau$, we will have distortions in the direction of the sagitta, of a few 100 μm, just a bit larger than could be ignored.

To monitor these distortions we will use in the ALEPH TPC a laser calibration system, which will provide 30 ionizing light rays similar to real tracks coming from the e^+e^- collision point. Any distortion of these straight tracks will be measured and used for the correction of the real tracks. It was Jack who noticed that for this calibration purpose only the straightness of the laser beams is important; neither their position nor their brightness needs to be known with precision.

The feasibility of this calibration system has been tested using a prototype (TPC 90) in a magnetic field up to 1.2 T. Laser ionization tracks 0.6 m long were drifted over 1.3 m, and their displacement was measured as a function of the magnetic field strength. Figures 2 and 3 [3] show that when the current is reversed in the magnet the antisymmetric displacements of a laser track (Δ) are proportional to $\omega\tau/(1 + \omega^2\tau^2)$, and the symmetric track displacements (S) are proportional to $\omega^2\tau^2/(1 + \omega^2\tau^2)$, as contained in (1). Assuming a homogeneous electric field it is possible to calculate with this equation the integral over the radial component of the magnetic field from the measured displacements. Figure 4 (from Ref. [3]) shows the perfect agreement of this calculation with the measured map of the magnetic field. The inhomogeneity in the ALEPH magnet will be an order of magnitude smaller than in the TPC 90 magnet.

The results of the tests described above have shown that with this procedure it is possible to compute and to correct the track distortions induced by the inhomogeneities of the fields, so that the accuracy of the track reconstruction is solely limited by ionization statistics, mechanical and electrical imperfections.

33

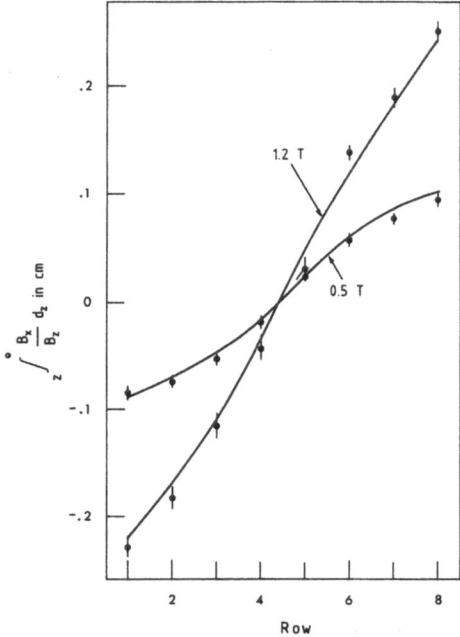

Fig.4 Measured distortions (points) induced by the magnetic field compared to those calculated using the field map (curves)

4. Statistical Variation of the Avalanche Position

The accuracy of the measurement of the electron swarm centre of gravity is affected by the following three effects, which are caused by the discontinuous nature of the ionization process.

a) The diffusion process over a drift distance L displaces each electron by a mean square value of 2DL/v. The diffusion constant D depends on the magnetic field approximately like

$$D(B) = D(0)/(1 + \omega^2\tau^2) \ . \tag{2}$$

Here lies another advantage of a high value of $\omega\tau$. Concerning the dependence of D(B) on the gas pressure p, we note that for large $\omega\tau$ D(B) is proportional to p, but for small $\omega\tau$ it is proportional to 1/p: large $\omega\tau$ and low pressure result in a small diffusion constant.

The electron cloud measured over one pad row consists of N_e electrons. Their centre position measured at right angles to the track has a variance of

$$2LD(B)/(v \ N_e) \ , \tag{3}$$

which is independent of p for large $\omega\tau$.

b) The angular wire effect arises when the track makes an angle α with the normal to the wire in the end plane (Fig. 5a). The ionization charge is then collected on the sense wire, not in one point but over a distance d tan α. The position of the centre of gravity of the charge has a variance of $(1/N_{eff})(d^2/12) \times \tan^2 \alpha$, where N_{eff} represents the effective number of fluctuating charges collected over the wire gap d;

N_{eff} may be smaller than the number of primary clusters created in d because of the cluster size variation, or it may approach the total number of electrons if, owing to strong diffusion, the electrons have lost the space correlation of their primary clustering.

The presence of the magnetic field makes this effect asymmetric and, on an average, larger: the electrons, as they are collected in the cylindrical field of the sense wire, travel across the magnetic field lines, and the $(\hat{E} \times \hat{B})$ term in (1) gives them a velocity component along the wire. The average angle under which they approach the wire is denoted by ψ (see Fig. 5b). Now the ionization charge is smeared out over a distance $d(\tan \alpha - \tan \psi)$, and the variance of the centre position along the wire is

$$(d^2/12\,N_{eff})(\tan \alpha - \tan \psi)^2 \ . \tag{4}$$

This influence of the magnetic field was discovered at TRIUMF [4]; in Fig. 6 we see the resolution measured there as a function of the angle α. It was large and asymmetric.

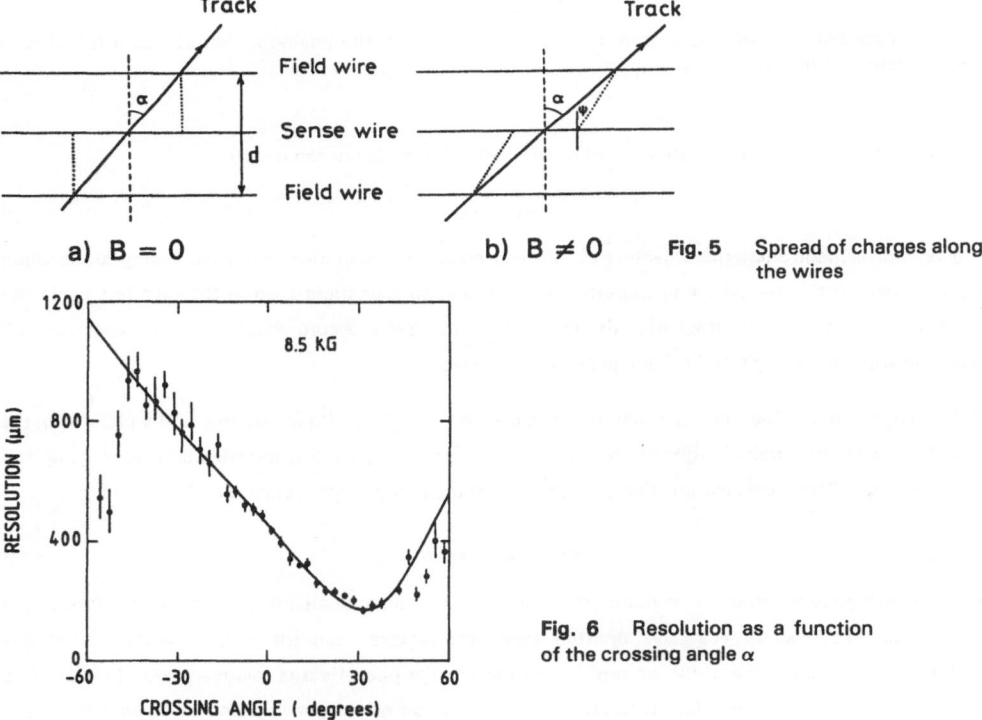

Fig. 5   Spread of charges along the wires

Fig. 6   Resolution as a function of the crossing angle α

When more than one wire contributes to the signal induced in a pad row, there is a statistical gain which is represented by a corresponding increase of N_{eff}. The easiest way to obtain the functional dependence of N_{eff} on the diffusion parameter and the pad length is a Monte Carlo simulation which contains the Landau distribution with its high-energy tail.

The result of such calculations for $\alpha = 0$ is drawn in Fig. 7. We see how the r.m.s. position accuracy σ, defined by both the wire (E × B) and diffusion effects, varies with the single electron diffusion s. A strong minimum is reached when s is about half the sense wire distance (2 mm).

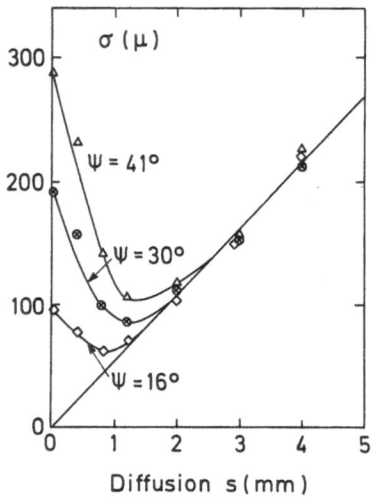

Fig. 7 Combined contributions from E × B and diffusion to the position accuracy as a function of single-electron diffusion

The gas pressure p has an influence of N_{eff} and on ψ. From (1) we can derive

$$\tan \psi = \langle \omega \tau \rangle \ . \tag{5}$$

The average is made along the trajectory of the electron approaching the sense wire. Owing to the much higher E field in this region τ and therefore $\langle \omega \tau \rangle$ is much smaller there than in the drift region. In the interest of a small E × B effect near the sense wire, the gas pressure should be high, since $\langle \omega \tau \rangle$ will decrease with increasing p and moreover N_{eff} will increase.

c) **The angular pad effect** arises when the track makes an angle β with the normal to the pad row in the end plane. The ionization charge is spread out over a distance h tan β along the pad row, h being the length of a pad. Therefore, the position of the centre of the charges has a variance of

$$(h^2/12 \ N_{eff}) \tan^2 \beta \ . \tag{6}$$

Here, the effective number N_{eff} of fluctuating charges has the same meaning as above; it refers to the gas sample thickness h. We have omitted statistical factors from the β-dependence of the gas thickness. Comparing (6) and (4) we notice that the angular pad effect is relatively large because h is several times as large as d. Also, when h increases, the pad effect goes up and the wire effect goes down. With the approximation that N_{eff} is proportional to h, (4) is proportional to 1/h, and (6) is proportional to h.

The greater part of the angular pad effect can be corrected [5, 6] when use is made of the measured pulse height of the wires. In practice it has been possible to correct a fraction $f_{corr} = 0.95$ of (6) in clean cases.

5. Errors from Electronic and Mechanical Sources

When the pad signals are amplified, digitized, and calibrated, additional errors may be introduced into the determination of the centre of the avalanche. Without going into detail we summarize the situation by stating that the error contributed by the electronics can be kept lower than 100 μm by suitable electronic design. The mechanical error of misalignment of the pads or sectors was estimated to be smaller than 50 μm. Both are independent of the track angle and the drift distance.

6. Experimental Results

Collecting all the statistical contributions from above we may parametrize the $r\phi$ resolution for the case where the track makes an angle α with the wire normal and the pad rows follow the wires (Berkeley geometry, $\alpha = \beta$):

$$\sigma^2 = \sigma_0^2 + \sigma_d^2 \, L/\cos \alpha + \sigma_1^2 \, (\tan \alpha - \tan \psi)^2 \cos \alpha + \sigma_2^2 \tan^2 \alpha \cos \alpha \quad , \tag{7}$$

where L is the drift length in metres. (Here we have included the statistical cos α factors describing the variation of the gas sample thickness, and the diffusion term contains the projection factor (1/cos α) applicable to the diffusion of an inclined line of electrons.

In a test chamber which reproduced the Berkeley geometry, the resolution was measured in an argon–methane mixture (90:10) at 1.5 T and a drift length of 3 cm, as a function of the angle α [6]. Curve A in Fig. 8 shows the result. After applying the angular pad correction we obtained curve B. A fit of (7) yielded

$\sigma_0 = (155 \pm 25)\,\mu$m

$\sigma_1 = (340 \pm 30)\,\mu$m

$\psi = (29 \pm 3)^\circ$

$\sigma_2 = (1000 \pm 50)\,\mu$m before correction

$\sigma_2 = (230 \pm 80)\,\mu$m after correction

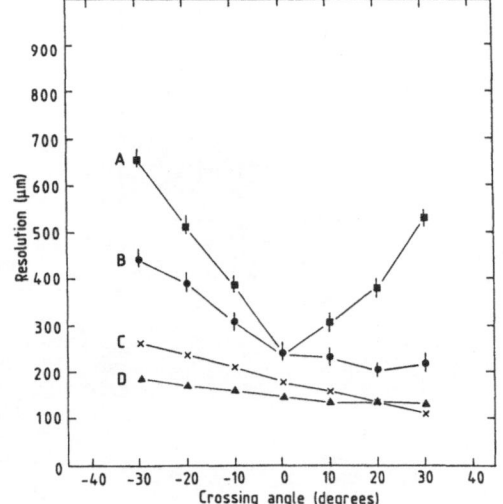

Fig. 8 Resolution as a function of the angle α at B = 1.5 T. Curve A: Berkeley geometry, no angular corrections; curve B: Berkeley geometry, with angular corrections; curve C: 30 mm radial pads, 4 cm drift; curve D: 30 mm radial pads, 220 cm drift

An independent measurement in TPC 90 tells us that under these conditions

$$\sigma_d \approx 80\,\mu\text{m} \quad .$$

These numbers implied that in this geometry it was essential to do the angular pad correction, and that the two angular effects were then comparable in size.

7. A New Geometry

Although the importance of the angular pad term in the resolution can be drastically reduced using the pulse heights of the wires, this is not really satisfying since these corrections can only be applied if the wire signals of the track are clean and not overlapped in time by the signals of other tracks. With the expected large track densities in jets at LEP energies this might be only rarely the case. Moreover, the experimental values of σ_1 and σ_2 (correction applied) show that in the case of the Berkeley geometry only a minor improvement of the resolution could still be obtained by increasing the pad length, because of the two competing angular effects.

A big step forward was made in 1983 when Jack, following an idea of Wenzel [7], proposed a new geometry for the end plate of the ALEPH TPC [8], based on circular rings of radial pads (Fig. 9). With this geometry the angular pad effect disappears for the high momentum tracks coming from the interaction point, which, having radial trajectories, are always at $\beta = 0°$ (Fig. 10). This then made it possible to increase the length of the pads to 30 mm, reducing the contributions of the angular wire effect and of the diffusion to the resolution. Collecting more charge also increased the ratio signal/electronic noise.

An open question was how much one could really improve the momentum resoution for stiff tracks and how much resolution would be lost for low momentum tracks, which have $\beta \neq 0$ owing to their bending in the magnetic field. It was again Jack who showed [9] that the angular wire effect is reduced by a factor 0.66 when the pad length is changed from 8 to 30 mm, proving that the rare cases of large energy release in a single ionization cluster, which could spoil the resolution, can easily be recognized

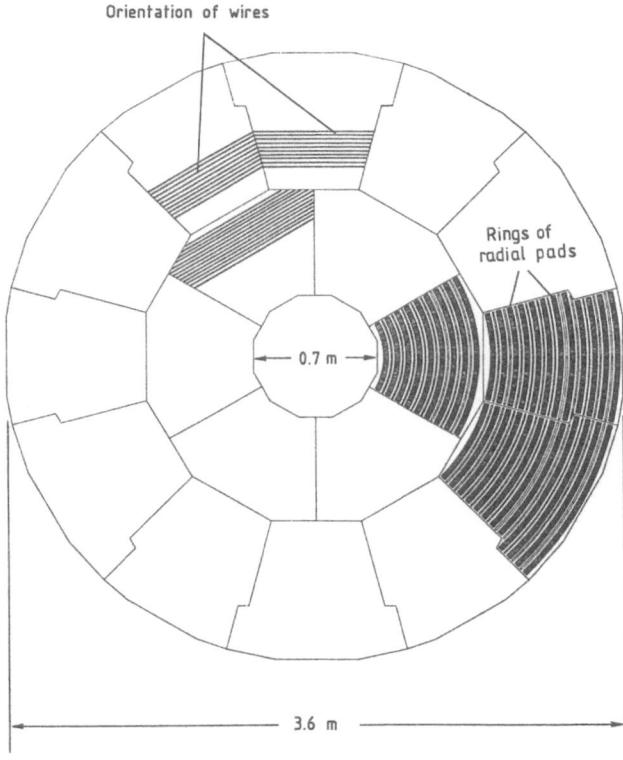

Orientation of wires

Rings of
radial pads

0.7 m

3.6 m

Fig. 9 End plate of the ALEPH TPC

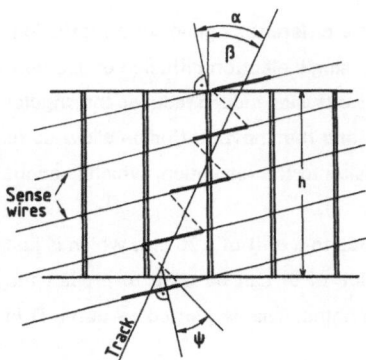

Fig. 10   Definition of angles

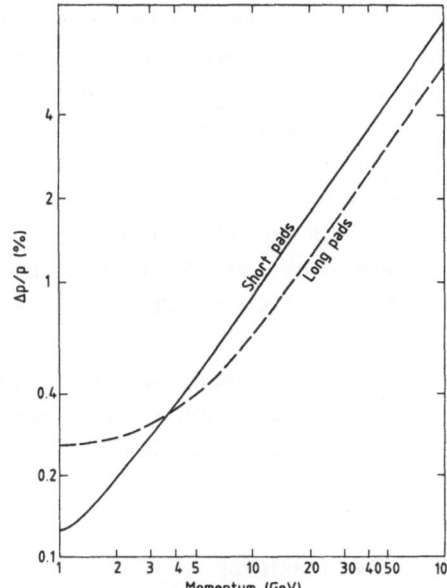

Fig. 11 Momentum resolution as a function of momentum for charged particles in the ALEPH detector

and disregarded. Figure 11, taken from his paper, shows the momentum resolution of the ALEPH detector as a function of the momentum of the track. Below 4 GeV, where 8 mm pads are better, the momentum resolution reaches the multiple scattering limit of 0.3% in dp/p, even for 30 mm pads.

8.   Experimental Evidence of the Advantages of the New Geometry

A measurement of the spatial resolution obtainable with long pads has been done recently with a small TPC equipped with four pad rows. The orientation of the sense wires could be changed with respect to the pads to simulate the local geometry of the ALEPH TPC. The chamber was operated with an argon–methane mixture of 90:10 in a magnetic field of 1.5 T. Particles of momentum larger than 15 GeV parallel to the pad axis ($\beta \simeq 0$) were measured.

The space resolution can be parametrized (with $\beta = 0$) as follows:

$$\sigma^2 = \sigma_0^2 + \sigma_d^2/L + \sigma_1^2 \cos^2 \alpha \, (\tan \alpha - \tan \psi)^2 \ . \tag{8}$$

The measured resolution σ is shown in curve C of Fig. 8 as a function of the angle α. A fit of (8) to these data gives

$$\sigma_0 = (104 \pm 1)\,\mu m$$
$$\sigma_1 = (219 \pm 3)\,\mu m$$
$$\psi = (32 \pm 0.5)° \ .$$

This measurement of ψ is in perfect agreement with the one mentioned earlier, and the ratio of the two results for σ_1 being 0.64 ± 0.06 also agrees very well with the expected value of 0.66 [9].

Another experiment was done with this chamber in order to study the combined influence of the angular effect and the diffusion on the spatial resolution at zero magnetic field ($\psi = 0$). The idea was to determine σ_1, which should not depend on the magnetic field, at a different single-electron diffusion. The average drift length was 4.5 cm, corresponding at B = 0 to a single-electron diffusion of 1.3 mm. Fitting the data with (8) gives $\sigma_1 = 125 \pm 5 \, \mu m$, indicating that the diffusion indeed reduces the angular wire effect. This measured value of σ_1, the measured value of ψ, and the known diffusion allow us to calculate the contribution of the angular wire effect and the diffusion to the resolution, which is about 100 μm, in perfect agreement with the Monte Carlo calculation (Fig. 7).

A single-electron diffusion of 1.3 mm at B = 1.5 T corresponds to a drift of 220 cm, which is just the maximum drift length in the ALEPH TPC. Therefore this value of σ_1 can be used to predict the resolution in the ALEPH TPC for stiff tracks at maximum drift lengths. This is plotted as curve D in Fig. 8.

Many ideas, good understanding, and careful work made the curve A of Fig. 8 come down to curve D. This would not have been possible without the enthusiastic atmosphere in the ALEPH Collaboration and the critical judgement of its spokesman.

References

1. D.R. Nygren: PEP 198 (1975).
   A.R. Clark et al.: PEP 4 Proposal (1976).
2. A. Barbaro-Galtieri: Proc. Int. Conf. on Instrumentation for Colliding Beam Physics, SLAC–250 (1982), p. 46.
3. S.R. Amendolia et al.: Nucl. Instrum. Methods **A235**, 296 (1985).
4. C.K. Hargrove et al.: Nucl. Instrum. Methods **219**, 461 (1984).
5. A. Barbaro-Galtieri: Tracking with the PEP-4 TPC, TPC–LBL–82–24 (1982).
6. S.R. Amendolia et al.: Nucl. Instrum. Methods **217**, 317 (1983).
7. W.A. Wenzel: Choice of pad geometry, TPC–LBL–78–48, BEV–3259 (1978).
8. J. Steinberger: Note on radially long pads and sector geometry, ALEPH–TPC note 95 (1983).
9. J. Steinberger: Ionization, pad length and TPC resolution, ALEPH note 146 (1985).

Forecast on the Future of Gaseous Detectors

G. Charpak

CERN, CH-1211 Genève 23, Switzerland

1. Introduction

Gaseous detectors in their primitive forms as single-wire cylindrical chambers or ionization chambers have played an immensely important role in the development of physics.

With the progress in the understanding of the properties of multiwire structures a great extension in their use was witnessed in the last 15 years. Exploiting the correlation between position and time, and the readout of the signals induced on multi-electrodes by the avalanches on wires, has resulted in a variety of detecting architectures adapted to very different experimental needs. It is this flexibility which has been one major reason for the rapid expansion in the use of drift and multiwire chambers. Needless to say, the considerable development of electronic techniques during that period played a great role.

This book is a good opportunity for me to mention that the enthusiastic support of Jack Steinberger when I started, in 1968 and 1969, to publicize the features of these new detectors, played a great role in their early adoption by several other experimental groups who made major discoveries with these instruments. Jack, in fact, undertook the first ambitious experiments making full use of the high rate capability and position resolution of multiwire chambers and requiring the construction of large surfaces, which at that time appeared as a risky undertaking. While we were making a parallel effort of research with the team building the 'Split Field Magnet' detector, Jack's group introduced original mechanical and electronic solutions which were widely adopted by many other groups who wisely copied their solutions.

While most groups using wire chambers or drift chambers, from the year 1970 on, were satisfied with the characteristics of wire-chamber structures as displayed in the original prototypes which we built in 1968 and 1969, others undertook a systematic research on various aspects of gaseous detectors, resulting in improvements, which, although often not revolutionary in their principle, have been of practical importance. In a non-exhaustive list let me mention the following steps, which are all referenced in a review article [1] or conference proceedings [2].

i) Determination of the accuracy limits of drift chambers with uniform guiding electrical fields at atmospheric pressure, which appears to be around 60 μm at 1 atm.

ii) Systematic study of high-pressure effects in gases. The limit of accuracy due to diffusion is pushed down to 20 μm at 4 atm. The real limit, however, appears to be coming from the geometry of the chambers, and at present for tracks parallel to the anode planes a resolution of 60 μm, at 1 atm, for 1 cm of drift seems a reachable practical goal for many high-accuracy vertex detectors under construction.

iii) At very low pressures, of the order of 1 Torr, a remarkable type of operation appears. Time resolutions below 100 ps are attained and huge gains are obtained before breakdown [3]. In fact, as shown by the work of A. Breskin and his group, the multiplication occurs in two steps: amplification in the region of uniform field, followed by amplification around the wires. While this mode of operation was developed for heavy-ion physics it is very useful for vacuum ultraviolet (VUV) photon detection since it can combine maximum efficiency for Cherenkov photons, or for scintillation photons from adequate crystals, with a negligible efficiency for charged particles.

iv) The centroid of the avalanches along the wires can be determined with a considerable accuracy and, for minimum ionizing particles orthogonal to the chambers, reaches 50 μm. This method finds its broadest application in the readout of the end caps of Time Projection Chambers (TPCs).

v) The existence of a new mode of operation, called limited Geiger or limited streamer, was found, with very important applications for large-surface detectors, since it permits a considerable increase in the gaseous gain and a corresponding reduction in the electronics cost.

vi) The conditions for multistep operation of gaseous detectors have been brought to light. The initial ionization may be amplified between parallel grids and a substantial fraction of the electrons of the avalanches can be partially transferred to following amplifying structures, which may consist of another gap between parallel electrodes or wire planes, or even spark chambers. This adds a new flexibility, with important consequences in some applications such as single-photoelectron counting or high rates.

vii) The mechanism of the avalanches around wires is fully clarified and it has been shown that it is possible to obtain, in some cases, the azimuth of the avalanches. This transforms the wire structure into a detector with a two-dimensional continuous response for X-rays and with a possibility, using the pulses induced by the avalanches initiated by the ionization clusters along a track, separated in time by a proper reduction of the drift fields, to overcome the geometry errors in drift chambers and reach the ultimate limit. This approach, suggested by Walenta, is so far not used in an experiment, but this limit shows why, despite the considerable efforts accomplished in the last ten years, it is likely that the gaseous detector may be superseded, for some specific applications, by solid-state detectors [1], which permit, in various arrangements, to reach accuracies of the order of 5 μm.

I have skipped a considerable amount of other improvements as well as original developments in the architecture of detectors such as the TPC, JADE, the UA1 visual detector, and high-density drift spaces [2]. I want to limit this contribution to a few lines of research which may be of importance for the future:

i) The detection of VUV photons with photo-ionizable vapours; applications for Cherenkov ring imaging; applications for the detection of photons from heavy scintillators.

ii) The optical imaging of photons emitted by avalanches in multistep parallel-plate counters of wire chambers permitting any complexity of particle configuration to be coped with.

iii) New structures with a high density of position-measuring wires, where the limited accuracy of gaseous detectors is partly overcome by having a great number of measuring points along a track.

2. Detection of UV Photons in Gaseous Detectors

It was the merit of J. Seguinot and T. Ypsilantis [2] to find the way to use wire chambers as VUV photon detectors. They showed that by adding vapours with a low ionization potential, wire chambers can detect efficiently and localize VUV photons within a spectrum range limited by the transparency of windows and the photo-ionization threshold. Figure 1 shows the properties of common windows and vapours which are typical in this application. This revolutionized the art of Cherenkov detectors, which were confined to the use of photomultipliers. The efficiency of gaseous detectors can be, at some wavelengths, superior to that of photocathodes, with the considerable advantages presented by wire chambers, namely high position resolution and large surfaces.

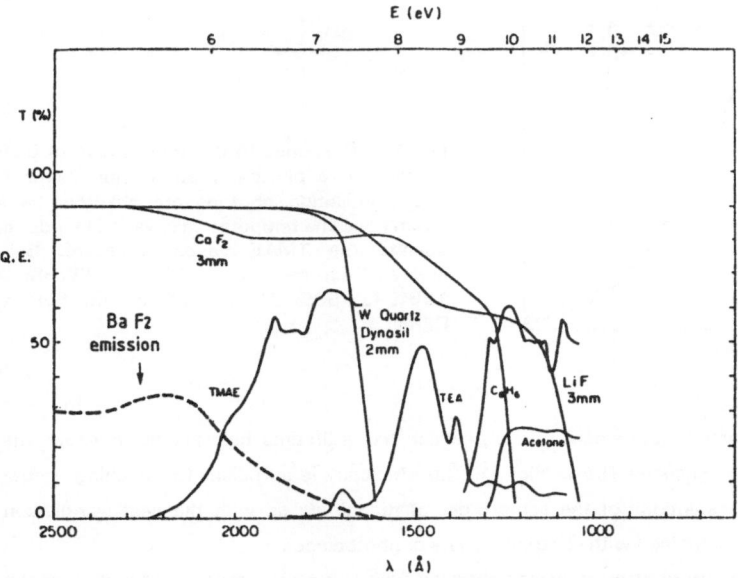

Fig. 1 Quantum efficiency of various vapours. Transparency of various windows. Spectrum of scintillation photon emission from BaF$_2$

The detection of photons can occur in various structures: drift chambers, multiwire chambers, multistep chambers. A surface of the order of 100 m^2 is foreseen for the DELPHI detector envisaged for the Large Electron Positron Storage Ring (LEP). The choice of the structure has to be adapted to such conditions as the expected counting rate or background, the knowledge or not of the production time for an event, and the multiplicity. Ring imaging Cherenkov counters practically insensitive to charged particles, owing to the very low pressure of the gas filling, are being developed [4]. They can stand high rates and high multiplicities. But the development which may in the future have the greatest importance is the combination of a heavy scintillator, BaF$_2$, with a multiwire chamber [5]. D. Anderson, while working in our group at CERN, has shown that the photons emitted by BaF$_2$ can be detected by a wire chamber containing tetrakis(dimethylamine)ethylene vapours. Although his starting point was an attempt to make a condensed liquid photocathode where the ionization potential is expected to be in the region of 4.1 eV, against 5.36 eV for the gaseous state, the happy fact is that the spectrum of the

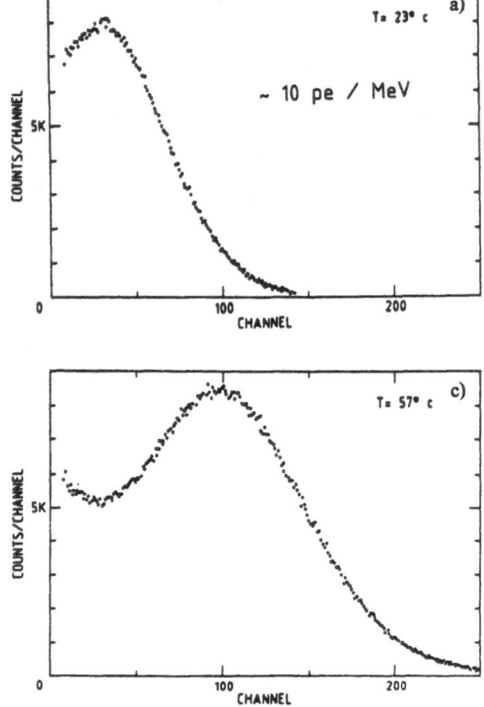

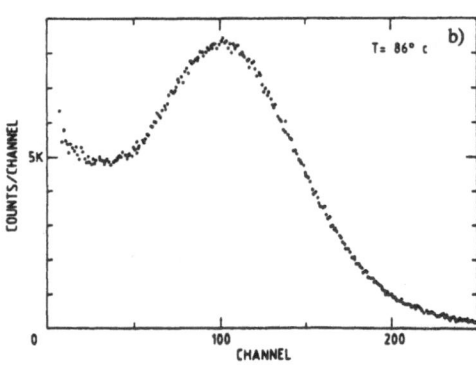

Fig. 2 Response to 0.6 MeV γ-rays of BaF$_2$ coupled to a photo-ionization wire chamber. The scintillation photons are absorbed in a 3 mm gap. The temperature is varied in order to change the TMAE vapour pressure. BaF$_2$ crystal: 5 cm thick, $\varnothing$ = 12.5 cm. FWHM = 125% (a), 90% (b), and 80% (c). (M. Suffert, CERN)

photons emitted by the fastest component of BaF$_2$, which has a lifetime below 1 ns, overlaps the spectral sensitivity curve of gaseous TMAE (Fig. 1). The efficiency is sufficient for reaching, in the region of 1 GeV, energy resolutions of the same order of magnitude as with the best scintillation crystals such as NaI or CsI, combined with photomultipliers or photodiodes.

Figure 2 shows the spectrum of pulse height obtained with 0.6 MeV γ-rays as a function of the temperature of a gaseous detector made of a 3 mm thickness of conversion gap, where the TMAE vapour is in equilibrium with the liquid. This spectrum illustrates why several groups are contemplating the use of BaF$_2$ crystals for medical imaging cameras, based on the positron annihilation radiation, where it will be possible to combine an efficiency close to 100%, with a time resolution of the order of 1 ns, and a localization accuracy depending on the crystal size, with all the advantages of wire chambers. Most existing cameras require hundreds or even thousands of photomultipliers.

In high-energy physics the advantage of calorimeters using the combination of BaF$_2$ crystals and wire chambers is that a much greater granularity can be reached than in the usual heavy crystal calorimeters, with the possibility of following neighbouring tracks or the pattern of shower development with an accuracy that could not be dreamed of with conventional methods of readout, based on photodiodes or photomultipliers.

Figure 3 illustrates some properties of such a calorimeter tested up to 200 MeV.

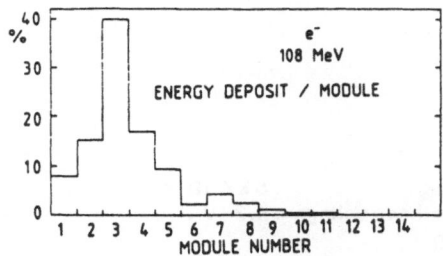 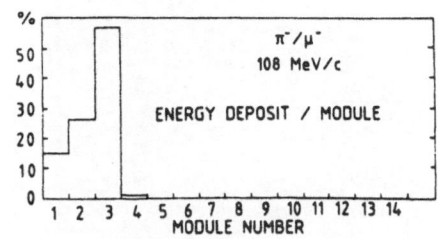

Fig. 3   Energy deposit for 108 MeV/c electrons and 108 MeV/c μ^- or π^- (Ref. [5])

3.   Optical Imaging of Photons from Avalanches in Gaseous Detectors

While the observation of photons from avalanches has for a long time given rise to a special class of gaseous detectors, the optical localization of these avalanches is a new approach with many intriguing applications. The first attempts were aimed at localizing the position of avalanches produced by X-rays in a parallel grid gap [6]. Tracks have also been observed in a high-pressure gap [7] combined with a gaseous amplification. In both cases the avalanches were produced against the fibre optics window of an image intensifier, limiting the imaging to very small surfaces. With the advent of multistep chambers a new area of application was opened since such structures permit higher gains than single-step parallel-plate or wire structures. R.S. Gilmore et al. [8] tried to image avalanches produced by single photons from Cherenkov radiation. This has the evident advantage of being compatible with any multiplicity. It demonstrated clearly that single-electron avalanches can be observed with a conventional lens optics followed by an image intensifier. This was obtained with a filling of Ar + acetone in a structure made of two successive amplifying parallel-grid gaps.

Repeating this approach we could not obtain enough intensity without reaching breakdown. We have then concentrated our efforts on finding structures and gases giving the maximum amount of light and compatible with large surfaces. We found that triethylamine (TEA) vapour is the ideal component so far and it copiously emits photons when mixed with any noble gas, in a region peaked at 300 nm, traversing easily a cheap plastic window made of Aclar. We also found that single-gap wire chambers are sufficient for an intense yield of light if thick wires of a diameter between 35 and 60 μm are used. These features permit the construction of very large surfaces with the following obvious applications.

i)   Double β-decay of ^{136}Xe.

ii)   Detection of VUV photons from Cherenkov radiators. In cases where the number of expected photoelectrons in a ring is considerable, as with heavy ions, this may be the only possible method.

iii)   Detection of the electrons from dense drift spaces similar to the method developed for the DELPHI hadron calorimeters. The advantage of optical readout is that it is compatible with any complexity of the ionization pattern.

It is possible to retrieve the information necessary for the three-dimensional reconstruction by measuring the arrival time of the pulses on wires, strips, and pads. The ambiguities plaguing the purely electronic retrieval are to a large extent eliminated by this method. It may be that the combination with a dense drift space can be contemplated for a future ideal proton decay detector of huge dimensions and optimum granularity. Figure 4 gives a few images obtained with cosmic rays in a drift space filled with gas or with a high-density structure.

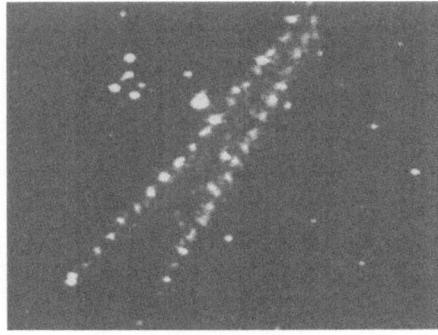

a)

Drift voltage	: 5 kV/10 cm
Anode–cathode voltage	: 4.2 kV
Gas filling	
Ar	: 47%
Isobutane	: 47%
TEA	: 6%

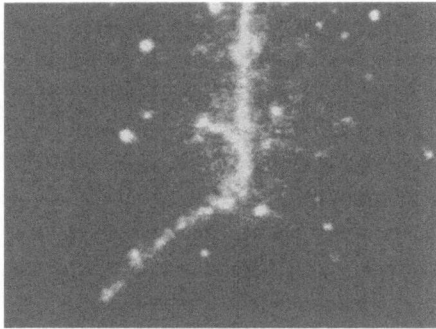

b)

Drift voltage	: 5 kV/10 cm
Anode–cathode voltage	: 4.2 kV
Gas filling	
Ar	: 47%
Isobutane	: 47%
TEA	: 6%

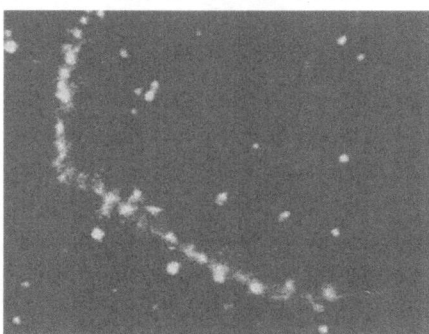

c)

Dense converter	: 2.5 cm
Drift voltage	: 1 kV/2.5 cm
Anode–cathode voltage	: 4.8 kV
Gas filling	
Henogal	: 40%
Isobutane	: 54%
TEA	: 6%

Fig.4　Images from cosmic rays and local radioactivity. a) and b) A drift volume of $10 \times 10 \times 10 \text{ cm}^3$ is filled with Ar + isobutane + TEA. c) Cosmic-ray event in a dense drift space made of copper grids with 3 mm square holes separated by insulating grids. Depth of drift: 2.5 cm. Gas filling: henogal + isobutane + TEA. The electrons drift to an end-cap wire chamber, with wires of $\varnothing = 35 \, \mu m$ and 4 mm spacing. (M. Suzuki et al., CERN, unpublished.) Images obtained with an UV-to-visible light converter (Proxitronic, BV 25329615) of 25 mm diameter, of gain 10, followed by a VARO electrostatic double stage of gain 5000, viewed by a Thomson video camera of sensitivity equivalent to 10^5 ASA

4. Multiwire Tubes

The reliability of detectors is a growing concern for the future since the luminosity of accelerators is going to increase and since each experiment is going to be of considerable complexity and cost, requiring — when one element fails — to stop the accelerator for repairs to the detector; gaseous detectors have particular reasons for ageing or failing. Leaving aside the nature of the gas, which is of paramount importance for the maximum possible integrated rate of a chamber and has been subject to many studies, the rupture of a wire is often a serious problem since it may wreck a whole wire plane or a sizeable fraction of a large-volume detector. This is one reason why tubes with a single wire are an attractive solution. In some cases, such as for large-surface muon counters, the matter introduced by the tube walls is of no relevance. The same is true for proton decay calorimeters. The only problem then is to bring down the cost of the tube production. For particle tracking, where a low density of matter is required, a step in this direction is observed at SLAC with the so-called MAC detector [9], where single thin tubes of aluminized mylar, of 7 mm diameter, are equipped with one amplifying wire. Measuring the drift-time gives surprisingly good accuracies and, in practice, for the whole system, an accuracy of 60 μm per point is obtained. We have in our group undertaken a more ambitious and more difficult approach [10]. In hexagonal carbon tubes of 0.4 mm thickness, 26 mm width, and 80 cm length, we fit 128 anode wires of 30 μm diameter and 378 cathode wires of 70 μm diameter. We thus have 128 independent measuring cells and we measure the drift-time of the ionization electrons, which have a maximum path length of only 1 mm. We can expect, a two-particle resolution of 500 μm. The spatial resolution per point should be around 100 μm, with about 12 measured points for a track segment of 26 mm length. The considerable advantage of this approach should be the reliability since the tubes can be readily changed and a few dead cells in a tube are of small importance for the system. We have traded many problems against a formidable mechanical problem which, after the first tests, does not

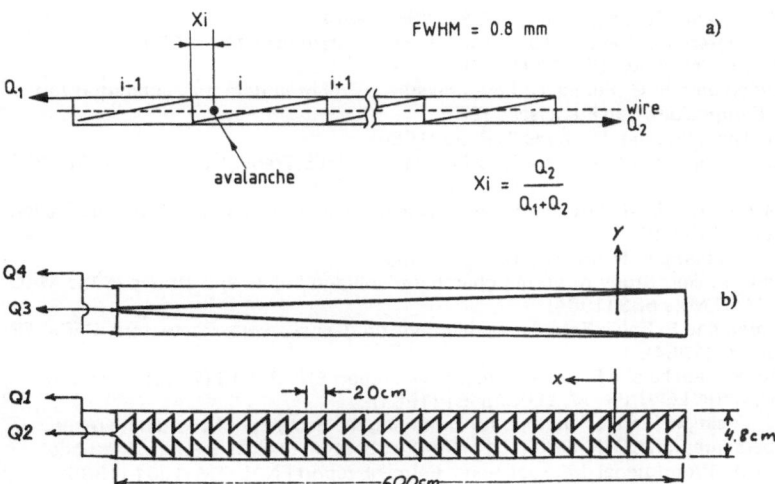

Fig. 5 Vernier methods for high-accuracy readout along long anode wires. The signals are induced on printed cathodes made of successive triangles (a); the ratio permits a high local accuracy, while gross methods permit the localization of the triangles (Refs. [13, 14]). Accuracies of 1 mm are obtained in tubes of 6 m length at Fermilab, with printed cathodes shaped as shown in (b)

seem to us impossible to master. Making detectors by using prefabricated tubes, easy to interchange, is a tempting solution.

Also worth mentioning for the future are a few simple developments which may be of importance. The maximum counting rate of wire chambers has been seriously increased by using rapid gases, made of CF_4 mixed with isobutane [11], and an anode–cathode gap smaller than 1 mm. A resolution time of 5 ns and a counting rate of $10^7/cm^2$ are obtained. This drastic reduction of gap is also now operational in a large calorimeter [12], where it permits a reduction in the longitudinal size.

The capability of gaseous detectors to offer the only alternative for high-accuracy detectors of a surface of the order of acres is illustrated by a recent simple progress. While long drift tubes with a single wire have for a long time been a popular solution for high-accuracy large-surface muon detectors, it has always been difficult to obtain simply the coordinate of avalanches along the wires which are so perfectly localized. Current division, in this respect, gives a very poor resolution for long tubes. By developing a 'vernier' method, based on pulses induced on triangular cathodes (Fig. 5), this problem is easily solved for moderate counting rates, and a resolution of 1 mm over a length of 6 m has been obtained for a detector being built at Fermilab [13, 14].

5. Conclusions

The few developments described in this article show that in the future gaseous detectors may well play an ever increasing role, especially in the detection and localization of VUV photons emitted by Cherenkov radiators or by a heavy scintillator such as BaF_2.

As particle tracking detectors a steady improvement in their reliability can still be expected and, except in the region of accuracy better than 50 μm, they provide, so far, an unmatched flexibility at low cost.

References

1. G. Charpak and F. Sauli: Ann. Rev. Nucl. Part. Sci. **34**, 285 (1984).
2. Proc. Wire Chamber Conference, Vienna, 1983 [Nucl. Instrum. Methods **217** (1983)].
3. A. Breskin: Nucl. Instrum. Methods **196**, 11 (1982).
4. S. Majewski, D. Anderson and G. Founakis: Low-pressure UV photon detector, submitted to the Nuclear Science Symposium, San Francisco, 1985.
5. D.F. Anderson et al.: Nucl. Instrum. Methods **228**, 33 (1984).
6. O.H.W. Siegmund, S. Clothier, J.L. Culhane and I.M. Mason: IEEE Trans. Nucl. Sci. **NS-30**, 350 (1983).
7. D.M. Potter: Optical readout of a high-resolution parallel plate avalanche chamber, Carnegie-Mellon University report CMU-HEP 83–14 (1983).
8. R.S. Gilmore et al.: Nucl. Instrum. Methods **206**, 189 (1983).
9. G.B. Chadwick and F. Muller: Study of vertex chamber resolution in jets and straws in the MAC geometry, SLAC MAC Note 683 (1984).
10. R. Bouclier, G. Charpak and F. Sauli: Some considerations on gaseous vertex detectors, CERN–EP Internal Report 84–03 (1984).
11. J. Fisher, A. Hrisoho, V. Radeka and P. Rehak: Brookhaven report BNL 35819 (1985).
12. OPAL Status Report, CERN LEPC/84–17, LEPC/m–51 (1984).
13. D.F. Anderson et al.: A simple 'vernier' method for improving the accuracy of coordinate readout in large wire chambers, preprint CERN–EP/84–01 (1984), submitted to Nucl. Instrum. Methods.
14. C. Brown et al.: E–740 – Proportional drift tube tests, Fermilab report FNAL–TM–1301 (1985).

The Involvement of Scientists in National Security Issues

S.D. Drell

Stanford Linear Accelerator Center, Stanford, CA 94305, USA

Dedicated to Jack Steinberger - a superb scientist and close colleague with whom I have enjoyed so many valuable discussions of physics through the years, and also more recently, on the subject of the essay.

Twenty-six years ago I was busy minding my own business--probably trying to understand a dispersion relation--when the then Executive Vice President of The Institute for Defense Analyses, Charles Townes, called early one evening to invite me to Washington to attend a briefing for a group of then bright and young physicists. As I remember, Charlie said this was an organization meeting planned to bring fresh and good young brainpower together to work on problems of national importance for our security in the still very young age of nuclear weapons, space, and intercontinental missiles. It was the genesis of JASON.

Unquestionably I was flattered, but I also was very dubious. I had no spare time. Doing my job as a physics professor, enjoying a family with two small children, and trying like hell to keep up with the exciting new data obtained by experimenters like Jack who had seen the light - and the original triangle anomaly - a decade earlier and had defected from the ranks of the theorists, as well as trying to keep up with all my very smart theoretical colleagues, already was requiring more than 24 hours a day. But the very <u>fact</u> of the   phone call from the very distinguished Vice President Townes, plus President Townes, plus his cogent arguments caused me to pause and think. So I talked with my colleague and close friend, Pief Panofsky, whose great personal commitment and enormously valuable contributions to U.S. national

security and sound policy I admired so much; and I thought about the value
and even the need for a new and younger group to add their talents to
those of the veterans of Los Alamos and the Manhattan Project, and the
MIT Radiation Laboratory who had already contributed so much. No one,
after all, was growing younger!

Encouraged and inspired by Pief and other similar models whose names
we all know, I went to the first JASON briefing that winter. I found the
problems interesting--at least some of them. I also found the arguments
for involvement compelling, and the conditions of broad access to
information, without industrial priority restraints and with very strong
staff support, attractive. So I agreed to taste the experience of being a
JASON, and I spent several months that summer working in one of the
temporary buildings on the hill at the Berkeley Rad Lab.

As luck would have it Mal Ruderman and I collaborated there on a most
interesting problem of calculating what happens in the upper atmosphere
following a high altitude nuclear explosion. We wanted to determine how
much NO was formed by the various collisions and reactions that occur in a
layer of hot air, because NO, a non-homopolar molecule with a dipole
moment, radiates in the IR region. That radiation was what we wanted to
calculate. This problem was of interest then as a possible countermeasure
to IR early warning satellites--it is of no less interest now as we
grapple with some of the technical realities of ABM, specifically target
discrimination and X-ray propagation.

The problem of IR radiation from hot air caused by a high altitude
nuclear explosion was fun physics--really fun--but, more than that, this
work introduced me to a number of technical issues and practical problems
that, today, are still central to thinking about nuclear weapons, arms
control, and improved national security. That first summer of work and
what I learned the first few years thereafter opened my eyes to a number

of issues that now seem more or less obvious, but were then quite new--at
least to me.

First there is the importance of implementing President Eisenhower's
goal of "open skies" with effective, reliable, and mutually acceptable and
non-provocative means. The better our early warning, the less trigger
happy or nervous we will be about sudden attacks--especially in time of
crisis. The better our reconnaissance and surveillance, the more
accurately we can gauge the threat posed by our opponents--particularly
the Soviet Union--and therefore the more rationally and responsibly we can
do our national defense planning. Moreover, and this I would emphasize,
the more we can find out, the more possibilities we can consider for
verifiable arms limitations. Remember that back around 1960 we were still
overflying with U-2's and the era of space-based sensors was just in its
infancy. In the 1950s we had spent huge sums on a very large, cumbersome,
and unnecessary air defense against a presumably enormous, but in reality
non-existent, Soviet bomber threat; and the fictitious missile gap played
center stage in the Presidential election of 1960. So much for the
virtues or bliss of ignorance. Open skies was and remains crucial! The
strategic arms negotiations of the 1970's and 1980's were made possible by
the development during the 1960's of the National Technical Means of
surveillance that gave us open skies.

Secondly, there was the need for making the best technical support
available in the Government, and especially to the White House, directly
to the President. Presidential decisions are fundamentally and most
importantly political ones, but in the area of national security and
military planning they do have crucial technical ingredients which define
what can and what cannot be done, and what should and what should not be
done, both for weapons systems and for policy, including arms control; and
the information available to the President on these technical ingredients

51

should be informed, accurate, and unbiased. Moreover, there is also need for the best projections as to what new technologies to bet on with priority support because of their future potential, and which ones to deemphasize or terminate because their promise is poor. President Eisenhower understood this need at the time of Sputnik when it became clear that we were facing a technologically astute adversary in the Soviet Union in an era of rapidly changing technology, particularly of nuclear weapons, missiles, and uses of space. He therefore activated a President's Science Advisory Committee (PSAC) to give him directly at Presidential level the best direct and balanced technical staff support on which he could rely with confidence.

It is no insult to, or lack of respect for, defense scientists with operational responsibilities in the government labs and executive departments and agencies to emphasize the importance of independent scientific staff advice to the leaders who have to make the decisions, including at the highest levels. When anyone has direct operational responsibilities and is very close to one part of a problem, it is very difficult to step back and see all aspects from a balanced viewpoint. Aren't we all familiar with the same problem back home in our own academic and research institutes? Don't we often find the strangest parallaxes creeping in to the ways we, or more frequently our colleagues, needless to say, view issues of central importance to our own institutions! We're all human, and a mechanism that can offer an independent, authoritative, balanced perspective--and which at the same time resists the temptations of micro-management--can be of immense value. This is particularly so when technologies are changing so rapidly--and when the gap is so great between the realities of science and technology and our government leaders. This circumstance was lamented by former British Prime Minister Harold Macmillan who wrote in his 1972 book Pointing the Way: "In all

52

these affairs Prime Ministers, Ministers of Defense and Cabinets are under a great handicap. The technicalities and uncertainties of the sophisticated weapons which they have to authorize are out of the range of normal experience. There is today a far greater gap between their own knowledge and the expert advice which they receive than there ever has been in the history of war."

Scientists have of course long been involved as experts on issues of weapons and war: Archimedes of Syracuse; Leonardo da Vinci as probably the greatest military scientist of his time in Milan; and Michelangelo as engineer-in-chief of the fortifications of Florence. But never has the gap been as great as it is now--nor has there been so precious little room for failure as in our present condition with "weapons of absolute destruction" in the words of C.P. Snow. This fact poses a special burden on Scientists!

And there were also the travails of JASON membership during the Vietnam years.

We have all learned by now of the dilemma faced by scientists who leave their ivory towers and laboratories, where we have been trained to study fixed and rational laws of nature, to enter as technical participants into an outer world with shifting and often apparently irrational laws of politics and society. On the grand scale, the sad tale of the fall of Robert Oppenheimer from hero to a personally devastated so-called security risk--and more recently the tragic saga of Andrei Sakharov--are well known and are part of our culture as physicists. But in our own much more limited ways we JASONs also tasted similar experiences during Vietnam in the early 1970s. Whether or not we wanted to, we had to settle in our own minds some of the deep reasons for JASON's existence and for our own personal involvement.

On many campuses, working for a Defense Department that was doing all
those awful things in Southeast Asia was de facto evidence of evil doings.
With varying intensity we were personally attacked—some even thrown out
of classrooms in Europe—for "supping with the devil". Some Jasons who
worked hardest felt bad, if not betrayed, by the uses made of their
contributions to the electronic battlefield in Vietnam—much as Sakharov
during the 1960s grew to oppose the nuclear weapons policies and continued
atmospheric testing by the Soviet government for which, twenty years
earlier, he had worked so hard and contributed so importantly in order to
insure that they could build the H-bomb. These experiences and unhappy
feelings were by no means universal during Vietnam, but all of us felt the
strain of having to keep part of our lives secret and off bounds to
discussions and questions while working with our students and colleagues
in the very open scientific world in which we enjoy the full and open
exchange of ideas and the public scrutiny that is traditional and
cherished in our purely scientific activities. This bifurcation in our
lives doesn't cause much of a strain and is not really an issue during
peaceful times, or times of national consensus—but when that consensus
breaks down it can be a very heavy burden.

During those days of confrontation and challenge it was ever more
clear to me how unhealthy it is for a society when its intellectuals and
its leaders are split into two warring camps: the intellectual-scientist-
academic critics on the outside, and the governmental decision makers on
the inside.

I emphasize this because we are perilously approaching that condition
once again in the current debate over nuclear arms control and Star Wars.

In his famous "Star Wars" speech of March 23, 1983, President Reagan
challenged the nuclear scientists "who gave us nuclear weapons" to "give
us the means or rendering these nuclear weapons impotent and obsolete."

54

We were told by the ardent Star Wars supporters, including the President, how much more moral it would be to replace deterrence based on the threat of nuclear annihilation. How do we respond to this challenge which has sparked a heated – and occasionally acrimonious debate within the scientific community?

President Reagan's call for a shift from deterrence based on the mutual fear of nuclear retaliation to reliance on defense against nuclear annihilation has strong appeal. The impulse to look to our weapons and armed forces to defend us, rather than threaten others, is a natural one and is deeply ingrained with long historical precedent in the era before nuclear weapons. Throughout history, military analysts have grappled with the role of offense versus defense. Many instances in the past can be cited where defense has proved decisive to the outcome of combat. However today, when we face the staggering destructive power of nuclear weapons, when just one single, relatively small nuclear bomb is a weapon of mass destruction, we must recognize that effective defenses must meet a very much higher standard of performance than at any previous time in the history of warfare. A 10 percent defense such as won the Battle of Britain – or even a 90 percent defense – against today's threat of almost 10,000 strategic nuclear warheads cannot protect a nation from nuclear annihilation. So profound has been the change in the destructiveness since the advent of nuclear weapons that simplistic historical comparisons can be, and often are, misleading. I cite this to emphasize that it is important to look at the practical side of the issue of defense versus offense under the shadow of current nuclear arsenals, now that the President's call for the **"means of rendering these nuclear weapons impotent and obsolete"** has rekindled a debate on defense that has been relatively dormant since the ABM Treaty of SALT I was ratified in 1972.

The ABM Treaty of 1972 is at the base of efforts to approach stability, avoid nuclear war, and achieve arms reductions. It is the formal recognition that mutual destruction **could not be escaped** if the superpowers were drawn by accident or design into nuclear war.

In its broadest dimension the ABM Treaty is an early milestone of the **political approach** to avoiding nuclear war and to providing hope and reassurance that, despite the East-West confrontation and the existence of nuclear weapons, nations and their leaders can act to keep nuclear war from erupting.

The technical and political realities of 1972 led the United States and the Soviet Union to endorse the basic principles of deterrence and stability and to negotiate limitations on ABM.

Recall what President Nixon said in explaining his decision on March 14, 1969, to forego a broad defense of the nation in favor of the limited Safeguard ABM system primarily to defend U.S. retaliatory forces:

"Although every instinct motivates me to provide the American people with complete protection against a major nuclear attack (emphasis added), **it is not now within our power to do so. The heaviest defense system we considered, one designed to protect our major cities, still could not prevent a catastrophic level of U.S. fatalities from a deliberate all-out Soviet attack. And it might look to an opponent like the prelude to an offensive strategy threatening the Soviet deterrent."**

But now in 1985 the whole approach has been challenged. The President's Star Wars initiative raises fundamental issues regarding U.S. strategic policy that go to the heart of the superpowers' strategic relationship.

This applies both to Star Wars I, which was advocated to escape from, or to transcend, deterrence by achieving a totally effective defense against strategic nuclear weapons as called for by the President; and also to the more modest goal of enhancing deterrence through an effective, but partial, nationwide defense that is now emphasized by the Defense Department. I am referring here to Star Wars II, the Strategic Defense Initiative (SDI) program as submitted to the Congress in April, 1985 by the Defense Department, which states:

> **"In pursuing strategic defenses, the U.S. goal has never been to eventually give up the policy of deterrence. With defenses, the U.S. seeks not to replace deterrence, but to enhance it."**

What has changed since the ABM Treaty was signed in 1972 to suggest that it is now practical to consider the prospect of nationwide defenses? Do the technical advances of the 13 years, as prodigious as they have been, offer us new and better choices for dealing with the nuclear threat in 1985?

There have indeed been tremendous advances in recent years in the technology pertinent to this problem that have removed some of the shortcomings of previous defense concepts. These include, in particular:

. sensors for finding, identifying, and following targets;

. the means of producing directed energy beams of high power that travel as accurate bullets at the speed of light; and

. the ability to gather promptly, process rapidly, and transmit reliably enormous quantities of data for the purpose of battle management.

While these developments represent advances in specific technologies pertinent to strategic defense, the challenge to build an effective nationwide defense involves much more than the solution of individual technical problems.

It is difficult for a responsible scientist to say flatly that a task is impossible to achieve by technical means without being accused of being a "naysayer." Indeed, many instances can be cited in which prominent scientists have concluded that a task is impossible only to be proved wrong by future discoveries. One should recognize, however, that the deployment of an effective nationwide defense is not a single technical achievement but the evolution of an extensive and exceedingly complex **system**; a system built together out of many links to form a defensive chain. Each link is crucial for protection against ballistic missiles. There must be sensors to provide early warning of an attack. There must be a command structure with the authority to make decisions on committing the defensive forces on exceedingly short notice and then to implement those decisions efficiently. This raises the grave problem of whether this chain of command, from warning to decision, must be **totally** automated, or may contain human links. The defensive system must also have sensors that acquire and track the enemy missiles and warheads, and then aim and fire the defensive devices, be they material interceptors such as chemical rockets or hypervelocity guns, or directed-energy devices. Sensors must also determine which of the attacking missiles and warheads have been destroyed in order to fight secondary or tertiary engagements successfully.

Furthermore, this system must work reliably in a hostile environment against a determined and uncooperative opponent, dedicated to defeating it with countermeasures to bypass, overcome, attack, or destroy it.

Because of these systems' issues, and in the face of countermeasures, there will still remain great operational problems in deploying nationwide defenses even if the expensive R&D program of the Administration achieves all of its ambitious technical goals. Furthermore, we must be able to maintain high confidence in such a system and its ability to accomplish the enormous task of battle management in a very short time, although it can never be tested under realistic conditions such as in an environment disturbed by nuclear explosions. To anyone who has experienced the turn-one of a high-energy accelerator, such as at SLAC, CERN, DESY or Fermilab, this appears to be not only an enormous, but a preposterous requirement. It calls to mind the claim of magic by Glendower in Shakespeare's HENRY IV, Part One, "I can call spirits from the vasty deep", to which Hotspur responds: "Why, so can I; or so can any man, but will they come, when you do call for them?"

One cannot compare these awesome requirements with those faced by the Apollo program to put a man on the moon - an analogy which has been drawn to point to other great challenges which have been met by science and technology. Putting a man on the moon was solely a technical challenge. The moon couldn't shoot back, or run away, or dispense moon decoys, or turn off its lights. Even so, recall how many lengthy holds we witnessed during the countdowns for the moon shots - or, for that matter, for the space shuttle flights. Such holds will not do for a defensive system which has barely seconds to spare, and no privilege of saying, "Hold your attack. I'm not ready. Please wait a few minutes—or hours!"

Supporters recognize these difficulties and propose to attain a highly effective nationwide defense by deploying a multi-layered system. Up until now, we have thought of defense in terms of intercepting the many attackers. - warheads and cruise missiles - as they arrive near their targets. In contrast with previous concepts of terminal defense, the crucial and necessary new ingredient of Star Wars is to be the first of three or four layers of defense that is designed to destroy most of the missiles over the enemy's own territory within seconds after they are launched, i.e., during their boost phase while their engines are still accelerating them into space. In the language of football, this layer corresponds to tackling the quarterback before he releases the pass rather than preventing the pass from being completed by covering the receivers. As stated in the 1985 report to Congress by the SDI office:

"Ability to effectively respond to an unconstrained threat is, therefore, strongly dependent on the viability of a boost-phase intercept system.... requiring 10 or more kills per second...."

Suffice it to say that as interesting or intriguing as the concept of boost-phase interception may be, its prospects, whatever they may be, lie well in the future, orders of magnitude beyond today's technology and sensitive to countermeasures available to the offense. In particular, a boost-phase defense must complete its mission in seconds. At most a few minutes are available to engage today's missiles during boost, but the technology to shorten the total burn time to less than one minute already exists. This means that the interceptors will have to be pre-deployed in space, or they will have to get there very, very quickly. If they are in space, their numbers must necessarily be very large, since only a small fraction of them will be on-station within range of Soviet launch areas at any one time. And they themselves will be more vulnerable than the missiles that are their targets, as they move in predictable orbits like ducks in a shooting gallery.

These disadvantages of space basing have suggested that we look to the possibility that the interceptors be based on the ground, poised to pop-up into space upon notification of enemy launch. The only practical device, compact and light enough for doing that, is the x-ray laser pumped by a nuclear explosive. There is no law of physics which says this cannot be done, but at present our physical understanding of the potential of such devices is much too primitive to draw any conclusions.

Moreover, we should remember that the laws of physics are not secret. They cannot be classified, and the possibilities of such lasers are well recognized by Soviet scientists. They have discussed these concepts in their literature as have American scientists in ours. As we well know from experience, they will build such devices if we do, probably not very

60

long after we start, and surely long before we have a full operational system of our own. It is ironic, but not surprising, that this widely heralded technology which first triggered so much interest in Star Wars is now being de-emphasized by the Administration's stated goal of developing a non-nuclear defensive system. In fact, its most practical application could well be as an anti-satellite weapon against an opponent's Star Wars defenses, should we go ahead.

Other schemes for accomplishing boost-phase intercept have been proposed, but all share one or both of the difficulties of having potentially vulnerable components - for example, large orbiting mirrors or particle accelerators - based in space; or having very little, if any time for accomplishing their intercept. In the face of such severe technical plus operational difficulties that must be mastered and cannot be avoided, **I see no prospect of building an effective nationwide defense now or of achieving one in the foreseeable future, unless the offensive threat is tightly constrained technically and greatly reduced numerically as a result of major progress in the U.S.-Soviet political process and dialogue.**

Although one occasionally hears excessive claims to the contrary, I know of no informed analysts who differ from the conclusion that we must have constraints on the offensive nuclear forces, if defensive deployments can hope to be effective. I agree with Dr. Richard DeLauer, Undersecretary of Defense for Research and Engineering during the first Reagan Administration, who said, **"With unconstrained proliferation, no defensive system will work."** It follows that we must achieve some success in the arms control negotiations before any defensive system can become workable.

The challenge of Star Wars, then, is not only a technical one, nor even an awesome operational one. It is also, perhaps foremost, a political challenge to succeed in limiting offensive forces while simultaneously developing defenses to stop them. But, is this possible or likely? Will the U.S. or the Soviet Union be willing to do this in a Star Wars environment? Would we be willing to accept limits or to consider reductions in the U.S. retaliatory capability, if we face the growing Soviet deployment of missile defenses which is sure to follow our own program? On the contrary, can't we expect a buildup of offensive weapons on each side as a counter to the other nation'Star Wars efforts, much as the U.S. undertook a major buildup of our offensive forces developing and deploying MIRVs in response to the initial limited deployment of Soviet anti-ballistic missile systems in the 1960s. The Russians followed suit, and now their large MIRVd ICBM force and its looming threat to the U.S. Minuteman silos has become the focus of instability in the 1980s. How can we hope to avoid a repetition of that pattern now? Only if we retain and strengthen our framework for arms control – and in particular for the ABM Treaty by resolving difficult compliance issues (viz. the large phased array radar that the Soviet Union is building near Krasnoyarsk in Siberia in apparent violation of the SALT I ABM Treaty).

The importance of this arms control framework is apparent as one looks ahead toward possible deployments of defensive systems. As Paul Nitze remarked in a speech to the Philadelphia World Affairs Council on February 20, 1985, if the Star Wars research program of the Administration is fully successful in developing new technologies that satisfy the two criteria of survivability and cost-effectiveness, then a decade or more from now, we can look forward to entering a transition period during which defensive weapons will begin to be deployed in conjunction with offensive forces. This, he said, would be a "tricky" enterprise:

> "We would see the transition period as a cooperative endeavor with
> the Soviets. Arms control would play a critical role."

62

Evidently, we will require an arms control regime in order to get anywhere with defensive deployments. Thus, the strategic defense programs in the coming decade must not contribute to dismantling our current treaty achievements.

In the face of these realities, what actions should the United States be taking now and during the coming decade to enhance our security and to reinforce stability, given the current state of the U.S.-Soviet competition and nuclear arsenals, and their likely trends? I recommend six actions which would meet our national security needs, including improving our prospects for success in arms negotiations:

1. Reaffirm and strengthen existing treaty commitments.

The ABM Treaty is of value for structuring our offensive forces and nothing that has happened since 1972 has changed its central pivotal role as a basis for common U.S.-Soviet efforts to avoid war. The ABM Treaty, of course, cannot be viewed in isolation of other issues in the U.S.-Soviet relationship. There are real concerns on both sides about each other's threatening behavior and these cannot be simply dismissed as unfounded or irrational. Both nations need some assurance, either through treaty or by displayed restraint, that their adversary is not working toward achieving a strategic advantage with nuclear weapons or a break-out from treaty provisions with their ongoing programs. As seen by the United States, the concerns emphasize the first strike potential by Soviet MIRVs versus our Minuteman silos and the worry of the Soviet break-out from the ABM Treaty through continuing developments like the Krasnoyarsk radar and improved interceptor missiles. As seen by the Soviet Union, the concern takes the form of fear of decapitation strikes against Soviet leadership - a concern enhanced by the deployment of Pershing IIs in Germany. It will be important for the United States and the Soviet Union to clarify, in future talks and in the Standing Consultative Commission, the distinction between the allowed research and the forbidden development and testing of ABM systems. This will require a clear and agreed definition of system components when dealing with the newer technologies. For example, the ABM Treaty forbids the development and testing of space-based and airborne components of defensive systems.

2. Organize and support a prudent, deliberate, and high-quality research program on defensive technology, within ABM Treaty limits.

The availability of such technology, together with the real possibility of American countermeasures, can contribute to discouraging the Soviets from deploying defensive systems in violation of the Treaty, minimizing the adverse effects of such a potential Soviet "breakout," if it should occur, and protecting us against technological surprises. We had such a program before the President's Star Wars speech first raised the prospect of transforming it from a hedge against future developments into one preparing for breakout from the ABM Treaty.

3. Avoid large-scale technology demonstrations.

It is now far too early for any program in strategic defense to consider technology demonstrations of types that could raise serious issue of compliance with the ABM Treaty. This was the explicit conclusion of a recent workshop at the Stanford Center for International Security and Arms Control, as endorsed by signatories that include supporters, as well as opponents, of the Strategic Defense Initiative.

Such demonstrations would be politically mischievous with regard to ABM Treaty compliance. They are also technologically unwarranted – the technology has very, very far to go before it can meet minimum systems criteria – and very likely will lead to cost overruns and to an emphasis on engineering design that would be far from optimal.

64

4. Form a strong "red team," that is, a team of devil's advocates to continually challenge the defense program concepts against potential Soviet countermeasures.

Any deployed defensive system will have to be effective against a determined opponent who can resort to a wide variety of countermeasures to defeat, deny, evade, destroy, or otherwise overpower the system. It will also have to be cost-effective in defeating the offense. The importance of these two demanding criteria were emphasized by Ambassador Nitze in his February 20 speech, referred to earlier, in which he said:

"The technologies must produce defensive systems that are survivable: if not, the defense would themselves be tempting targets for a first strike. This would decrease, rather than enhance, stability."

"New defensive systems must also be cost-effective at the margin — that is they must be cheap enough to add additional defensive capability so that the other side has no incentive to add additional offensive capability to overcome the defense. If this criteria is not met, the defensive systems could encourage a proliferation of countermeasures and additional offensive weapons to overcome deployed defenses, instead of a redirection of effort from offense to defense."

If we go ahead with development and deployment of a Star Wars defense that fails to satisfy these two requirements, we will be guilty not only of folly, but of a dangerous folly. Wasting money is bad enough. Wasting human talent and technological muscle is worse. But decreasing stability and intensifying the arms build-up should be totally unthinkable. In view of the nature and importance of its task, the "red team" should have a charger independent of the management of the defensive weapons R&D.

5. Enact legislation in Congress with the explicit provision that the research and technology program must proceed by means that are fully consistent with the ABM Treaty.

65

An appropriation of roughly $1.5 to $2 billion per year - up from the current level of $1.4 billion this year but well below the $3.7 billion requested by the Administration for next year - is fully adequate for a strong program that makes appropriate choices among priorities in developing new technologies.

6. Form a senior-level, independent oversight panel that would examine the work on defensive R&D and report at the highest levels of government, to the Congress as well as to the Administration.

This panel of experts and former public officials in defense and arms control fields should maintain a sufficiently detailed knowledge of the technical and programmatic status of the work to advise the government on the setting of priorities. It should also conduct periodic program reviews to ensure that the work remains in close harmony with our overall military, security, and arms control goals, and strategic policy.

I emphasize the importance of this last recommendation to create an oversight panel. The Administration and the Pentagon SDI office have stated their intentions to conduct the R&D on defensive weapons **"in a manner fully consistent with all U.S. Treaty obligations."** However, the 1985 Report on SDI submitted to Congress in April has the strong flavor of what SALT I negotiator Ambassador Gerard Smith has called **"an anticipatory breach of contract."**

The 1985 Report exploits inevitable verbal ambiguities in the 1972 Treaty to pursue the new technologies. For example, Article V of the Treaty forbids development, testing, or deployment of **"ABM systems or components"** which are space-based or air-based. In the 1972 ratification hearings before the Senate Armed Services Committee, Ambassador Smith defined the prohibitions on development as starting when **"field testing is initiated on either a prototype or breadboard model."** At issue is how far one can go in disaggregating components into subcomponents in order to

66

claim compliance. Is an airborne or space-based sensor okay to test against a warhead, so long as it lacks a direct communication link to the ground? Is it legal for an airborne or spaceborne missile interceptor to be tested against a satellite, just not against an ICBM warhead? Or may it be tested against a warhead, so long as it doesn't carry its own target-tracking system? How can one develop space-based high velocity interceptors (kinetic kill vehicles or railguns)?

The answers to these and many other similar questions will soon be required. The U.S. Congress must make certain that the answers it gets can be defended in the Court of World Opinion, as well in the Standing Consultative Commission, where issues of treaty compliance are addressed and must be resolved. The Pentagon report gives me no confidence that satisfactory answers to these questions will be provided by this Administration, in its rush towards strategic defense.

To summarize, a deliberately paced U.S. research effort on strategic defense, organized and pursued with clearly stated goals and means, abiding by the ABM Treaty, and in accord with these six recommended actions, should serve our security well at reasonable cost - not very different from what we have been spending in the recent past. It should also give the Soviet Union - itself engaged in R&D on strategic defenses - no real reason to feel the need for additional offensive or defensive forces in response. A U.S. effort with better planning, explanation, and diplomacy might moderate Soviet concerns, and their perceptions, which now regard the U.S. program as threatening their deterrent capability, menacing strategic stability, and defeating the prospects for arms control. It would also be of enormous value for the Soviets to take some reciprocal steps to explain some of their troubling actions, including in particular discussion of the Krasnoyarsk radar. More openness in this dialogue would be immensely valuable.

The impact of such clearly defined and constrained activities on the ABM Treaty should be nil, in the near- and mid-terms. Genuine success in creating a new approach to effective nationwide strategic defense would indeed present a challenge to the ABM Treaty. If this should occur, it would do so well into the future, and regardless of which side would succeed in developing the defensive system. In such an eventuality, the U.S. would need to conceive and negotiate a new arms control regime that would successfully integrate offensive and defensive forces.

But above all, we all should step back and view the threat of nuclear weapons in proper perspective. This threat is not solely or even primarily a technical problem. The difficulty is much deeper and the solution much more radical than that which can be achieved by continually calling on the next stage of technology to provide a new Maginot Line.

In the current debate over Star Wars we scientists can make important contributions. Not only do we bring valuable physical insights and understanding as to what science can do; equally important, we can help clarify what are the limits to the promise of technological fixes as potential cures for our nuclear dangers.

Shortly after the atom bombs were dropped on Hiroshima and Nagasaki within days of Japan's capitulation, Robert Oppenheimer emphasized the futility of looking to science and technology to cure the dangers of nuclear weapons. In a transmittal letter to Secretary War Henry Stimson of a report of the Scientific Panel of what was known as the "Interim Committee," on the scope and program of future work in the field of atomic energy, Oppenheimer said that:

> "We have been unable to devise or propose effective military counter-
> measures for atomic weapons. Although we realize that future work
> may reveal possibilities at present obscure to us, it is our firm
> opinion that no military countermeasures will be found which will be

adequately effective in preventing delivery of atomic weapons."
"We believe that the safety of this nation - as opposed to its
ability to inflict damage on an enemy power - cannot lie wholly or
even primarily in its scientific or technical prowess. It can be
based only on making future wars impossible. It is our unanimous
and urgent recommendation to you that, despite the present incomplete
exploitation of technical possibilities in this field, all steps be
taken, all necessary international arrangements be made, to this one
end."

The destructive power of nuclear weapons is so great that we
require a radically new way of thinking in dealing with them. In our
field of physics, the advent of quantum mechanics presented not just a
technical problem of matrix algebra to physicists 60 years ago. It also
demanded a whole new way of thinking; and physicists who couldn't master
it sank in the wake of rapidly advancing progress. The advent of nuclear
weapons of mass destruction - fission bombs 40 years ago and thermonuclear
weapons 30 years ago - presents all mankind with more than a technical
problem of rescaling warfare. It demands that we develop ways of thinking
about weapons and national security; and new means for resolving our
differences and conflicts. A nuclear war would be nothing short of
suicide.

Our challenge is thus not just to make a better laser or computer.
For the U.S. and the Soviet Union, above all, as possessors of close to 99
percent of the world's nuclear weapons, the challenge is to get serious -
really serious - and committed to improving our political, diplomatic, and
human relations. We must face our common danger: not each other, but
nuclear weapons - ours, theirs, and those of the increasing number of
nuclear-weapons-capable nations. Avoiding a nuclear holocaust is a deep
moral obligation we bear to generations yet unborn.

Neutral Currents

F. Dydak

CERN, CH-1211 Genève 23, Switzerland

C. Geweniger

Institut für Hochenergiephysik der Universität,
D-6900 Heidelberg, Fed. Rep. of Germany

1. Introduction

Neutral Current (NC) phenomena are an appropriate subject to be included in the Festschrift dedicated to Jack Steinberger. His initiative, his insight into the secrets of good experimentation, and last but not least his driving force have left unmistakable footprints in the field. And this right from the moment on when he and his team joined the club of neutrino scatterers as early as 1977. From then on, for almost a decade of continuously refined experimentation, important results emerged and still keep emerging.

The authors have had the privilege of working under Jack's leadership on the subject of NCs for a good fraction of their life. In the hope of understanding a little of what they are talking about, they aim at reviewing the developments of NC phenomena from a somewhat personal point of view. They also wish to touch upon the likely developments in the foreseeable future. This appears most appropriate the more Jack engages himself already in the next generation of electroweak experiments.

2. The beginnings of Neutral Currents

The early history of NCs is not a brilliant chapter of experimental high-energy physics. The only place where one could hope to see weak NC interactions, without being swamped by the electromagnetic interaction, was in neutrino scattering. However, the experimentalists did their best to discuss away all hints of neutrino-induced NC interactions in the early experiments, as "neutron-induced interactions". To see an example of both the importance which was assigned to the study of neutrino scattering on the one hand, and of the reluctance to accept NC scattering events on the other hand, let us take a look at Fig. 1. The photograph, taken in 1964, shows G. Bernardini, the then leader of the neutrino team at CERN, presenting a long list of results in the CERN Main Auditorium, including a (wrong!) upper limit of NC to CC (Charged Current) scattering.

Certainly, the smallness of the then used neutrino detectors made them almost transparent for neutrons, rendering a proof for the existence of neutrino NC scattering next to impossible. We suspect, however, that besides this real difficulty, the experimentalists had understood too well the then valid dogma: "Weak interactions are charged-current interactions", since the

Fig. 1   G. Bernardini giving a talk on results from early CERN neutrino experiments

absence of NCs had been experimentally shown with high precision in flavour changing decays like $K^+ \longrightarrow \pi^+ e^- e^+$ [1].

It needed a strong push from theorists to make experimentalists re-think the matter. After t'Hooft's proof [2] of the renormalizability of the "Weinberg model", which became known later as the Glashow-Salam-Weinberg (GSW) theory of electroweak interactions [3], the existence of weak NCs was strongly favoured as a natural consequence of the proposed unification of weak and electromagnetic interactions. In addition to the charged intermediate bosons $W^{\pm}$, the neutral intermediate boson Z^0 was postulated to mediate weak NC processes such as $\nu_\mu + e^- \longrightarrow \nu_\mu + e^-$ (elastic neutrino-electron scattering) or $\nu_\mu + N \longrightarrow \nu_\mu + X$ (inelastic neutrino-nucleon scattering).

The newly postulated NC interaction was flavour-conserving. The absence of flavour-changing NCs had found a convincing theoretical explanation by the postulated existence of the fourth, charm, quark, in the frame of the Glashow-Iliopoulos-Maiani scheme [4]. Did flavour-conserving NCs exist (as required by theory) or not (as claimed by early experiments)?

The breakthrough came with the observation of a single-electron event, interpreted as a $\nu_\mu + e^- \longrightarrow \nu_\mu + e^-$ event [5], in 1973. The event was observed in the heavy-liquid bubble chamber Gargamelle (GGM), exposed at CERN to an intense muon-neutrino beam. Shortly afterwards (still in 1973), the claim was made, by the GGM team, of the observation of NC neutrino-nucleon scattering [6]. The claim was based on the observed flat distribution of event vertices along the beam direction, whereas one expected a rather exponential fall-off for neutron-induced background.

For many years to come, this discovery of the GGM team, headed by P. Musset, was considered the most important result obtained at CERN. Not everybody was convinced right from the beginning. The authors remember Jack being sceptical. It must be one of the rare occasions where his judgement proved wrong.

At the time of this discovery at CERN, the 400 GeV Fermilab accelerator had begun operation, with two neutrino experiments set up to take data. They were quick in confirming the GGM result [7,8].

With these and a few more experimental results, the existence of semileptonic NC scattering was established beyond any doubt as early as 1975. For example, we recall in Fig. 2 the event-length distribution in iron, as measured by the Caltech-FNAL Collaboration [8]. The distribution shows at short event-lengths an unmistakable peak, attributed to NC events, on top of a background of CC events having typically long event lengths because of the penetrating final-state muon.

From the outset, the Weinberg model was a serious contender to explain the properties of the new interaction. In its minimal version, it has just one free parameter, the electroweak mixing parameter $\sin^2\theta_W$, to be determined by experiment. The discriminating power of the first round of experiments in 1974/75 among different models was weak; however, very soon

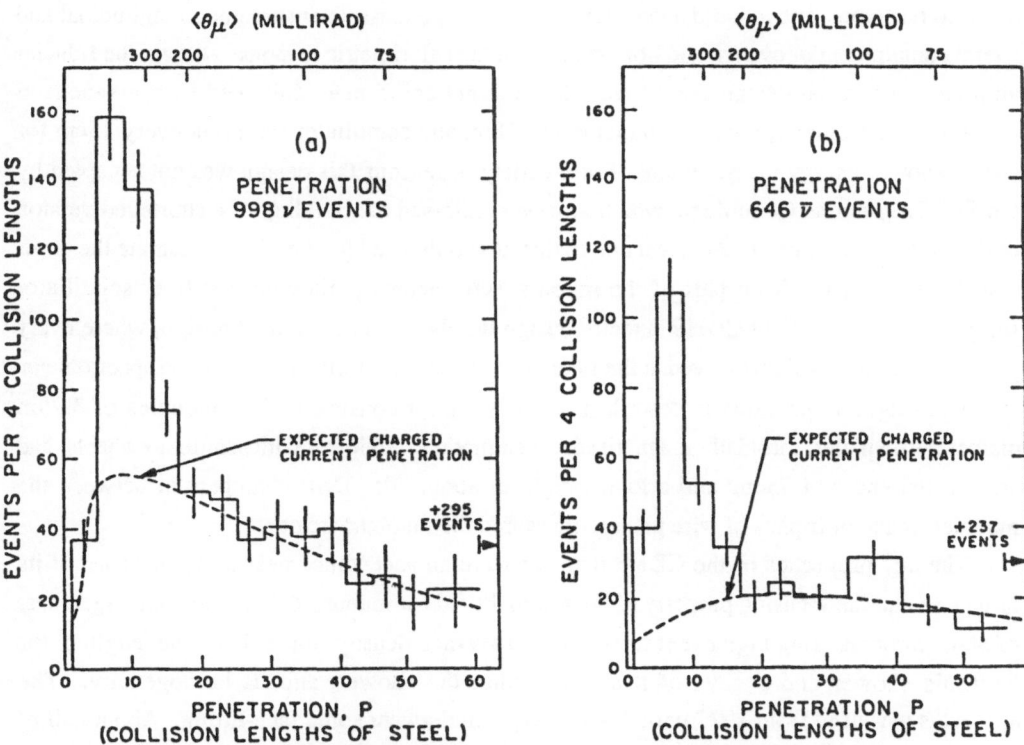

Fig. 2   Event-length in iron of neutrino and antineutrino events, as measured by the 1975 Caltech-FNAL experiment [8]

73

experiments tended to favour the Weinberg model, with a mixing parameter $\sin^2\theta_w = 0.4 \pm 0.1$ [9].

Would the Weinberg model withstand a detailed experimental scrutiny? If so, what was the precise value of the electroweak mixing parameter?

3. The first precision experiments

3.1 The CDHS vFe scattering experiment at CERN

Towards the end of 1976, the CERN SPS and the West Area Neutrino Facility became operational. The first neutrino beam to be exploited was a Narrow-Band Beam (NBB) with 200 GeV parent momentum, very good sign selection, and well known energy spectrum.

At the same time, the neutrino detector of the CERN-Dortmund-Heidelberg-Saclay (CDHS) Collaboration, led by Jack Steinberger, was completed and ready to take data. Since this detector was incisive in the further development of neutrino physics, it might be worth recalling its main characteristics.

As nearly as the authors can tell, the idea of building at CERN a large and dense calorimetric device together with an iron-core magnetic spectrometer for final-state muons, was due to Jack. His basic design idea, influenced by the strengths and weaknesses of the FNAL neutrino detectors, was to build a detector with very large mass, to ensure good longitudinal and lateral containment of events, and homogeneity in the calorimetric response all over the fiducial volume. The initial design comprised a large target calorimeter followed by a magnetized iron-core spectrometer, with an iron length of 18m, and scintillator sampling every 15cm for coarse shower energy measurement. For a variety of reasons this design was not accepted by the SPS Programme Committee, which however endorsed the building of a shortened version of the muon-spectrometer. As a reaction to this, it was decided to effectively integrate the target calorimeter into the front part of the muon spectrometer by allowing for finer scintillator sampling. Thus the final CDHS detector design was born: an integrated design, where every part of the fiducial volume served at the same time as target, calorimeter and muon spectrometer.

The original apparatus is described in Ref.[10]. It consisted of 19 modules of 3.75m diameter iron plates, toroidally magnetized. Each module, equipped with scintillator sheets, had an iron thickness of 75cm, and a total weight of about 70t. Drift chambers in between the modules, made of triplets of wire planes, measured the muon trajectory.

The key properties of the CDHS detector were an acceptance near unity, because of its large size and the focusing property of the toroidal field for muons, its large fiducial target mass of about 500t ensuring high event rates, its high average density minimizing the length of the hadronic shower and decays of π and K within the shower, and its homogeneity. The apparatus was used from 1977 until 1980 when it underwent some face-lifting. About half of the detector was replaced by modules with finer sampling, and x-y positioning of the scintillator sheets to allow a better localization of the position of hadronic showers. It was this latter feature which was specifically aimed at a better systematic measurement of NC events.

The good functioning of the CERN SPS was no less crucial to the success of the CDHS detector. The SPS delivered, right from the beginning, 400 GeV protons with ever increasing intensity, and great reliability. SPS running efficiencies below 80% were the exception.

Semileptonic NC events were trivially seen as muon-less events in the CDHS detector. The event-length method, pioneered by the Caltech-FNAL Collaboration, was used (and is still used) to separate NC from CC candidates. Already in the first series of NBB exposures early in 1977, about 8000 ν and 2200 $\bar{\nu}$ NC events were collected, enabling a much more precise determination of the NC to CC ratio than was possible before. The experimental progress may perhaps best be visualized by the length distribution shown in Fig. 3 as compared to the one in Fig. 2. The results for the NC to CC ratios for hadronic energies $E_H > 12$ GeV were [11]:

$$R_\nu = 0.293 \pm 0.010$$
$$R_{\bar{\nu}} = 0.35 \ \pm 0.03.$$

These ratios were consistent with the GSW theory predictions, with a value of the mixing parameter $\sin^2\theta_W = 0.24 \pm 0.02$, obtained in the frame of the Quark-Parton Model (QPM) description of the nucleon. Radiative corrections were not yet an issue at that time.

The low value of $\sin^2\theta_W$ was a major surprise since the then commonly accepted value was much higher. In his rapporteur's talk at the 1977 Lepton-Photon Symposium at Hamburg, P. Musset [12] chose to calculate a world average for $\sin^2\theta_W$ "without CDHS", $\langle\sin^2\theta_W\rangle = 0.31 \pm 0.03$, as opposed to the low CDHS result, $\sin^2\theta_W = 0.24 \pm 0.02$.

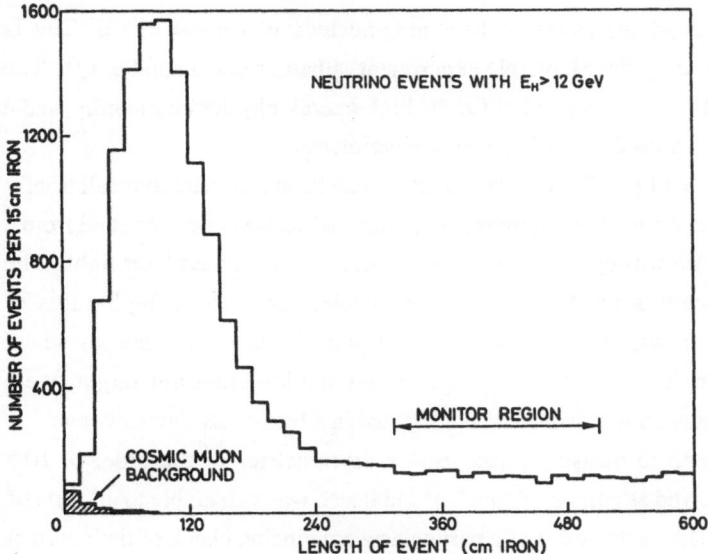

Fig. 3   Event-length in iron of neutrino events, as measured by the 1977 CDHS experiment [11]

75

The data also allowed a meaningful test of the Lorentz structure of the NC interaction. This structure is directly measured in the "inelasticity" or y-distribution ($y = E_H/E_\nu$). Comparing the NC y-distribution with the CC y-distribution it was shown [13] that they were very similar. This demonstrated that the hadronic NC for scattering off an isoscalar target is dominated by V-A. A small admixture of V+A was seen, with a significance of 4 σ, in support of the prediction of the GSW theory for $\sin^2\theta_W = 0.24$.

3.2 The eD scattering experiment at SLAC

While neutrino-quark scattering experiments were busy measuring the properties of the NC interactions, many years of hard work at SLAC came to fruition in 1978: in the scattering of longitudinally polarized electrons off unpolarized deuterium nuclei a parity-violating asymmetry was observed by a SLAC-Yale Collaboration [14]. Since the electromagnetic interaction conserves parity, the parity-violating asymmetry was attributed to the weak NC interaction. This experiment demonstrated the interference of the weak and electromagnetic currents, just as predicted by the GSW theory. Moreover, it gave a value of $\sin^2\theta_W = 0.224 \pm 0.02$, compatible with the one found by CDHS in neutrino scattering. This constituted a major triumph for the GSW theory, and the henceforth accepted value of the electroweak mixing parameter was $\sin^2\theta_W \approx 0.23$ [15].

The SLAC experiment gave its result at the right moment. Two atomic physics experiments, carried out at Seattle (USA) and Oxford (UK), which measured parity-violating effects in optical transitions of heavy nuclei, failed to see such effects [16,17] at the level predicted by the GSW theory. This was disturbing, and caused a fair amount of concern.

The parity-violation experiments in heavy atoms measure the weak NC interaction between electrons of the atomic shell, and quarks inside the atomic nucleus: $e^- + q \longrightarrow e^- + q$. This is the same reaction as measured by the SLAC eD experiment, albeit at much smaller Q^2. The asymmetry measured at SLAC was a great relief for the high-energy physics community, and a relaxed attitude was taken vis-à-vis the atomic physics experiments.

Because of the importance of the SLAC experiment, it may be appropriate to recall briefly the ingenious set-up of this experiment. Longitudinally polarized electrons were emitted from a gallium arsenide crystal, which was optically pumped by circularly polarized laser light. The electrons were then accelerated in the Stanford Linear Accelerator, with negligible loss of polarization. The polarization was reversed on a pulse-to-pulse basis. The electrons which were scattered under an angle of 4° from an unpolarized liquid-deuterium target, were momentum-analyzed in a magnetic spectrometer, and detected in a lead-glass shower counter.

The experiment attempted to measure cross-section asymmetries of the order of 10^{-4} (because of the small Q^2 involved as compared to m^2_Z), and hence was vulnerable to all sorts of systematic effects. Figure 4 shows the result of a particularly convincing check of their correct understanding of the experiment: the measured asymmetry varies as a function of the beam

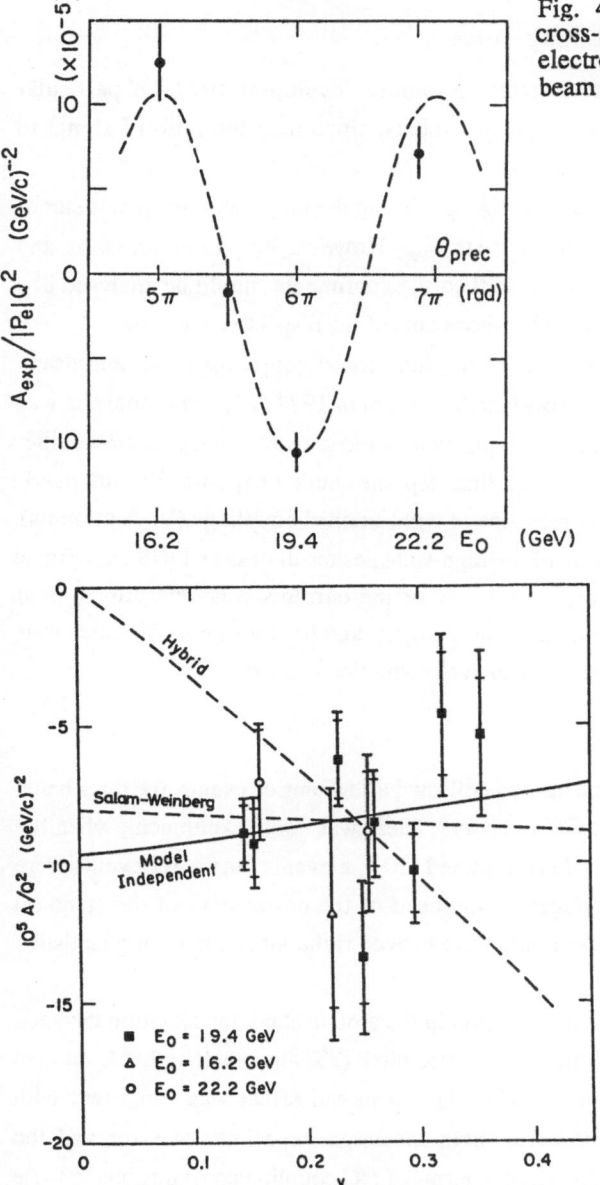

Fig. 4   Asymmetry in the scattering cross-section of longitudinally polarized electrons on deuterium as a function of the beam energy [14]

Fig. 5   Asymmetry in the scattering cross-section of longitudinally polarized electrons on deuterium as a function of the inelasticity y [18]

energy, just as predicted by the g-2 precession of the electron spin in the beam transport elements.

The asymmetry in the scattering of polarized electrons on deuterium as a function of the inelasticity y [18] is shown in Fig. 5. As can be seen, also the measured differential asymmetry is consistent with the prediction of the GSW theory.

4. Towards the glory of 1979

4.1 The NC coupling constants of the hadronic current

The investigation of NC phenomena became a rapidly developing field. Of particular importance were neutrino-nucleon scattering experiments, since here the ratio of signal to background was the most favourable.

Experimental results were usually presented by specifying the only unknown parameter of the minimal GSW theory, the mixing parameter $\sin^2\theta_W$. However, in particular, Hung and Sakurai [19] pointed out that for a real test of the theory, experiments should be analysed in a model-independent way, in terms of the coupling constants of the respective currents.

A first complete analysis of the strengths of the four chiral couplings of up and down quarks, assuming V and A currents only, was given by Seghal in 1977 [20]. This analysis was followed by others [21], where, in addition, attempts were made to resolve the sign ambiguities of the coupling constants. In these analyses, in a first step, the counter experiments with nearly isoscalar targets were used to test for the presence of right-handed couplings (V+A currents). These tests became positive with the advent of the high-statistics result of the CDHS experiment in 1977 [13]. In a second step the isospin structure of the currents was determined using bubble-chamber results with hydrogen and deuterium targets, and finally sign ambiguities were resolved with the help of exclusive and semi-inclusive production processes.

4.2 Other NC experiments

Neutrino-electron scattering experiments had collected a few tens of events, the rates being in agreement with expectation from the GSW theory. There was some excitement, when the GGM Bubble Chamber Collaboration in 1978 reported 10 $\nu_\mu e$ events where 1.7 events were expected[22]. But most people, being already convinced of the correctness of the standard model, took a "wait-and-see" attitude. And they were proved right, since increasing statistics washed the anomaly away.

As for experiments searching for parity violation in the atomic shell, the situation changed in 1978. Parity violation in bismuth was found at Novosibirsk [23] at a much higher level than the limits set by the earlier experiments [16,17]. The observed effect was consistent with expectations from the GSW theory; however, both the large experimental error and the theoretical difficulties in understanding the result in terms of NC coupling constants, made these rather exotic experiments less interesting from a quantitative point of view.

In the middle of 1979, the experimental situation was reasonably well under control, and gave overwhelming support to the GSW theory.

4.3 Global fits

The programme of model-independent determination of the NC coupling constants finally led to global fits to all available experimental data, comprising altogether some 25 experiments.

All but the early experiments on parity violation in heavy atoms were in reasonable agreement with the predictions of the GSW theory. A value of $\sin^2\theta_W = 0.230 \pm 0.015$ virtually satisfied all existing data [24].

A closer inspection of the global fits shows, however, that they were really dominated by the precise CDHS νN, and by the SLAC eD results. The agreement of most other experiments was rather qualitative, and in general better than one could expect on the basis of the quoted errors. Why did not even one of the 25 experiments show a (perfectly acceptable) 3σ deviation? As a consequence, one of the problems of such a global fit to all experimental data is the much too good χ^2 of the fit.

The Swedish Academy of Sciences did not worry about such minor things. They awarded the 1979 Nobel Prize of Physics to S.L. Glashow, A. Salam and S. Weinberg, the fathers of the successful GSW theory of the electroweak interaction.

5. The request for higher precision

5.1 SU(5) and all that

The breakthrough in the formulation of a unified gauge theory of weak and electromagnetic interactions based on $SU(2)_L \times U(1)$ was almost immediately followed by the formulation of QCD based on $SU(3)_C$. These theoretical ideas, supported by the discovery of NCs and charm, called for a "grand unification" of strong and electroweak interactions. The most popular model became the minimal SU(5) model of Georgi and Glashow [25]. This model predicted relations between the three coupling constants of the standard model $SU(3)_C \times SU(2)_L \times U(1)$, which may be expressed in terms of the fine structure constant α, the strong coupling constant α_s, and $\sin^2\theta_W$. In particular, the value $\sin^2\theta_W = 3/8$ at some very large mass scale and the decay of the proton were predicted. For a comparison with measurements at the scale of the W boson mass, radiative corrections had to be applied to obtain $\sin^2\theta_W$ at the W mass scale. Was this latter $\sin^2\theta_W$ consistent with the $\sin^2\theta_W$ determined in lepton-scattering experiments? In order to assess this question properly, the need to calculate the one-loop electroweak radiative corrections for the lepton-scattering data became evident. At the same time, the experimentalists felt challenged to reduce the experimental uncertainties as much as possible, in order to make the comparison of their results with Grand Unified Theory predictions significant.

A first solution of the problem of electroweak radiative corrections was given in 1981 by Sirlin and Marciano [26]: they obtained $\sin^2\theta_W = 0.215 \pm 0.014$, determined from neutrino-nucleon scattering experiments, in good agreement with the SU(5) prediction $\sin^2\theta_W = 0.215 \pm 0.006$. It is amusing to note that the eD scattering experiment, after radiative corrections, yielded the same numerical result: $\sin^2\theta_W = 0.215 \pm 0.015$ [27].

5.2 Measuring higher-order electroweak effects

With time passing by, and experiments failing to see proton decay thus putting the minimal SU(5) model into severe trouble, a second aspect of the electroweak radiative corrections to the results from lepton-scattering experiments aroused a lot of interest: the correct prediction of the $W^{\pm}$ and Z° masses, especially on the eve of their planned discovery at the CERN $p\bar{p}$ Collider.

Assuming that there is no "New Physics" going on, different experimental results should be described by a single parameter, $\sin^2\theta_W$. However, when experiments become more precise, experiment-dependent higher-order corrections are expected to come in:

$$(\sin^2\theta_W)_{exp} = \sin^2\theta_W + \Delta_{exp},$$

where $\sin^2\theta_W$ is a universal parameter. The correction Δ_{exp} has to be calculated for each particular process.

Having measured $\sin^2\theta_W$ the masses m_W and m_Z may be predicted in the Born approximation using the relations

$$\sin^2\theta_W = \frac{\pi\alpha}{\sqrt{2}\, G_F\, m_W^2}$$

$$\sin^2\theta_W = 1 - \frac{m_W^2}{m_Z^2}\ .$$

Using the second relation as a definition for the renormalized parameter $\sin^2\theta_W$, where m_W and m_Z are the physical masses of W and Z, respectively, then the first relation receives higher order corrections:

$$\sin^2\theta_W = \frac{\pi\alpha}{\sqrt{2}\, G_F\, m_W^2} \; \frac{1}{1 - \Delta r},$$

where $\Delta r = 0.0696 \pm 0.0020$, for three fermion families, $m_{Higgs} = m_Z$, and $m_{top} = 36$ GeV/c^2 [28]. The resulting renormalization of the W mass is sizeable: 2.8 GeV/c. Notice that here one relates the W mass with the renormalized $\sin^2\theta_W$, which in itself is obtained from, for example, neutrino-nucleon scattering after electroweak radiative corrections of about 5% [26,27].

Many theorists have contributed to the evaluation of electroweak radiative corrections. For a detailed list of references the reader is referred to Ref.[29].

6. Recent experimental progress

6.1 The discovery of W and Z

For a number of years, the $W^{\pm}$ and Z° were the most well-established but unobserved particles. It needed the transformation of the CERN SPS into a $p\bar{p}$ collider, the provision of a high-intensity $\bar{p}$ source, and a huge experimental effort to record and analyse the data from high-energy $p\bar{p}$ collisions in order to establish experimentally the existence of $W^{\pm}$ and Z^{0}.

The idea goes back to 1976 when Rubbia, together with Cline and McIntyre [30], proposed to convert the CERN SPS into a $p\bar{p}$ collider. Its modification was proposed even before it started operation as a 400 GeV proton accelerator at the very end of 1976.

The first $p\bar{p}$ collisions were observed in July 1982. One year later, enough statistics were accumulated by the UA1 and UA2 detectors to allow the discovery of the $W^{\pm}$ [31,32], followed shortly afterwards by the discovery of the Z^{0} [33,34].

Nobody was really surprised that the $W^{\pm}$ and Z^{0} particles had been found experimentally. They were found even precisely at their predicted mass values, based on the value of $\sin^2\theta_W$ obtained from neutrino- and electron-nucleon scattering experiments. This was another triumph of the GSW theory, and after the 1979 Nobel Prize awards, this achievement was crowned with the 1984 Nobel Prize for Physics awarded to C. Rubbia and S. van der Meer.

Today's best values for the mass of the $W^{\pm}$ and Z^{0} particles are [35,36]:

UA1: $m_W = 83.5 \pm 1.1$ (stat.) ± 2.7 (syst.) GeV/c^2
            $m_Z = 93.0 \pm 1.4$ (stat.) ± 3.0 (syst.) GeV/c^2

UA2: $m_W = 81.2 \pm 1.1$ (stat.) ± 1.3 (syst.) GeV/c^2
            $m_Z = 92.5 \pm 1.3$ (stat.) ± 1.5 (syst.) GeV/c^2 .

From these mass values, one readily derives the following values of $\sin^2\theta_W$:

UA1: $\sin^2\theta_W = 0.214 \pm 0.006$ (stat.) ± 0.015 (syst.)
UA2: $\sin^2\theta_W = 0.226 \pm 0.005$ (stat.) ± 0.008 (syst.).

The results are well consistent with the ones from neutrino-scattering experiments. From the statistical point of view, the neutrino experiments cannot compete in the long run. From the point of view of systematic accuracy, the present UA1 and UA2 experiments are limited by the uncertainty of the absolute calibration of their calorimeters. As long as this obstacle is not removed the neutrino experiments are able to give $\sin^2\theta_W$ with comparable or even better precision than the UA experiments.

6.2 Neutrino-electron scattering

The scattering of neutrinos off electrons is of particular interest, since one is concerned with an elastic collision of point-like leptons. No uncertainties due to the internal structure of extended hadrons arise. However, the price one has to pay for this simplicity is high: the cross-section is lower than the neutrino-nucleon cross-section by a factor of the order of

$m_N/m_e = 2000$, rendering both the accumulation of large statistics and the separation from background a formidable experimental challenge.

Scattering of v_e and $\bar{v}_e$ off electrons, though theoretically very interesting due to the unique possibility of studying the interference of NC and CC, has just been observed with rates being consistent with the expectations [37,38].

Scattering of v_μ and $\bar{v}_\mu$ off electrons, proceeding via NC only, has been studied extensively at high-energy accelerators. From the kinematical constraint for the energy and angle of the final-state electron,

$$E_e\theta_e^2 < 2m_e,$$

the experimental signature is a high-energy electron emitted under very small angle in the forward direction, together with the absence of hadronic activity. Although bubble chambers played an important role at the beginning, the field is now dominated by electronic detectors. The best results have so far been obtained by the CHARM collaboration [39] and a BNL experiment [40]. The angular distributions of $v_\mu e$ and $\bar{v}_\mu e$ candidate events in both experiments are shown in Fig. 6. It is evident from the figure that the separation of signal and background, in particular in the case of the CHARM experiment, is a somewhat difficult task.

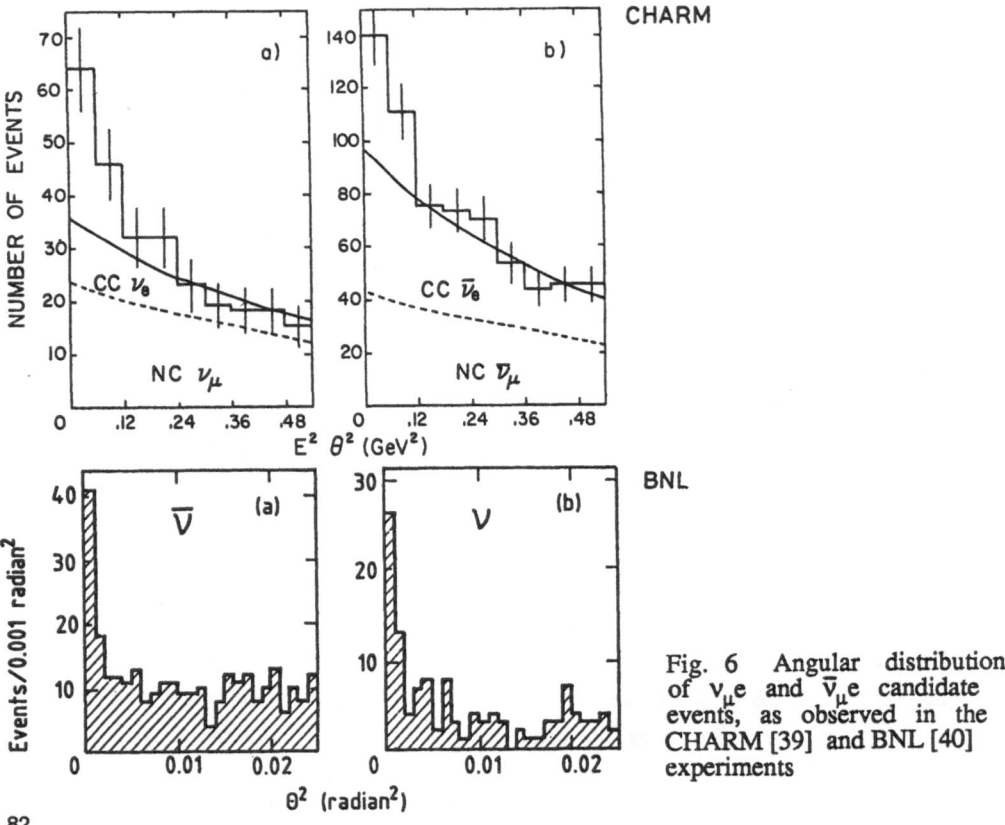

Fig. 6 Angular distribution of $v_\mu e$ and $\bar{v}_\mu e$ candidate events, as observed in the CHARM [39] and BNL [40] experiments

The value of $\sin^2\theta_W$ may be determined from the absolute scattering cross-sections. However, as was first noticed by the CHARM collaboration, a better quantity to use is the cross-section ratio $\sigma(\nu_\mu e)/\sigma(\bar{\nu}_\mu e)$ which is strongly dependent on $\sin^2\theta_W$ around $\sin^2\theta_W \sim 0.2$. Based on the expectation that many systematic errors would cancel in the cross-section ratio, one may achieve a systematically more precise measurement of $\sin^2\theta_W$ via this ratio. The present world average is [41]:

$$\sin^2\theta_W = 0.212 \pm 0.021 \text{ (stat.)} \pm 0.009 \text{ (syst.)}.$$

This value, albeit less precise, is well compatible with the more precise determinations of $\sin^2\theta_W$ discussed above.

6.3 Charged lepton-quark scattering

Analogously to the eD scattering experiment at SLAC, the BCDMS experiment at CERN was able to measure a cross-section asymmetry of the scattering of left-handed μ^+ and right-handed μ^- off unpolarized carbon nuclei [42]:

$$B = \frac{\sigma^+_L - \sigma^-_R}{\sigma^+_L + \sigma^-_R} \quad .$$

Owing to the large Q^2 up to $100\ \text{GeV}^2$, obtainable in high-energy muon beams, the asymmetry B is of the order of 1%, much higher than in the SLAC eD experiment. Yet the experiment was very difficult, since it required the understanding of scattering cross-sections measured in running periods with different beam polarities.

Figure 7 shows the obtained B asymmetry, which is expected to depend linearly on Q^2, for 120 and 200 GeV/c beam momenta. The data are corrected for electroweak radiative effects. They support the GSW theory for a value of the mixing parameter

$$\sin^2\theta_W = 0.23 \pm 0.07 \text{ (stat.)} \pm 0.04 \text{ (syst.)}.$$

Charged lepton-quark scattering is also studied, at even higher values of Q^2 (albeit time-like rather than space-like as above), in e^+e^- annihilation into quarks, at PETRA and PEP. In the process $e^+e^- \longrightarrow f^+f^-$, where f denotes a quark or a lepton, one is at present energies sensitive to the effect of electroweak interference. It manifests itself in a $\cos\theta$ term in the angular distribution of the final-state fermions on top of a $(1 + \cos^2\theta)$ distribution, giving rise to a forward-backward asymmetry A. This is given, to a good approximation, by

$$A = -\frac{3}{2} \frac{a_e\, a_f}{Q_f} \chi \quad .$$

Q_f denotes the electric charge of the final-state fermion f, a_e and a_f the axial-vector coupling

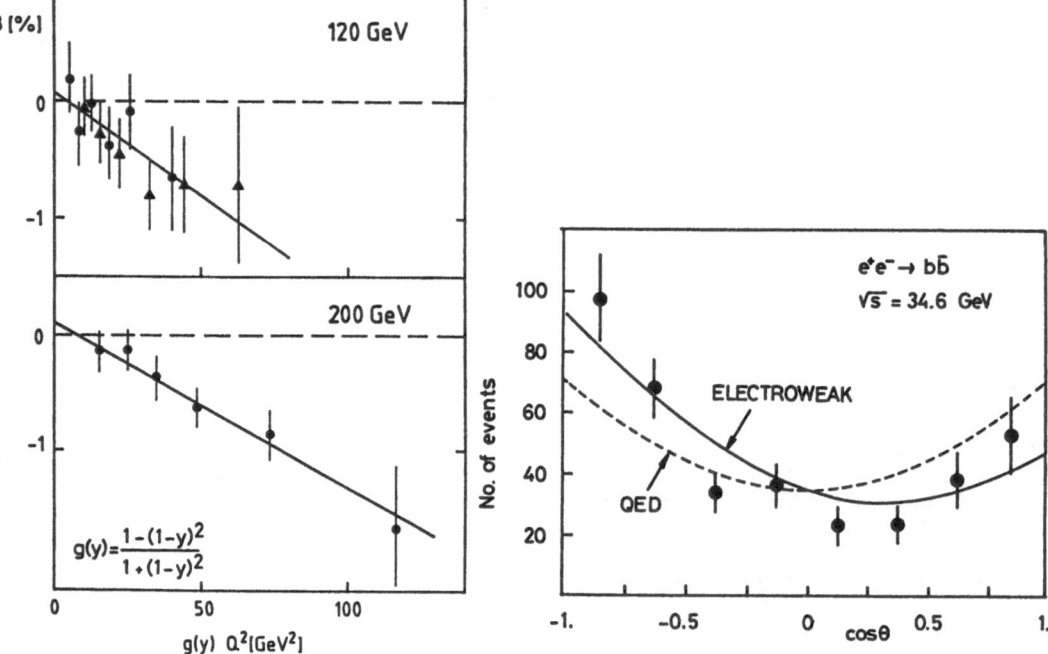

Fig. 7   Cross-section asymmetry for the scattering of μ^+_L and μ^-_R off carbon nuclei, as observed in the BCDMS experiment [42]

Fig. 8   Forward-backward asymmetry of $e^+e^- \longrightarrow b\bar{b}$, as observed by the JADE Collaboration [43]

constants of the electron and the fermion, and

$$\chi \cong gs \frac{1}{s/m_Z^2 - 1} \qquad \text{with}$$

$$g = \frac{G_F}{8\pi\alpha\sqrt{2}} \cong 4.5 \times 10^{-5} \text{ GeV}^{-2} \quad .$$

Measuring the forward-backward asymmetry of a given quark flavour is a difficult task. The JADE Collaboration [43] has succeeded in selecting a relatively pure sample of 306 $e^+e^- \longrightarrow b\bar{b}$ events out of 24 000 hadronic events, and measured the $b\bar{b}$ asymmetry (see Fig. 8):

$$A_{b\bar{b}} = -22.8 \pm 6.5\%.$$

From $A_{b\bar{b}}$ they deduce an axial-vector coupling constant

$$a_b = -0.90 \pm 0.26,$$

consistent with the GSW assignment $a_b = -1$.

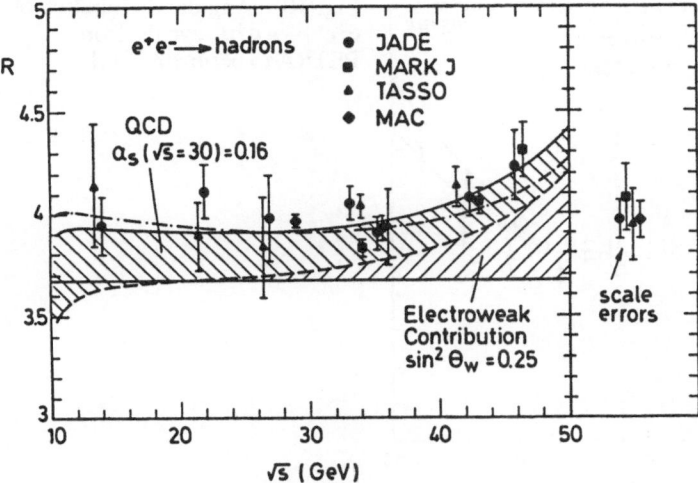

Fig. 9   Measurements of R at PETRA and PEP [44]

At energies higher than accessible at PETRA and PEP, a dramatic rise of the e^+e^- annihilation cross-section is expected. From a compilation of recent data [44] on R, the ratio of the hadronic to the point-like QED cross-section, one might infer the onset of the Z^o resonance at the highest PETRA energies (see Fig. 9).

6.4 e^+e^- annihilation into charged leptons

The forward-backward asymmetry of the process $e^+e^- \longrightarrow \mu^+\mu^-$ has been measured by various collaborations at PETRA and PEP. A compilation [44] of angular distributions of $e^+e^- \longrightarrow \mu^+\mu^-$ events, and the results for the asymmetry, are shown in Figs. 10 and 11. In first approximation, the measured points pretty much follow the GSW theory. However, the data points have a tendency to fall below the expectation for $m_Z = 93$ GeV. The average value of the axial-vector coupling constant is

$$a_\mu = -1.13 \pm 0.07 \quad .$$

This is the most precise determination of a single coupling constant. It differs by 2σ from the GSW assignment $a_\mu = -1$.

The measurements of the asymmetry for $e^+e^- \longrightarrow \tau^+\tau^-$ are also shown in Fig. 11. In contrast to the muonic case, here the data points have rather a tendency to lie above the expectation for $m_Z = 93$ GeV. Altogether, involving μ-τ universality and combining $\mu^+\mu^-$ and $\tau^+\tau^-$ data, the agreement with the GSW theory is good.

The asymmetry at PETRA/PEP energies is essentially given by the axial-vector coupling constants of the participating fermions, which are independent of $\sin^2\theta_W$. How do the PETRA/PEP groups obtain a measurement of $\sin^2\theta_W$ from the asymmetry?

85

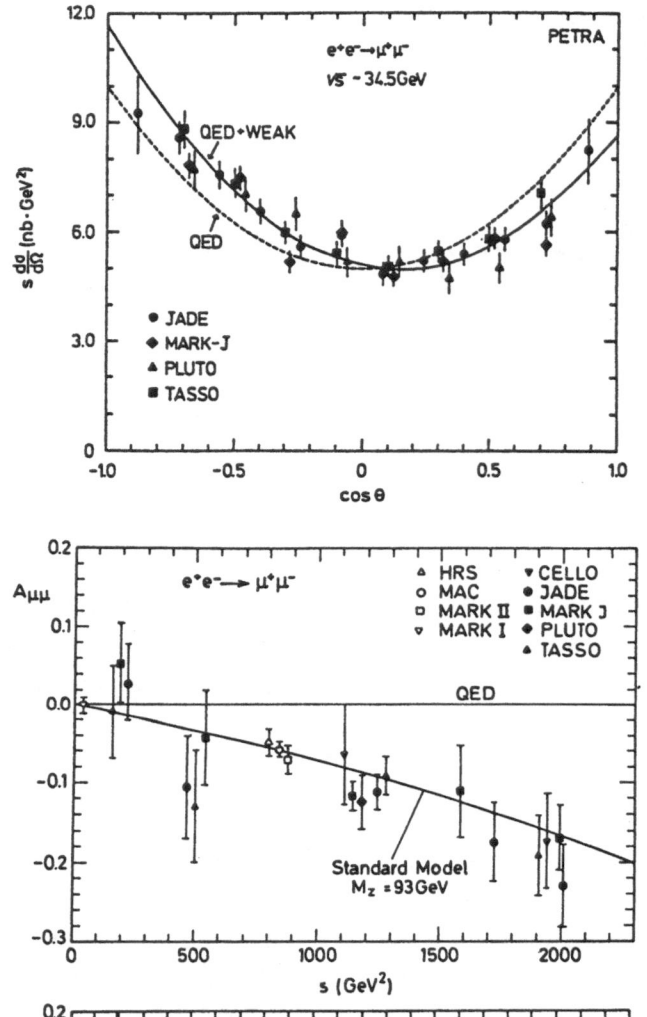

Fig. 10 Angular distribution of $e^+e^- \longrightarrow \mu^+\mu^-$ events from PETRA experiments [44]

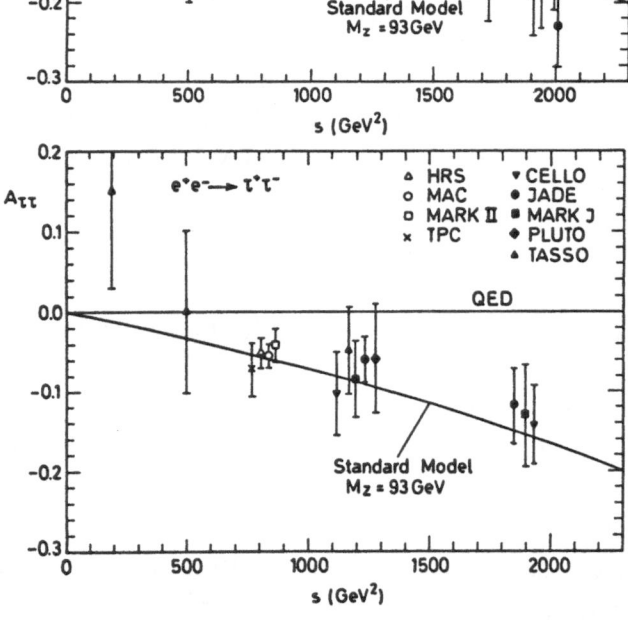

Fig. 11 Asymmetry measurements of $e^+e^- \longrightarrow \mu^+\mu^-$ and $e^+e^- \longrightarrow \tau^+\tau^-$ [44]

The dependence comes from the factor χ in the expression for the asymmetry A given in Section 6.3. Expressing g in terms of m_Z and $\sin^2\theta_W$ one obtains

$$\chi = \frac{1}{16 \sin^2\theta_W \cos^2\theta_W} \; \frac{s/m_Z^2}{s/m_Z^2 - 1}$$

which is dependent on $\sin^2\theta_W$. Determined this way, the coupling strength of the weak NC interaction is measured by $\sin^2\theta_W$, whereas all other determinations of $\sin^2\theta_W$ assume the same coupling strength for NC and CC, and rather determine $\sin^2\theta_W$ from the fermion coupling constants. Of course, both determinations of $\sin^2\theta_W$ coincide in the minimal GSW theory.

The value of $\sin^2\theta_W$ is [44]:

$$\sin^2\theta_W = 0.21 \pm 0.019 \pm 0.013 \; ,$$

entirely consistent with other determinations. Here the overall experimental error and the error due to the uncertainty of m_Z are given in turn. The specific interest in this particular value is the high Q^2 and the purely leptonic character of the data.

6.5 Semileptonic neutrino scattering off nucleons

Measuring the ratio of NC to CC scattering of neutrinos off heavy nuclei (nearly isoscalar targets) still provides the most precise way of determining $\sin^2\theta_W$. However, in order to extract $\sin^2\theta_W$, recourse is taken to the QPM description of the nucleon structure. In this approximation, in a world of u and d quarks only,

$$R_\nu = \left(\frac{NC}{CC}\right)_\nu = \frac{1}{2} - \sin^2\theta_W + \frac{5}{9} \sin^4\theta_W \; (1+r) \; ,$$

$$R_{\bar\nu} = \left(\frac{NC}{CC}\right)_{\bar\nu} = \frac{1}{2} - \sin^2\theta_W + \frac{5}{9} \sin^4\theta_W \; (1+\frac{1}{r}) \; , \qquad \text{with} \qquad r = \frac{CC_{\bar\nu}}{CC_\nu} \; .$$

These formulae have been derived by Llewellyn-Smith [45] by assuming isospin invariance only, allowing the determination of $\sin^2\theta_W$ in a largely model-independent fashion. The QPM is then needed only for minor corrections such as the scattering off the strange sea and the onset of charm production.

The sensitivity of R_ν on $\sin^2\theta_W$ is sizeably larger than that of $R_{\bar\nu}$, for $\sin^2\theta_W = 0.22$. Beam time is thus better invested in neutrino running.

The measurement of the NC to CC ratio has been a key issue for neutrino experiments ever since the discovery of NCs. The more precise results have been obtained by electronic detectors, because they can accumulate higher statistics than bubble chambers. There are four large counter experiments, two at FNAL and two at CERN. Their recent results are quoted in Table 1.

Table 1 Results for R_ν and $\sin^2\theta_W$ from semileptonic neutrino scattering.

Collaboration	Target Material	R_ν	$\sin^2\theta_W = 1 - m_W^2/m_Z^2$
CCCFR [46]	Iron		0.242 ± 0.012
CDHS [47]	Iron	0.301 ± 0.007	0.227 ± 0.012
CHARM [48]	Marble	0.320 ± 0.010	0.209 ± 0.016
FMM [49]	Sand		0.246 ± 0.018

It should be noted that the reported values of R_ν are not directly comparable since the experiments employed different cut-offs in hadronic energy, used different beam energies, and have target materials with different deviations from isoscalarity. However, the quoted values of $\sin^2\theta_W$ are directly comparable. They are compatible. The weighted average of the four experiments is $\sin^2\theta_W = 0.230 \pm 0.007$. The authors, however, are reluctant to accept the small error of this weighted average from experimental results which are all plagued by systematic rather than statistical errors.

On top of the quoted experimental error on $\sin^2\theta_W$, an additional "theoretical" error of about 0.006 arises from uncertainties in the determination of $\sin^2\theta_W$ due to the incomplete knowledge of QPM parameters. A detailed analysis of these errors can be found, for instance, in Ref.[47].

6.6 The 1984 CERN experiments

At the 1982 SPS Fixed-Target Workshop, Llewellyn-Smith argued that there is no obvious theoretical obstacle to measuring $\sin^2\theta_W$ to a precision of ± 0.005 arising from uncertainties of the hadronic structure of the nucleon [45]. Since a high-precision measurement of $\sin^2\theta_W$ was considered an important issue, the CDHS Collaboration launched into this programme. The main requirements to obtaining higher precision were firstly significantly larger event statistics, and secondly careful control of the so-called Wide-Band Beam (WBB) background in a NBB.

A design of a NBB was made by Grant and Maugain [50] with 160 GeV/c parent momentum, aimed at a flux larger by a factor of two as compared to the conventional 200 GeV CERN NBB which had been used before. Several upgrades were made in the beam line, the most important of which was the installation of a remotely-controlled total absorbing dump at the end of the parent hadron beam, just before the beginning of the decay tunnel. This way a precise experimental subtraction of parent decays occurring before the decay tunnel (the WBB background) could be achieved. This almost trivial new installation enabled a sizeable improvement in precision. The layout of the new beam line is shown in Fig. 12.

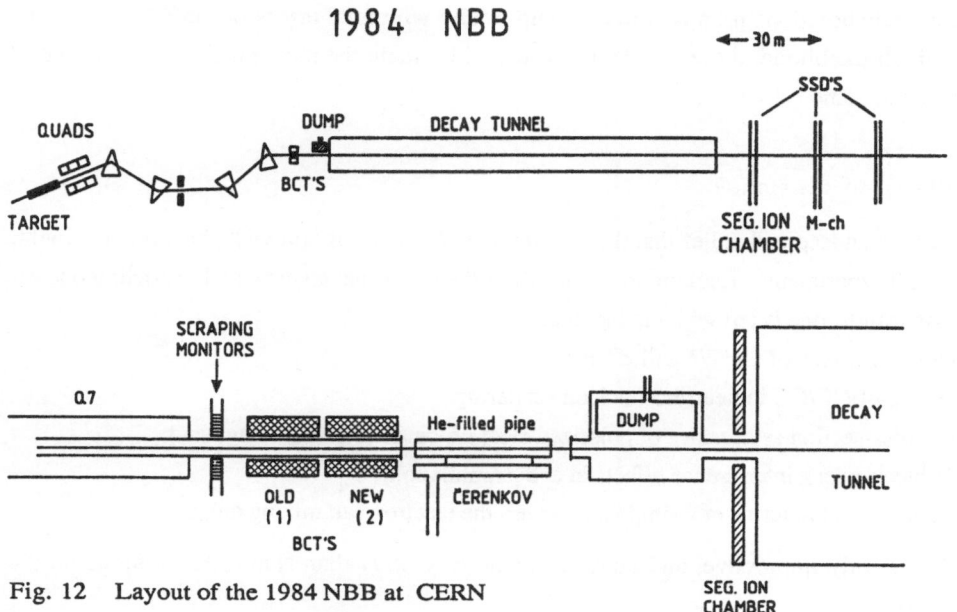

Fig. 12   Layout of the 1984 NBB at CERN

The CDHS detector had undergone an upgrade in 1980/81, which helped in the attempt to measure NC/CC more precisely. The hodoscope structure of the scintillator sheets enabled a precise location of the shower vertices independently of the existence of a reconstructed muon and provided a more uniform calorimetric response across the fiducial volume. After the proposal of the CDHS Collaboration had been approved, the CHARM Collaboration joined in the enterprise.

The year 1984 saw a brilliantly working CERN SPS, delivering a total of 6×10^{18} protons of 450 GeV/c momentum onto the neutrino target. The beam line worked according to expectation.

Both the CDHS and CHARM Collaborations reported preliminary results of their analyses at the 1985 International Conference on High-Energy Physics at Bari:

CDHS [51]: $R_\nu = 0.306 \pm 0.004$
$\sin^2\theta_W = 0.218 \pm 0.007$

CHARM [52]: $R_\nu = 0.317 \pm 0.006$
$\sin^2\theta_W = 0.215 \pm 0.010$

The same remarks as made in Section 6.5 apply to these results. They constitute the most precise values of $\sin^2\theta_W$ at present. They are well compatible with each other, which is a non-trivial happening in view of several occasions in the past where the CDHS and CHARM Collaborations had conflicting results.

Both collaborations promised to come up finally with total errors on $\sin^2\theta_W$ as small as 0.005. In all likelihood, these results on $\sin^2\theta_W$ will remain the most precise ones for two to three years to come.

7. Where do we stand?

There is widespread belief that the minimal GSW theory is "proved", beyond any doubt, by about 30 experiments. Looking more closely at the experimental data, and restricting oneself to precise results, one is left with the fact that

(i)   the mass values of the $W^\pm$ and Z^o bosons,

(ii) the rate of NC/CC in neutrino-nucleon scattering,

(iii) the cross-section asymmetry of polarized electron-deuterium scattering, and

(iv) the electroweak interference effects in e^+e^- annihilation,

can be understood in terms of a single parameter, the electroweak mixing angle.

This is truly impressive, and there seems no way other than Nature knowing about the GSW theory.

But: how well has the GSW theory been really tested experimentally?

The equality of $\sin^2\theta_W$ from the $W^\pm$ and Z^o mass on the one hand, and from νN scattering on the other hand, is tested to 5% precision. The best measured single coupling constant is a_μ, measured to 6% precision. The hadronic NC coupling constants are after ten years of neutrino experimentation still rather crudely known. For example, the predicted presence of V+A currents in νN scattering has never been proved beyond any doubt. Its significance has mainly been obtained from measured rates, rather than from shapes of y-distributions.

Most of the 30 supporting experiments give moral support only to the GSW theory, and do not allow stringent quantitative checks to be made. Moreover, one might wonder about the independence of all these experimental results. Remember that all experiments before 1977 favoured $\sin^2\theta_W \sim 0.33$, until CDHS claimed a much lower value.

Fortunately, significant experimental progress is in sight. The CHARM II experiment at CERN aims at measuring $\sin^2\theta_W$ to a precision of ± 0.005 from the purely leptonic process of $\nu_\mu e$ scattering. With sizeably more luminosity at the CERN $p\bar{p}$ Collider and at the Tevatron I at FNAL, in conjunction with better calorimetric devices than are available at present, the W mass will be measured with a precision of 100 - 200 MeV/c^2. The SLC machine at SLAC, and LEP at CERN are designed to study electroweak physics. They will permit a precise measurement of all leptonic NC coupling constants, and the determination of $\sin^2\theta_W$ from various processes with a precision of the order of 0.001. The question of the Higgs particle will be addressed at SLC and LEP. In a second stage, LEP II with $\sqrt{s} \sim 200$ GeV, aims at measuring the interactions of intermediate bosons with each other.

At the time of the SLAC eD experiment, the asymmetry due to the weak-electromagnetic interference was of the order of 10^{-4}. In 1990, HERA is planned to start operation, colliding 820 GeV protons with 30 GeV electrons, with Q^2 larger than m_Z^2. Weak-electromagnetic interference effects will then show up at the order of unity, and will constitute the background for the physics of tomorrow.

Acknowledgements

It is a pleasure to thank our colleague J. Wotschack for a critical reading of the manuscript. Many thanks are due to Mrs. A. Mazzari and Mrs. J.M. Rabbinowitz for their careful and patient editing work.

References

[1]   U. Camerini et al., Phys. Rev. Lett. 13 (1964) 318.

[2]   G. t'Hooft, Nucl. Phys. B35 (1971) 167.

[3]   S.L. Glashow, Nucl. Phys. 22 (1961) 579.
        S. Weinberg, Phys. Rev. Lett. 19 (1967) 1264; Phys. Rev. D5 (1972) 1412.
        A. Salam, Proc. 8th Nobel Symposium, Aspenäsgarden, 1968.
        (Almquist and Wiksell, Stockholm, 1968), p. 367.

[4]   S.L. Glashow, J. Iliopoulos and L. Maiani, Phys. Rev. D2 (1970) 1285.

[5]   F.J. Hasert et al., Phys. Lett. 46B (1973) 121.

[6]   F.J. Hasert et al., Phys. Lett. 46B (1973) 138.

[7]   A. Benvenuti et al., Phys. Rev. Lett. 32 (1974) 800.

[8]   B.C. Barish et al., Phys. Rev. Lett. 34 (1975) 538.

[9]   H. Faissner, Proc. Neutrino '75, Balatonfüred, Hungary, 1975
        (eds. A. Frenkel and G. Marx, Budapest, 1975), Vol. I, p. 116.

[10]   M. Holder et al., Nucl. Instrum. Methods 148 (1978) 235.

[11]   M. Holder et al., Phys. Lett. 71B (1977) 222.

[12]   P. Musset, Proc. Int. Symposium on Lepton and Photon Interactions, Hamburg, 1977
        (ed. F. Gutbrod, DESY, Hamburg 1977), p. 785.

[13]   M. Holder et al., Phys. Lett. 72B (1977) 254.

[14]   C.Y. Prescott et al., Phys. Lett. 77B (1978) 347.

[15]   C. Baltay, Proc. Int. Conf. on High Energy Physics, Tokyo 1978 (Physical Society of
        Japan, Tokyo 1979), p. 882.

[16]   L.L. Lewis et al., Phys. Rev. Lett. 39 (1977) 795.

[17]   P.E.G. Baird et al., Phys. Rev. Lett. 39 (1977) 798.

[18] C.Y. Prescott et al., Phys. Lett. 84B (1979) 524.

[19] P.Q. Hung and J.J. Sakurai, Phys. Lett. 63B (1976) 295.

[20] L.M. Seghal, Phys. Lett. 71B (1977) 99.

[21] For a list of references see J.E. Kim et al, Rev. Mod. Phys. 53 (1981) 211.

[22] P. Alibran et al., Phys. Lett. 74B (1978) 422.

[23] L.M. Barkov and M.S. Zolotorev, Proc. "Neutrinos-78", West Lafayette, 1978
(ed. E.C. Fowler, Purdue University, West Lafayette, 1978), p. 423.

[24] P. Langacker et al., Proc. "Neutrino-79", Bergen, 1979
(eds. A. Haatuft and C. Jarlskog, Bergen, 1979) Vol. 1, p. 276.
I. Liede and M. Ross, ibid., p. 309.

[25] H. Georgi and S.L. Glashow, Phys. Rev. Lett. 32 (1974) 438.

[26] A. Sirlin and W.J. Marciano, Nucl. Phys. B189 (1981) 442.

[27] J.F. Wheater and C.H. Llewellyn-Smith, Nucl. Phys. B208 (1982) 27.

[28] W.J. Marciano and A. Sirlin, Phys. Rev. D29 (1984) 945.

[29] M.A. Bég and A. Sirlin, Phys. Rep. 88 (1982) 1.

[30] C. Rubbia, P. McIntyre and D. Cline, Proc. Int. Neutrino Conf., Aachen 1976 (Vieweg, Braunschweig, 1977), p. 683.

[31] G. Arnison et al., Phys. Lett. 122B (1983) 103.

[32] M. Banner et al., Phys. Lett. 122B (1983) 476.

[33] G. Arnison et al., Phys. Lett. 126B (1983) 398.

[34] P. Bagnaia et al., Phys. Lett. 129B (1983) 130.

[35] G. Arnison et al., preprint CERN-EP/85-185 (1985).

[36] J.A. Appel et al., preprint CERN-EP/85-166 (1985).

[37] F. Reines, H.S. Gurr and H.W. Sobel, Phys. Rev. Lett. 37 (1976) 315.

[38] R.C. Allen et al., Phys. Rev. Lett. 55 (1985) 2401.

[39] M. Jonker et al., Phys. Lett. 117B (1982) 272:
F. Bergsma et al., Phys. Lett. 147B (1984) 481.

[40] L.A. Ahrens et al., Phys. Rev. Lett. 51 (1984) 1514 and 55 (1985) 1814.

[41] G. Barbiellini and C. Santoni, preprint CERN-EP/85-117 (1985).

[42] A. Argento et al., Phys. Lett. 120B (1983) 245.

[43] W. Bartel et al., Phys. Lett. 146B (1984) 437.

[44] B. Naroska, Talk given at "Physics in Collision V", Autun, France, 1985; preprint DESY 85-090 (1985).

[45] C.H. Llewellyn-Smith, Nucl. Phys._B228 (1983) 205.

[46] P.G. Reutens et al., Phys. Lett. 152B (1985) 404.

[47] H. Abramowicz et al., Z. Phys. C28 (1985) 51.

[48] M. Jonker et al., Phys. Lett. B99 (1981) 265.

[49] D. Bogert et al., preprint FERMILAB-Conf-85/107-E (1985).

[50] A. Grant and J.M. Maugain, Internal Report CERN/EF/BEAM 83-2 (1983).

[51] A. Blondel, Talk given at the Int. Conf. on High Energy Physics, Bari, 1985.

[52] R. Pain, Talk given at the Int. Conf. on High Energy Physics, Bari, 1985; preprint CERN-EP/85-113 (1985).

How Well Do We Know the Parton Distributions in the Nucleon from Deep Inelastic Scattering Experiments?

F. Eisele

DESY, D-2000 Hamburg 52, Fed. Rep. of Germany

R. Turlay

CEN, Saclay, France

Introduction

The discovery of pointlike partons inside the nucleon by the inelastic electron scattering experiments at SLAC (68/69) has been a milestone for the development of our present understanding of the building blocks of matter and their fundamental interactions as summarized by the standard model. Afterwards early neutrino experiments were able to identify the partons as quarks. With the increase of the energy scale of scattering experiments at the super proton synchrotrons, the ISR, the $e^+ - e^-$ storage rings and recently the proton antiproton collider, the quark and gluon substructure of the hadrons became visible in many other so-called hard scattering processes. The QCD-inspired quark parton model is able to describe these processes surprisingly well and to relate them using for all of them the same parton distributions for incoming hadrons and the same fragmentation functions. The genuine hard scattering process is given by the elastic scattering of pointlike partons and/or leptons via the weak, electromagnetic or strong interaction. Well known examples are: the Drell-Yan process of high mass lepton pair production and recently the production of W and Z^0 vector bosons, heavy quark production in photon and hadron collissions, single hard photon production and high E_T jet production in hadron-hadron collisions. All these processes have in common that they require for their description the knowledge of the parton distributions of the initial nuclei or nucleon e.g. the quark and antiquark distributions separated for all flavours and the gluon distribution. It is our understanding that one of the most important accomplishments of neutrino experiments has been the provision of a reliable set of parton distributions in the nucleon.

We have undertaken to write this article to show our respect and friendship for Jack with whom we had the chance and pleasure to work together for so many years. The subject of this article seems to us specially suited for the occasion since the study of the nucleon structure has been for Jack the main interest of neutrino physics at the SPS and the field where he has invested most of his work. It is interesting to look back to an interview which he gave to the CERN Courrier of March 1978. One can read: " A veteran at the neutrino game, having been a member of the team which first studied high energy neutrino interactions at Brookhaven back in 1962, Steinberger prefers instead to overlook transient trends and concentrate on what in his view is the central objective of neutrino physics: The study of the structure of the nucleon." We should remember what was the situation in 1978 to understand how this statement looks now profound and wise. At this time the high energy physics community was hunting for big discoveries. After a ephemoral "high y anomaly" effect, it was the multilepton events, quite spectacular, which were exciting physicists. It is quite impressive among all this excitation that Jack was looking and leading towards the most important contribution of neutrino physics at the SPS: the study of the nucleon structure at the parton level. It will become apparent in the course of this article that this work was moreover very successfull. The task is nearly completed to provide a reliable and consistent set of parton distributions.

Experimental determination of parton distributions

In principle all hard scattering processes provide information on the parton distributions. Deep inelastic scattering of neutrinos and charged leptons however play a unique role for the quark and anti-quark distributions since the pointlike leptons serve as ideal probe with well known interactions and the scattering processes occur at the parton level in most of the available kinematic range. Deep inelastic neutrino experiments have also determined the gluon distribution by a QCD-analysis of the Q^2-dependence of the structure functions, which is an indirect method but based on safe predictions. A direct measurement of the gluon distribution is accessible only to hadronic interactions and substantial progress has recently been made at the $\bar{p}p$-collider as will be discussed below.

Let's shortly summarize the strategy to obtain a consistent set of parton distributions including their Q^2-dependence. The key role is reserved for neutrino experiments since they are able to separate valence and sea quark distributions, to measure the strange sea distribution and to determine the longitudinal structure function q_L with good sensitivity over a large Q^2-range. This however is only true for experiments on heavy targets like iron where only the sum of up and down quark distributions can be determined. The separation of up and down flavours e.g. the determination of $xu_V(x,Q^2), xd_V(x,Q^2), x\bar{u}(x,Q^2), x\bar{d}(x,Q^2)$ is achieved by low statistics neutrino experiments on deuterium and hydrogen targets. The combination of all neutrino measurements could give a complete set of parton distributions with rather good precision. Unfortunately this is not immediately possible since the muon experiments at CERN and e-A scattering experiments at SLAC have demonstrated that parton distributions are different for free nucleons compared to nucleons which are tightly bound in heavy nuclei. This so-called " EMC-effect " (European Muon Collaboration : Aubert et al. 1982) is a major source of uncertainties in our knowledge of parton distributions at present as will be discussed below. The gluon distribution is not directly accessible to lepton scattering experiments since gluons are inert to leptons. These experiments determine however the momentum fraction of the gluons via the energy-momentum sum rule. Neutrino experiments are able to determine also the shape of the gluon distribution by a combined analysis of the scaling violations of the structure functions F2 , xF3 and $\bar{q}^{\nu}$ within the framework of QCD. Though this analysis is indirect and model dependent in contrast to the determination of the quark distributions it is nevertheless regarded as one of the more reliable ways to learn something about the gluon distribution.

Experimental input: main improvements

Parametrisations of parton distributions have been provided in the past by various authors and have received wide application. Their decomposition of the flavour content of the nucleons has suffered however from inconsistencies of experimental data and also from uncertainties about the impact of the EMC-effect on our knowledge of parton distributions. The experimental inconsistencies have been to a large extent due to i) uncertainties of normalisation mainly for neutrino experiments ii) insufficient experimental information on the longitudinal structure function q_L and iii) uncertainties in the magnitude and energy dependence of the strange sea component as seen by neutrino experiments. These defects have been cured by new experimental results and at the same time our knowledge of the EMC-effect has improved such that a substantial improvement of our knowledge of parton distributions will be possible in the near future. We discuss the main improvements below and afterwards give some results of a preliminary analysis which illustrate changes with respect to older parametrisations and their impact on hard scattering processes.

● *normalisation for neutrino experiments*

The measurements of total neutrino and antineutrino cross-sections had led to experimental discrepancies in the past. Recently these could be resolved by the CDHSW collaboration which, by two independent methods, was able to confirm the cross-section measurements of the CCFRR collaboration and at the same time shed light on the origin of the experimental discrepancies. As a result the cross-sections are now well known which allows to give precise values for the momentum fractions of quarks and gluons in the nucleon. The present world averages (ignoring measurements prior to 1981 with known defects) are:

$$\sigma^{\nu}/E_{\nu} = (.680 \pm .014) \cdot 10^{-38} [cm^2 GeV^{-1}]$$
$$\sigma^{\bar{\nu}}/E_{\nu} = (.326 \pm .008) \cdot 10^{-38} [cm^2 GeV^{-1}]$$

$$\int_0^1 F2(x,\nu)dx = .485 \pm .015$$

$$\int_0^1 xF3(x,\nu)dx = .332 \pm .010$$

$$\int_0^1 xG(x,\nu)dx = .515 \pm .016 \ .$$

These averages are valid for an average energy transfer $\langle\nu\rangle= 50$ GeV and change with ν.

● *measurement of the longitudinal structure function q_L*

From the experimental point of view it is a real challenge to measure q_L and published results have been very unsatisfactory so far. Also the knowledge of q_L is essential to determine the sea components in the nucleon which are very strongly correlated to it. A breakthrough has been recently achieved by the CDHSW collaboration based on unprecedented statistics and, more important, a new method specific to neutrino experiments which gives much higher sensitivity to q_L . Their measurement is shown in figure 1 as a function of x together with the QCD prediction for q_L. For the first time these data give convincing evidence for a strong x-dependence whereas a Q^2-dependence is not observed. This new measurement gives a satisfactory experimental basis to use the QCD prediction for q_L to determine the flavour composition.

● *momentum fraction and energy dependence of the strange sea*

The strange sea distribution can be measured only by neutrino experiments via single charm production which in the case of antineutrinos is mainly due to $s \rightarrow c$ transitions whereas for neutrinos this transition contributes about 50%. Since scattering off the strange sea leads to charm production in about 95% of all cases its contribution to neutrino scattering is suppressed by the charm threshold effect which has to be corrected in order to obtain the strange sea component. Single charm production has been studied with high statistics by CDHS via opposite sign dimuon events with the following results (Abramowicz et al. 82): i) The shape of the strange sea agrees with the shape measured for $x(\bar{u}+\bar{d})$ ii) There is experimental evidence for

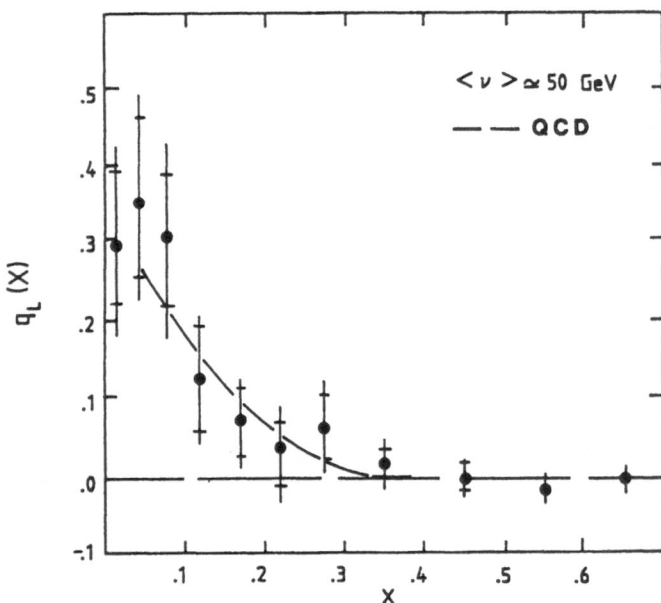

Figure 1 : *Measurement of the longitudinal structure $q_L(x)$ by the CDHSW collaboration (1985, preliminary) compared to the QCD-prediction*

a threshold suppression in agreement with the slow-rescaling model iii) The sea is not SU(3) symmetric but the strange quark carries only $(57 \pm 10)\%$ of the momentum fraction of a light quark for an average momentum transfer $\langle \nu \rangle$ of 30 GeV. (This latter result makes use of the Kobayashi-Maskawa mixing matrix elements as determined for three quark generations.)

- *separation of up and down flavours*

A complete separation of $u_V, d_V, \bar{u}, \bar{d}$ distributions has been acheived by the WA25 bubble chamber experiment at CERN which uses neutrino and antineutrino interactions on deuterium. Their results are partially displayed in figure 2 (Allasia et al. 84). This experiment provides the following information which is confirmed by other neutrino experiments using hydrogen and also by charged lepton experiments: i) The sea distributions for up and down quarks agree both in shape and in magnitude ii) the valence distributions differ in shape and are approximately related by $d_V(x) = .57 \cdot (1-x) \cdot u_V(x)$ in the

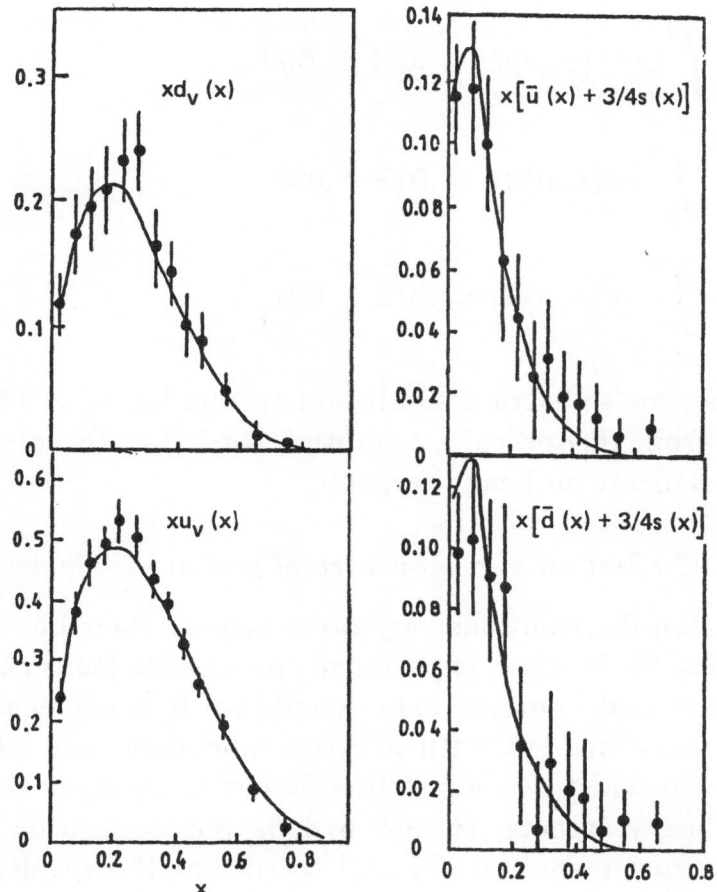

Figure 2 : *Measurements of up and down quark distributions on free nucleons by the WA25 νD_2 experiment. The solid lines are the predictions of the parametrisation discussed in this article*

range $0 \le x \le .8$. iii) the Q^2-dependence of the up and down valence distributions is in good agreement with the scaling violations as found by neutrino and muon experiments on heavy targets (Allasia et al.85).

The momentum fractions for an average energy transfer $\langle \nu \rangle = 25$ GeV are:

$$\int_0^1 x u_V(x, \nu) dx = .245 \pm .010$$

$$\int_0^1 x d_V(x, \nu) dx = .087 \pm .006$$

$$\int_0^1 x \bar{u}(x, \nu) dx = .019 \pm .005$$

$$\int_0^1 x \bar{d}(x, \nu) dx = .018 \pm .004 \quad.$$

The statistics is too low to extract significant results for q_L . This input has to come from theoretical assumptions or it has to be obtained from measurements on heavy targets .

- *impact of the EMC-effect on our knowledge of parton distributions*

Applications of parton distributions very often require their knowledge for free protons. Since muon experiments on free nucleons cannot separate flavours and neutrino experiments on free nuclei are limited by statistics we are left with a rather unsatisfactory situation. One approach which we will follow below is to start with neutrino data on heavy targets, to add u/d flavour separation as given by neutrino experiments on D_2 and to correct the resulting parton distributions using the measurements of charged lepton scattering experiments on the ratio of the structure function F2(A) on a heavy target to $F2(D_2)$ on deuterium. Recent results are shown in figure 3 (Feltesse 85). The following conclusions can be drawn: i) there is a clear and well established effect in the intermediate x-region $.25 \leq x \leq .60$ which shows no Q^2- dependence and which is known to rise approximatly like $\ln(A)$ with the atomic number of the target (Bodek et al. 83) ii) charged lepton scattering experiments give discrepant results in the small x region (sea region).

Three neutrino experiments have actually published sea ratios. The results given are ($U = \int_0^1 xu(x) dx$ etc.):

$$(\bar{U} + \bar{D} + 2S)_{Ne} / (\bar{U} + \bar{D} + 2S)_{D_2} = 0.91 \pm .07 \; (Cooper \; A \; M \; et \; al. \; 84)$$

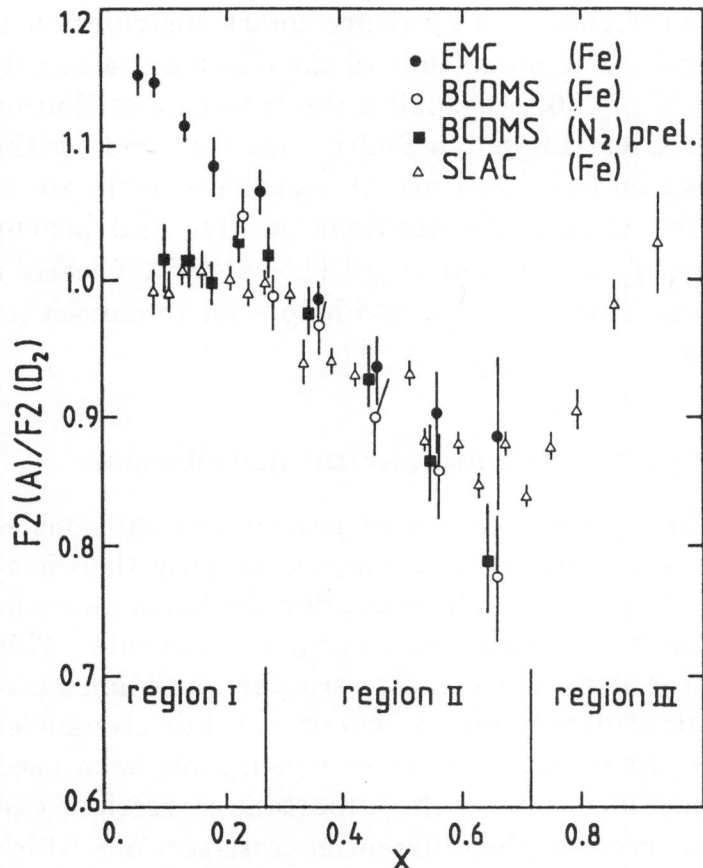

Figure 3 : *Ratio of the structure function F2(A) on heavy targets to F2(D₂) on deuterium versus x as measured by three charged lepton experiments (Feltesse 85)*

$$(\bar{U} + \bar{D} + 2S)_{Ne} / (2\bar{D} + 2S)_{H_2} = 0.95 \pm .16 \; (Parker \; M \; A \; et \; al. \; 84)$$

$$(\bar{U} + \bar{D} + 2S)_{Fe} / (2\bar{D} + 2S)_{H_2} = 1.10 \pm .12 \; (Abramowicz \; et \; al. \; 84).$$

Thus neutrino experiments give no evidence for a difference of the sea component and we can derive a conservative upper limit of this difference of $\leq 15\%$.

The uncertainty of the parton distributions due to the EMC-effect can be summarized as follows: The sea component differs by at most 15%between free nucleons and bound nucleons and there is at present no evidence for any difference. The valence quark distribution is harder for a free nucleon. The magnitude of the effect is reasonably well measured for $.25 \leq x \leq .6$. At small x the valence distribution is constrained by the Gross-Llewellyn-Smith sum rule and by the neutrino measurements on free nucleons. A reasonable estimate at present is to relate the valence distributions for free and heavily bound nucleons by $q_V(x)_{free} \approx q_V(x)_{bound} * (1 - x)^{.2 \pm .1}$. Clearly a theoretical model of the EMC- effect would help a lot to correct for it in a more reliable way.

Example of a strategy to separate parton distributions

The strategy to obtain a consistent set of parton distributions is to some extent a matter of taste. Several authors (mostly theorists) have used the approach to give a simultaneous description of all available data of both muon and neutrino scattering experiments. This immediately leads to problems since these structure function measurements are inconsistent partly due to experimental discrepancies mostly however due to different assumptions which have been used for their determination. Since some of the important corrections can only be done if one has access to the differential cross-sections which are generally not published, this problem cannot be easily overcome.

We advocate and follow a different approach which uses only very few selected data sets to keep systematic problems as small as possible. In our approach the main experimental information is provided by the CDHS structure function measurements on an iron target which is by far the most precise and complete set measured over a large Q^2- range. Measurements include $xF3(x,Q^2)$, $\bar{q}^{\bar{\nu}}(x,Q^2)$, $q_L(x,Q^2)$, $F2(x,Q^2)$ and the measurement of the shape and momentum fraction of the strange sea $xs(x)$. This heavy target data is complemented by the structure function measurements of the WA25 neutrino experiment on deuterium which has by far the best data to separate the up and down quark distributions on free nucleons.

The Q^2-dependence of the structure functions is well described by QCD and this is also the only reliable basis to make extrapolations of present measurements to higher energy machines. In a first step we therefore determine the scale parameter Λ (in leading order) and at the same time the shape of the gluon distribution $xG(x,Q_0^2)$ from a combined QCD-analysis of F2 and $\bar{q}^\nu$ (Abramowicz et al. 83). In this step we use $q_L(x)$ as predicted by QCD in agreement with recent measurements and the slow rescaling correction for the strange sea. In a second step we determine simple parametrisations for all quark and antiquark flavours at a representative value $Q_0^2 \approx 5 \; GeV^2/c^2$. This is done by a combined fit to xF3 (x,Q^2), $d_V(x)/u_V(x)$, $\bar{q}^\nu(x,Q^2)$ and $2S/(\bar{U} + \bar{D})$ where the Altarelli-Parisi evolution equations are used to interpolate all measurements to the same reference value Q_0^2. The sea quark distributions are obtained under the assumption $\bar{u}(x) = \bar{d}(x) = \alpha \star s(x)$ in this analysis in agreement with the measurements as discussed above. This completes the flavour separation apart from corrections related to the A-dependence of the parton distributions (EMC-effect). Given the present uncertainties in the understanding of the effect and it's rather small magnitude we believe that it is adequate to make ad hoc corrections to the valence distributions at this step (e.g. multiply these distributions by $(1-x)^{-.2\pm.1}$ for free nucleon distributions) and possibly also corrections for the magnitude of the sea distributions if they should turn out to be significant in future measurements. Finally, in order to apply the parton distributions at different values of Q^2 we have to make an evolution using the Altarelli-Parisi equations.

We have performed a preliminary analysis along these lines using published CDHS data (Brummel 84). In order to facilitate applications we have chosen to provide parametrisations which include the Q^2-dependence up to the highest energies of the next generation of colliders. This has been done in practice by numerical integration of the Altarelli-Parisi equations (using Q^2/Λ^2 as evolution parameter) and subsequent ad hoc parametrisation of the numerical results in (x,Q^2) adequate for computer applications. This preliminary analysis ignores the EMC-effect and is still based on a cross-section which is 15% lower than the new world average. A more refined analysis is underway which will be based on the new structure function measurements of CDHSW based on much higher statistics (more than

1 million events for both neutrino and antineutrino measurements). We are confident that these new results will provide a firm and reliable basis for further work which will not be superseded in the forseeable future. The limitations and necessary improvements will be summarized at the end.

Illustration of parametrisations and examples of application

Figure 4 shows the shape of all parton distributions for an average energy transfer of 30 GeV (representative for present measurements) and for 50000 GeV which is about the maximum value obtainable at a future e-p collider. It can be seen that sea quark and gluon distributions at small x (x≤ .1) change dramatically and are dominated by the evolution not by their present magnitude. All parton distributions at large x (x≥.2)on the other hand are strongly coupled to their present shape and magnitude and to the magnitude of the scale parameter Λ. Figure 2 which gives the free target measurements of

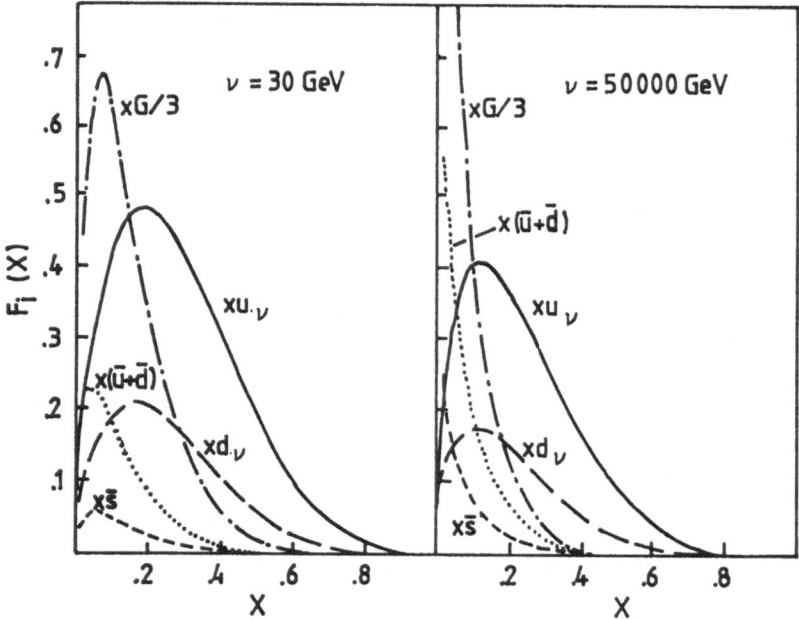

Figure 4 : *The shape of all parton distributions for two values of the energy transfer ν as given by the parametrisation described in this article*

WA25 shows the predictions of the parametrisations (solid lines) at the relevant value of the energy transfer. It should be noted that the magnitude and shape of all distributions are determined by heavy target data apart from the ratio $d_V(x)/u_V(x)$. The good agreement illustrates again the smallness of the EMC-effect and justifies the procedure.

Finally we give two examples for application to hard scattering processes:

- *Muon pair production in proton nucleon scattering*

Figure 5 shows measured differential cross-sections as measured in p-N scattering by the CFS-collaboration (Ito et al. 81) and in p-p scattering at the ISR (Antreyasyan et al 80). The QPM-predictions based on our parametrisations are shown as solid and dashed-dotted curves, the ratio of measured cross-section to this prediction - the famous k-factor - is shown in the lower part of the figure. The QPM prediction is lower by an average value of k$\approx$ 1.8 but shows a very similar x-dependence. The importance of a good knowledge of parton distributions for these processes is just to get a reliable determination of k(x) which has to be explained by theory. (The conjecture is that k(x) is not equal to 1.0 due to calculable QCD corrections to the QPM diagram). For the given example the knowledge of the antiquark distributions for free and bound nucleons is of prime importance. Our parametrisations differ from older ones mainly in that the sea quark distributions are smaller due to a better understanding of the scaling violations and because the large values of $q_L(x)$ at small x imply smaller values of $\bar{q}(x)$. As a result, the k-factors come out larger than in the first publications. Similar conclusions have recently been reached by new experiments (NA3, NA10) which have made use of the same parametrisations. Finally it should be noted that the k-factor for the data shown in figure 5 is very sensitive to a difference of the sea quark distributions for free nucleons compared to bound nucleons. A 15% difference, still compatible with present data, would increase k(x) by 15% in average and would induce a strong x-dependence.

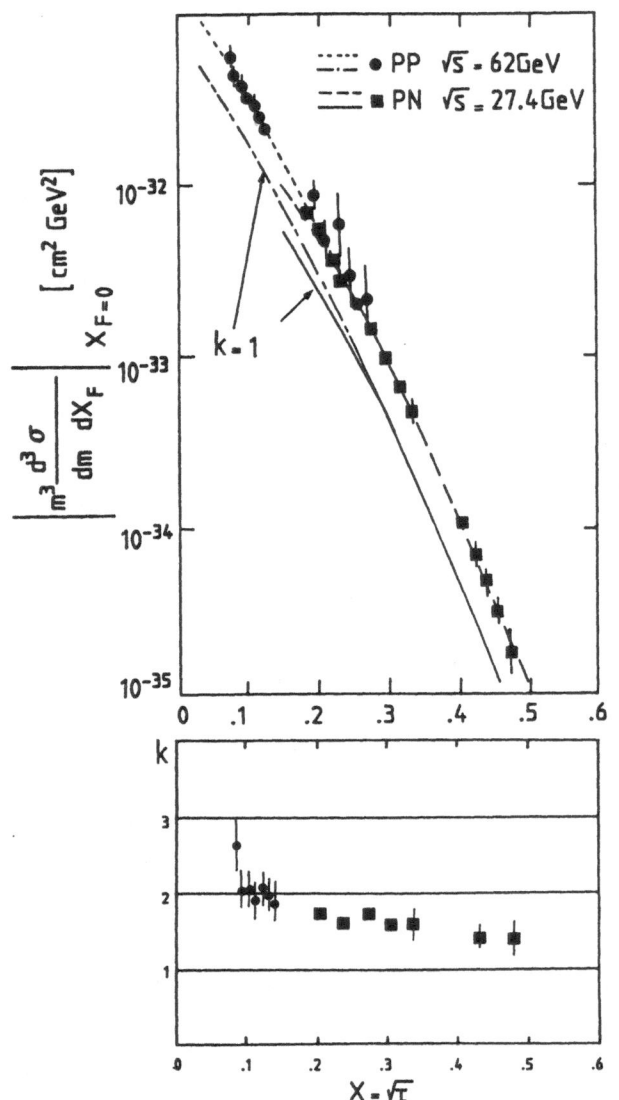

Figure 5 : *Differential cross-section for muon pair production in proton nucleon scattering outside the resonance regions compared to the QPM pedictions based on the parametrisations of this article (solid lines). The dashed lines give the predictions multiplied by a k-factor of 2.0. The lower part of the figure shows this factor directly: k=measured cross-section/QPM*

• *Jet-Jet cross-sections at the $\bar{p}p$-collider*

Hard scattering processes at the $\bar{p}p$-collider provide the best direct access to the gluon distribution at present. The interpretation of the Jet-Jet cross-sections is however only possible in the framework of the QCD-inspired parton model and involves the assumption that leading order diagrams describe the data up to a **constant** factor k. Figure 6 shows recent results from the UA1 experiment (Arnison et al. 84) for the structure function $xG(x)+4/9\star(q(x) + \bar{q}(x))$ com-

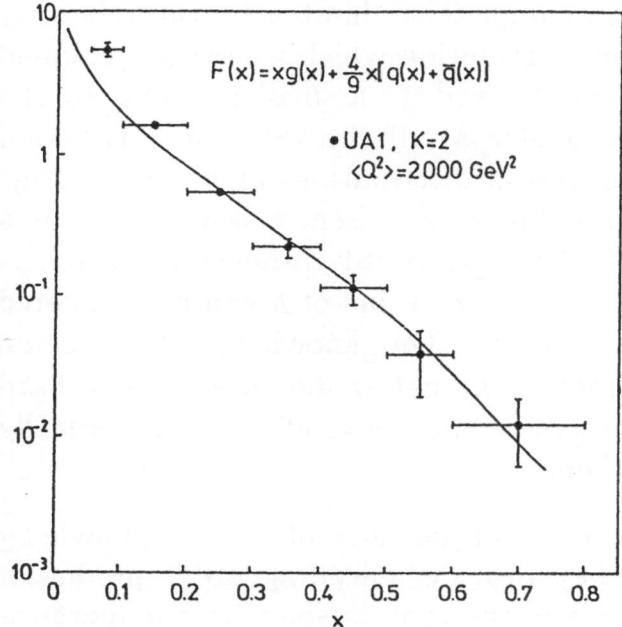

Figure 6 : *The structure function $F(x)=xG(x)+4/9*(q(x) + \bar{q}(x))$ versus x as derived by the UA1 collaboration from high E_T Jet-Jet cross-sections in $\bar{p}p$-collisions. The solid curve gives the parametrisations for $F(x)$ evoluted to a value of $Q^2=2000$ GeV^2 and multiplied by $k=2.0$*

pared to the prediction of our parametrisation evoluted to a value of $Q^2=2000$ GeV^2/c^2, which is supposed to be a relevant energy scale for this data . The agreement is satisfactory giving us confidence that the QCD-inspired parton model works well. We may also conclude, though not too convincingly, that our parametrisation of the gluon distribution is adequate.

Summary and conclusions

Deep inelastic scattering experiments have provided a reliable basis for our knowledge of quark and antiquark distributions for all flavours. Remaining uncertainties are mainly due to the poor measurement and understanding of the A-dependence (EMC-effect). Improvements would be welcome there and are possible. The application of parton distributions at high Q^2 is limited by the relatively poor knowledge of the gluon distribution which is strongly coupled to the sea distributions at small x and, to a smaller extent, by the uncertainty of the scale parameter Λ. If the value of Λ is varied between .1 and .4 GeV, the parton distributions at large x change by about 20% for an an extrapolation of present measurements by a factor more than 1000 in Q^2. This uncertainty is however not inherent to the parametrisations since every value of Λ can be simulated by changing the evolution scale Q^2. Our knowledge of the gluon distribution can only be improved by better measurements of hard hadron-hadron scattering processes and above all by a theoretically satisfactory description of them.

This brings us back to the main importance of a good knowledge of parton distributions: They provide a leading order prediction for hard scattering processes and the comparison with the measurements gives experimental constraints for the theoretical description of higher order corrections and the relevant energy scale for these processes. A simple and surprising result so far is that all hard scattering processes seem to be well described by the QCD- inspired QPM up to a nearly **constant** factor.

In summary, deep inelastic scattering experiments at the SPS have nearly completed their task to explore the nucleon structure at the parton level. Intermediate experimental problems have been resolved such that these experiments, which have been a major part of the SPS experimental program, will be able to provide a reliable and consistent set of parton distributions in the nucleon which constitute one of the basic ingredients of the standard model. We will have to rely on these measurements for a long time since they cannot be repeated with similar accuracy and since it will be hardly possible to supersede them in the forseeable future.

REFERENCES

Abramowicz H et al. 1982, Z. Phys. C15, 19
Abramowicz H et al. 1983, Z. Phys. C17, 283

Allasia D et al. 1984, Phys. Lett. 135B, 231

Allasia D et al. 1985, Z. Phys. C28, 321

Antreyasyan D et al. 1980, Phys. Rev. Lett. 45, 863

Arnison G et al. 1984, Phys. Lett. 136B, 294

Aubert J J et al. 1983, Phys Lett. 123B, 275

Bodek A et al. 1983, Phys Rev. Lett. 50, 1431 and 51, 534

Brummel H D 1984. Diplomarbeit, Inst. Phys. Dortmund. This reference contains explicit parametrisations, comparisons with other parametrisations and with other data.

Cooper A M et al. 1984, Phys. Lett. 141B, 133

Feltesse J 1985, Review talk at the Europhysics Conference on High Energy Physics Bari, 18-24 July 1985, DPHPE 85-09.

Ito A S et al. 1981, Phys. Rev. D23, 604

Physics at LEP

J. Ellis
CERN, CH-1211 Genève 23, Switzerland

1. Introduction

Particle physics is now at a very interesting stage. We have the Standard Model that describes all existing data very well, but could not possibly be an ultimate Theory of Everything. Opinions are divided as to the way to advance. What type of experiment will be the first to find a discrepancy with the Standard Model — a precision measurement checking a refined prediction, or the dramatic brute force discovery of a new particle? Which type of theoretical idea may best guide us — unified gauge theories, composite models of quarks and leptons, attempts to understand the origin of particle masses, or something else? None of us knows, though each of us has his prejudices. In such a situation, every one of us must gamble on which line to follow next. Jack's bet is on a large experiment at LEP. In this contribution I will describe why this seems to me a very promising gamble, and how the ALEPH experiment may be able to go beyond the Standard Model and discriminate between rival theoretical ideas. I have my own bet as to which theories are ripe to be investigated, and these prejudices will emerge clearly from my presentation. They are reflected in the choice of LEP physics made in this article. A comprehensive preview of all LEP physics is now being prepared [1], and I have selected a few topics which interest me particularly, and will perhaps appeal to Jack.

LEP will not be the first accelerator to explore energies up to 200 GeV in the centre of mass, but it will be much cleaner than its predecessors, the $p\bar{p}$ colliders at CERN and FNAL. The simplicity of the initial state and the absence of spectator debris in the final state make e^+e^- annihilation events comparatively precise experimental probes. Clearly precision tests of the Standard Model, using millions of Z^0 events, will be possible at LEP. But the clean experimental environment will also make possible searches for weakly interacting particles with subtle signatures which would be missed at $p\bar{p}$ colliders. Thus both types of experiment mentioned in the first paragraph — precision tests and new particle searches — will be possible at LEP.

Jack has been active in both kinds of experiment, and here I only mention one of each type. On the one hand, data gathered and analysed by the WA1 Collaboration have contributed greatly to the experimental elucidation of the Standard Model, notably its neutral currents and QCD as probed by scaling violations. On the other hand, Jack is a co-discoverer of the second neutrino. He will be able to pursue both lines with the LEP experiment, for example by making precision measurements at the Z^0 peak, searching for the Higgs boson, and counting the total number of neutrino species.

My own interests lie more with search experiments than with precision experiments, and this preference is reflected in subsequent sections. Section 2 discusses some of the various theoretical ideas which can be tested at LEP, and here I emphasize the importance and timeliness of attempting to understand the origins of particle masses. Some approaches to this problem imply the existence of scalar Higgs bosons [2], and of supersymmetric particles with spin 0 and $\frac{1}{2}$ [3]. Section 3 of this contribution describes ways to look for a Higgs boson at LEP, while Section 4 discusses possible searches for supersymmetric particles. Precision measurements at the Z^0 peak are described elsewhere in this volume [4], and the most recent comprehensive report on LEP physics [1] has already been mentioned. Section 5 of this article contains a few concluding thoughts about the future at LEP.

2. Some theoretical ideas

Although the Standard Model is very successful, it cannot possibly be the final Theory of Everything. It contains many free parameters, including three distinct gauge coupling constants, has no explanation or understanding of the number of fundamental quarks and leptons and their weak mixing angles, and the Higgs mechanism for generating the weak boson masses is experimentally untested as well as theoretically unsatisfactory. Three main problem areas can be distinguished, namely those of unification, flavour and mass, and there are ideas on the market for solving each of them.

One can try [5] to unify all the fundamental particle interactions in a Grand Unified Theory (GUT) with a single gauge coupling whose symmetry is broken at some very large energy scale. To test this type of theory directly, one should look for some new type of interaction such as baryon decay, or some new superheavy particle such as a monopole. Searches for these phenomena have been rather unsuccessful so far, and LEP is not appropriate to continue the search. However, there are some indirect tests of GUTs which can be made at LEP. For example, GUTs predict [6] a value for the electroweak mixing angle $\sin^2 \theta_W \sim 0.22$ — which is consistent with the results from neutrino–hadron scattering and elsewhere — and many GUTs cannot [7] accommodate more than three generations of quarks and leptons. As discussed elsewhere [4] in this volume, LEP will be able to make the most precise measurement of $\sin^2 \theta_W$ at the Z^0 peak, and measurements there and at slightly higher energies will be able to count the number of neutrino species and hence the number of generations. These important measurements will not be discussed further in this contribution.

A favoured approach to the flavour problem is to make the quarks and leptons composite, and perhaps the $W^\pm$ and Z^0 bosons as well. We know from precision experiments at low energies that the scale of compositeness must be at least 1 TeV, and in some cases much higher. There is no clear upper bound on the scale of compositeness, and hence no guarantee that it will produce any effects observable at LEP. A detailed study of possible manifestations of compositeness at LEP is available elsewhere [8], and I will not discuss them here. Constructing a theoretically consistent composite model is very difficult, and most models on the market are not satisfactory [9]. Moreover, the idea of 'another layer of the onion' does not appeal to me as very original or exciting.

Elementary particle masses in the Standard Model are generated by couplings to the Higgs boson, and at least one physical, neutral Higgs particle with mass O(10 GeV to 1 TeV) must exist [2]. However, no sign of this mythical scalar has ever been seen. This Higgs mechanism is theoretically unsatisfactory, since the mass of the Higgs particle itself is very unstable and tends [3] to acquire large corrections in many different theoretical frameworks. For example, radiative corrections in the Standard Model generate large effects unless they are cut off by some new physics at a relatively low-energy scale $\Lambda = O(m_W/\sqrt{\alpha}) = O(1$ TeV). Corrections to the Higgs boson mass in GUTs and other unified theories are even more unwieldy. One way to cut the Standard Model off with new physics is to make the Higgs boson composite [10] at an energy scale ≤ 1 TeV. If one extends [11] this theory to give masses to the quarks and leptons as well as to the $W^\pm$ and Z^0, it tends to predict [12] many new particles with masses ≤ 100 GeV which could be seen at LEP. Unfortunately, other predictions [13] of such composite Higgs models tend to conflict with upper bounds on flavour-changing neutral currents, notably from experiments on the K^0-$\bar{K}^0$ system. An alternative candidate for the new physics which cuts off the Standard Model at an energy ≤ 1 TeV is supersymmetry [3]. This theory predicts many new particles which must have masses ≤ 1 TeV if they are to do their job of protecting the Higgs boson mass. Many specific models further predict the existence of at least some supersymmetric particles with masses ≤ 100 GeV which could be produced and detected at LEP.

In my opinion, the best prospects for early experimental tests of these different ideas is offered by the mass problem. It sets a (relatively) low energy scale for its own solution, and predicts many new particles — Higgs bosons, perhaps supersymmetric particles — which could be detected at LEP. Accordingly, these are the subjects of the next two sections of this article.

3. Higgs bosons

In the minimal version of the Standard Model, there is a [2] single physical neutral Higgs boson H^0 with well-defined couplings to fundamental fermions and gauge bosons:

$$g_{H^0 f\bar{f}} = 2^{1/4} \sqrt{G_F} \, m_f$$

$$g_{H^0 VV} = 2^{5/4} \sqrt{G_F} \, m_V^2 \quad . \tag{1}$$

Unfortunately, the mass of this single Higgs boson could have any value between $O(10)$ GeV and $O(1)$ TeV. However, the known couplings (1) enable one to predict its principal decay modes quite unambiguously:

$$\Gamma(H^0 \to b\bar{b} : c\bar{c} : \tau^+\tau^- : \mu^+\mu^-) \propto 1 : \frac{m_c^2}{m_b^2} : \frac{m_\tau^2}{3m_b^2} : \frac{m_\mu^2}{3m_b^2} \tag{2}$$

$$\approx 1 : \frac{1}{10} : \frac{1}{30} : 10^{-4}$$

and also to predict its production cross-sections.

A very promising production mechanism is toponium $\zeta \to H^0 + \gamma$ decay [14], which is expected in leading order to have the branching ratio:

$$\frac{\Gamma(\zeta \to H^0 + \gamma)}{\Gamma(\zeta \to \gamma^* \to e^+ e^-)} = \frac{G_F m_\zeta^2}{4\pi\sqrt{2}\alpha} \left(1 - \frac{m_H^2}{m_\zeta^2}\right) \quad . \tag{3}$$

QCD radiative corrections [15] reduce the branching ratio relative to this leading-order formula by $O(50)\%$, while γ-Z^0 interference affects the denominator by a factor which varies significantly for $m_\zeta =$

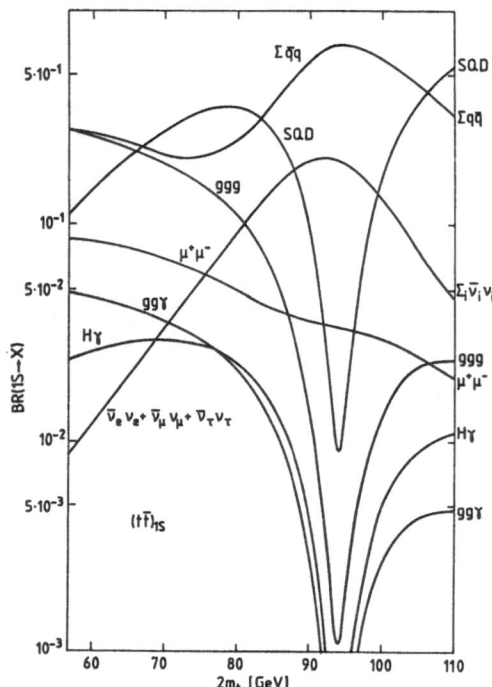

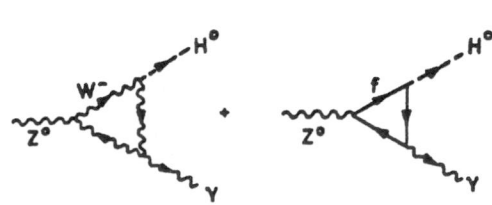

Fig. 1 Important branching ratios [16] for toponium ζ decay, as functions of m_ζ near m_{Z^0}

Fig. 2 Boson loop diagrams [18] contributing to $Z^0 \rightarrow H^0 + \gamma$ decay

$O(m_{Z^0})$. The variation of the principal branching ratios of toponium [16], including $B(\zeta \rightarrow H^0 + \gamma)$, are shown in Fig. 1. If $m_\zeta \sim 70$ to 90 GeV as suggested by indications for the t-quark from some UA1 data [17], the decay $\zeta \rightarrow H^0 + \gamma$ should be easily observable for $m_{H^0} \lesssim (0.8$ to $0.9) \times m_\zeta$, and the backgrounds appear small.

Another production mechanism for the H^0 which has been considered [18] is $Z^0 \rightarrow H^0 + \gamma$. There is no tree-level amplitude for this process, but it can arise from loops of fermions and bosons, as indicated in Fig. 2. These could be regarded as electroweak analogues of the fermion loops for $\pi^0 \rightarrow 2\gamma$ decay calculated [19] by Jack back in 1949. In our case, the W loops of Fig. 2 yield an amplitude

$$A = -\left[4.56 + 0.25 \left(m_{H^0}^2 / m_{W^\pm}^2\right)\right] \qquad (4)$$

which, when substituted into

$$\frac{\Gamma(Z^0 \rightarrow H^0 + \gamma)}{\Gamma(Z^0 \rightarrow \mu^+\mu^-)} = \frac{\alpha^2}{\pi^2 \sin^2 \theta_W} \left(\frac{E_\gamma}{m_{Z^0}}\right)^3 \frac{A^2}{\left[1 + (1 - 4 \sin^2 \theta_W)^2\right]} \qquad (5)$$

yields

$$\frac{\Gamma(Z^0 \rightarrow H^0 + \gamma)}{\Gamma(Z^0 \rightarrow \mu^+\mu^-)} \approx 6.3 \times 10^{-5} \left[1 - \left(\frac{m_{H^0}^2}{m_{Z^0}^2}\right)\right]^3 \left[1 + 0.14 \left(\frac{m_{H^0}^2}{m_{Z^0}^2}\right)\right] . \qquad (6)$$

This corresponds to a branching ratio in Z^0 decay of order 10^{-6} as seen in Fig. 3, which is probably too

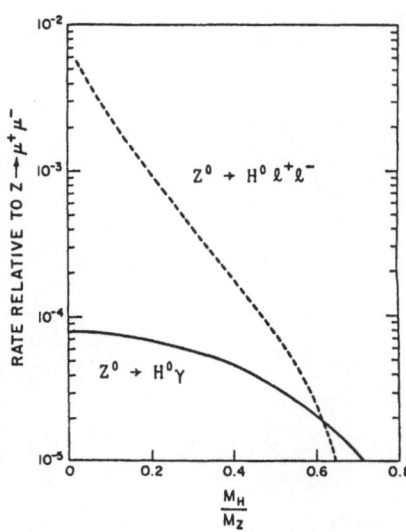

Fig. 3 Branching ratios for $Z^0 \rightarrow H^0 + \gamma$ [18] and $Z^0 \rightarrow H^0 + \ell^+\ell^-$ [20]

small to be measured, particularly since the backgrounds are not negligible. This is therefore not a promising process for hunting the Higgs.

Better prospects are offered by the decay $Z^0 \rightarrow H^0 + \ell^+\ell^-$, which exploits the $Z^0 Z^0 H^0$ vertex in (1). Here one calculates [20]

$$\frac{1}{\Gamma(Z^0 \rightarrow \mu^+\mu^-)} \frac{d\Gamma(Z^0 \rightarrow H^0 + \ell^+\ell^-)}{dx_{H^0}} = \frac{\alpha}{4\pi \sin^2 \theta_w \cos^2 \theta_w} \times$$

$$\frac{1 - x_{H^0} + (x_{H^0}^2/12) + 2/3 \, (m_{H^0}^2/m_{Z^0}^2)}{\left[x_{H^0} - (m_{H^0}^2/m_{Z^0}^2)\right]^2} \times \left[x_{H^0}^2 - \frac{4m_{H^0}^2}{m_{Z^0}^2}\right]^{1/2} \tag{7}$$

where $x_{H^0} \equiv 2E_{H^0}/m_{Z^0}$ has the kinematic range

$$\frac{2m_{H^0}^2}{m_{Z^0}} \leq x_{H^0} \leq 1 + (m_{H^0}^2/m_{Z^0}^2) \ . \tag{8}$$

As shown in Fig. 4, formula (7) gives an $\ell^+\ell^-$ invariant mass spectrum peaked at high masses, as close as possible to the virtual Z^{0*} pole in $Z^0 \rightarrow H^0 + (Z^{0*} \rightarrow \ell^+\ell^-)$. Integrating (7) one obtains the ratio of decay rates $\Gamma(Z^0 \rightarrow H^0\ell^+\ell^-)/\Gamma(Z^0 \rightarrow \mu^+\mu^-)$ shown in Fig. 3. It is higher than the corresponding ratio for the $Z^0 \rightarrow H^0 + \gamma$ decay mode, if $m_{H^0} \lesssim 50$ GeV. Also the backgrounds from higher-order QED are smaller. Therefore this process may be observable for $m_{H^0} \lesssim 40$ GeV. A closely related process is the reaction $Z^0 \rightarrow H^0 + (Z^{0*} \rightarrow \nu\bar{\nu})$, which has a rate about six times higher than (7). Here the signature is less clean than that for $Z^0 \rightarrow H^0 + \ell^+\ell^-$, but it still seems possible to detect this reaction if $m_{H^0} \lesssim 40$ GeV.

Another promising process for Higgs detection is $e^+e^- \rightarrow Z^{0*} \rightarrow Z^0 + H^0$, which has the cross-section [21]

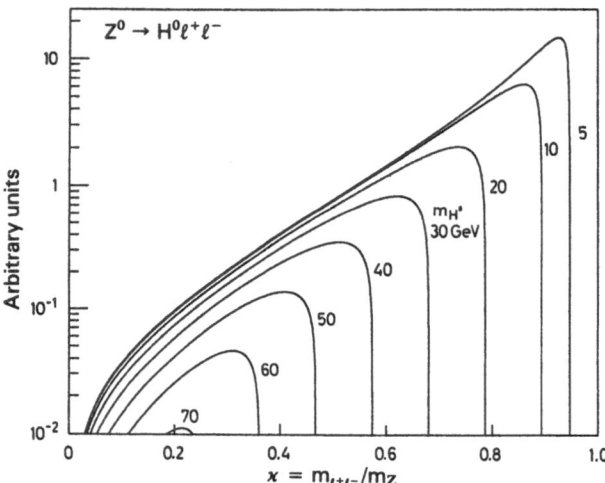

Fig. 4 Spectrum of $(\ell^+\ell^-)$ invariant masses in $Z^0 \to H^0 + \ell^+\ell^-$ decay

$$\sigma(e^+e^- \to Z^{0*} \to Z^0 + H^0) = \frac{\pi\alpha^2}{24} \left(\frac{2\kappa}{\sqrt{s}}\right) \frac{\kappa^2 + 3m_Z^2}{(s - m_Z^2)^2} \frac{(1 - 4\sin^2\theta_w + 8\sin^4\theta_w)}{\sin^4\theta_w (1 - \sin^2\theta_w)^2} \tag{9}$$

where κ is the final-state momentum. As shown in Fig. 5, this has a large enough cross-section to be seen for $m_{H^0} \lesssim 100$ GeV at $\sqrt{s} = 200$ GeV. How easy it would be to pick out this process depends on the mass of the H^0. If $m_{H^0} < m_{W^\pm}$, it might be feasible to use the dominant decays $Z^0 \to q\bar{q}$ jets, $H^0 \to q\bar{q}$ jets, but this would suffer a very large background if $m_{H^0} \sim m_{W^\pm}$ or m_{Z^0}. The final states $Z^0 \to \ell^+\ell^-$ would be the cleanest to search for, with very little background, but in these cases one must pay a factor of O(1/30) in rate. The final states $Z^0 \to \nu\bar{\nu}$ would be intermediate in rate, and might also be

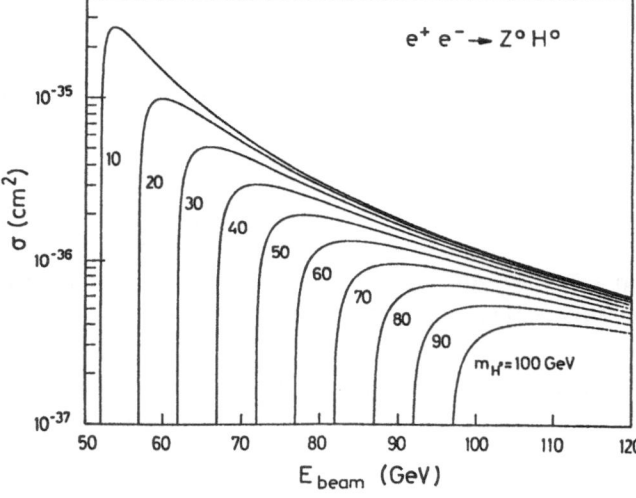

Fig. 5 Cross-sections [21] for $e^+e^- \to Z^0 + H^0$

detectable given good enough measurements of hadronic jets. The process $e^+e^- \rightarrow Z^{0*} \rightarrow Z^0 + H^0$ is the most promising way to look for $m_{H^0} \lesssim 100$ GeV at LEP.

In all electroweak models except the minimal Standard Model, there are physical charged Higgs bosons $H^\pm$ as well as more neutral Higgses. It appears possible [2] to detect the reaction $e^+e^- \rightarrow H^+H^-$ if $m_{H^\pm} < m_{W^\pm}$.

In my opinion LEP has the best prospects of any existing or planned accelerator to find a neutral Higgs boson H^0. The importance of finding a Higgs boson has been underlined in Section 2. It seems to me that this search for the Higgs boson is sufficient to justify the construction of LEP, and that extending the search up to masses $O(100)$ GeV is ample justification for taking $\sqrt{s}$ to 200 GeV at LEP II.

4. Supersymmetric particles

The supersymmetry stategy [3] for stabilizing the mass of the Higgs boson involves doubling the particle spectrum by introducing a new set of particles with the same internal quantum numbers (Q_{em}, B, L, colour, ...) as conventional particles, but with spins differing by $\frac{1}{2}$ a unit as listed in Table 1. Their identical quantum numbers mean they have identical couplings, up to trivial spin factors. Then the radiative corrections which previously gave large shifts in the Higgs boson mass squared are now tamed. The boson and fermion loops almost cancel each other out, and would do so exactly if the particles and their supersymmetric partners had exactly the same masses. This is of course experimentally excluded, with lower bounds [22] on sparticle masses ranging from

$$m_{\tilde{\ell}^\pm}, \ m_{\tilde{q}} \gtrsim 20 \text{ GeV}: \quad e^+e^- \text{ annihilation} \tag{10}$$

to

$$m_{\tilde{g}}, \ m_{\tilde{q}} \gtrsim 50 \text{ GeV}: \quad p\bar{p} \text{ collisions} . \tag{11}$$

In theories [3] with broken supersymmetry, it is the differences in boson and fermion masses squared $m_B^2 - m_F^2$ which provide an effective new physics cut-off that must be less then $O(1)$ TeV. Therefore the sparticles should weigh less than $O(1)$ TeV, and may be accessible to LEP. Here I will just mention the simplest examples of sparticle pair production, namely $e^+e^- \rightarrow \tilde{e}^+\tilde{e}^-$ and $e^+e^- \rightarrow \tilde{q}\bar{\tilde{q}}$.

Table 1 Supersymmetric particles

Particle	Spin	Sparticle	Spin
Quark $q_{L,R}$	1/2, 1/2	squark $\tilde{q}_{L,R}$	0, 0
Lepton $\ell_{L,R}$	1/2, 1/2	slepton $\tilde{\ell}_{L,R}$	0, 0
Photon γ	1	photino $\tilde{\gamma}$	1/2
Gluon g	1	gluino $\tilde{g}$	1/2
$W^\pm$	1	Wino $\tilde{W}^\pm$	1/2
Z^0	1	Zino $\tilde{Z}^0$	1/2
Higgs H	0	shiggs $\tilde{H}$	1/2

In most supersymmetric models there is a new quantum number R which is multiplicatively conserved, and takes the values: + 1 for all particles, − 1 for all sparticles. R conservation means that sparticles must always be produced in pairs, that heavier sparticles decay into lighter ones, and that the lightest sparticle (LSP) must be absolutely stable. This LSP should therefore present in the Universe today as a relic from the Big Bang, and the absence [23] of anomalous heavy isotopes then suggests [24] that the LSP should have neither strong nor electromagnetic interactions. Every heavier sparticle eventually decays into the LSP, which can only have weak interactions, and so escapes from experimental apparatus much like a neutrino. Thus the signature for sparticles is missing energy–momentum, and a favoured [24] candidate for the LSP is the photino $\tilde{\gamma}$.

The cross-section for $e^+e^- \to \tilde{e}^+\tilde{e}^-$, including direct channel γ exchange and the crossed channel exchange of a massless photino $\tilde{\gamma}$ is [25]:

$$\frac{d\sigma(e^+e^- \to \tilde{e}^+\tilde{e}^-)}{d\cos\theta} = \frac{\pi\alpha^2\beta^3 \sin^2\theta}{4s}\left[1 + \left(1 - \frac{4\kappa}{1 - 2\beta\cos\theta + \beta^2}\right)^2\right].\qquad(12)$$

This cross-section is large enough to be observable for $m_{\tilde{e}}$ almost as large as the beam energy. The decays $\tilde{e}^\pm \to e^\pm \tilde{\gamma}$ would give missing energy of order the beam energy, in general with significant missing momentum in the transverse and longitudinal directions as well. The only significant backgrounds are from higher-order QED such as $e^+e^- \to e^+e^-\gamma$ where the γ is not detected. A hermetic detector, which can veto on the presence of a γ down to angles a few degrees from the beam-pipe, can easily [3] remove this background, as shown in Fig. 6. Thus LEP should be able to probe much higher $\tilde{e}$ masses than present-day e^+e^- machines.

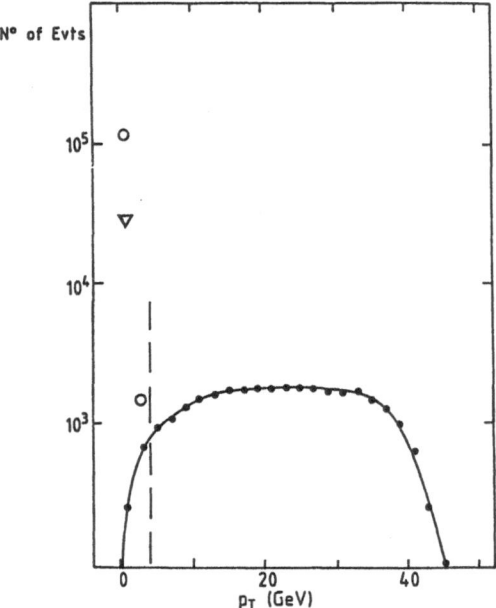

Fig. 6   Missing-p_T signature for $e^+e^- \to \tilde{e}^+\tilde{e}^-$ ($\bullet$), compared with QED backgrounds ($\triangledown$, $\circ$)

The cross-section for $e^+e^- \to \tilde{q}\bar{\tilde{q}}$ is large on the Z^0 pole if $m_{\tilde{q}} < m_{Z^0}/2$. This process is quite easy to observe if $\tilde{q} \to q\tilde{\gamma}$ decays dominate, but more difficult to see if the $\tilde{q}$ have a more complicated decay chain, e.g. $\tilde{q} \to q\tilde{\gamma}$ followed by $\tilde{\gamma} \to q\bar{q}\tilde{\gamma}$. In the former case, $e^+e^- \to \tilde{q}\bar{\tilde{q}}$ gives two-jet events with large missing energy, while in the second case the final states tend to have more jets and a smaller fraction of missing energy and are hence more difficult to detect. Table 2 shows event rates for $e^+e^- \to Z^0 \to \tilde{q}\bar{\tilde{q}} \to q\tilde{\gamma}\bar{q}\tilde{\gamma}$ after simple topological cuts which suffice to beat down the background from conventional $e^+e^- \to q\bar{q}$ events. This is also possible with more complicated $\tilde{q}$ decay modes, though not so easily.

It should be possible at LEP to detect all supersymmetric particles which couple to the photon or Z^0 and have masses ≤ 100 GeV.

Table 2 Number of events from $\tilde{q} \to q + \tilde{\gamma}$ at the Z^0 peak

Sphericity	Missing energy (GeV) 25	30	35	40
0.2	16 500	16 000	14 500	11 900
0.3	13 000	12 600	11 300	9 200
0.4	9 000	8 700	7 700	6 200

5. Conclusions

In this contribution I have given a brief survey of some of the most interesting new particle searches which can be performed at LEP. Particular emphasis has been given to the particles (Higgs bosons, supersymmetry) whose existence is motivated by attempts to understand the masses of the elementary particles. There are many other aspects to LEP physics, such as precision measurements on the Z^0 peak, toponium, QCD studies, and looking at the reaction $e^+e^- \to W^+W^-$. However, I think that these few examples are sufficient to show that LEP physics will be very interesting, to Jack and to the rest of us.

References

1. Physics at LEP, CERN Report in preparation; for previous reviews of LEP physics, see CERN Reports 76–18 (1976) and 79–01 (1979).
2. J. Ellis, M.K. Gaillard and D.V. Nanopoulos, Nucl. Phys. **B106**, 292 (1976);
   M.K. Gaillard, Commun. Nucl. Part. Phys. **8**, 31 (1978);
   ECFA/LEP Exotic Particles working group: G. Barbiellini et al., DESY preprint 79/27 (1979).
3. H.E. Haber and G.L. Kane, Phys. Rep. **117C**, 75 (1985);
   J. Ellis, CERN preprint TH–4255 (1985).
4. See also G. Altarelli et al., contribution to ref. [1].
5. H. Georgi and S.L. Glashow, Phys. Lett. **32**, 438 (1974).
6. H. Georgi, H.R. Quinn and S. Weinberg, Phys. Rev. Lett. **33**, 451 (1974);
   M.S. Chanowitz, J. Ellis and M.K. Gaillard, Nucl. Phys. **B128**, 506 (1977);

A.J. Buras, J. Ellis, M.K. Gaillard and D.V. Nanopoulos, Nucl. Phys. **B135**, 66 (1978).

7. D.V. Nanopoulos and D.A. Ross, Nucl. Phys. **B157**, 273 (1979).

8. R. Peccei et al., contribution to ref. [1].

9. M.E. Peskin, Proc. Int. Symp. on Lepton and Photon Interactions at High Energies, Kyoto (1985).

10. S. Weinberg, Phys. Rev. **D13**, 974 (1976) and **D19**, 1277 (1979);
    L. Susskind, Phys. Rev. **D20**, 2619 (1979).

11. S. Dimopoulos and L. Susskind, Nucl. Phys. **B155**, 237 (1979);
    E. Eichten and K.D. Lane, Phys. Lett. **90B**, 125 (1980).

12. J. Ellis, Proc. 1981 SLAC Summer Institute on Particle Physics, ed. A. Mosher (SLAC–215, Stanford, 1982).

13. S. Dimopoulos and J. Ellis, Nucl. Phys. **B182**, 505 (1981).

14. F.A. Wilczek, Phys. Rev. Lett. **39**, 1304 (1977).

15. M.I. Vysotsky, Phys. Lett. **97B**, 159 (1980).

16. W. Buchmüller et al., contribution to ref. [1].

17. UA1 Collaboration, G. Arnison et al., Phys. Lett. **147B**, 493 (1984).

18. R.N. Cahn, M.S. Chanowitz and N. Fleishon, Phys. Lett. **82B**, 113 (1979).

19. J. Steinberger, Phys. Rev. **76**, 1180 (1949).

20. J.D. Bjorken, Proc. 1976 SLAC Summer Institute on Particle Physics (SLAC–198, Stanford, 1977).

21. B.W. Lee, C. Quigg and H. Thacker, Phys. Rev. **D16**, 1519 (1977).

22. S. Komamiya, Proc. Int. Symp. on Lepton and Photon Interactions at High Energies, Kyoto (1985).

23. P.F. Smith et al., Nucl. Phys. **B206**, 333 (1982).

24. J. Ellis, J.S. Hagelin, D.V. Nanopoulos, K.A. Olive and M. Srednicki, Nucl. Phys. **B238**, 453 (1984).

25. G. Farrar and P. Fayet, Phys. Lett. **89B**, 191 (1980).

Future Prospects for Elementary Particle Physics: High and Low Energy Probes

M.K. Gaillard

Lawrence Berkeley Laboratory and Department of Physics, University of California, Berkeley, CA 94720, USA

The tremendous progress that has been achieved over the last two decades in our understanding of the elementary particle interactions poses some troubling questions for the future of the field. The standard model for the strong, weak and electromagnetic interactions continues to enjoy verification by experiment. Periodic hints of discrepancies that generate temporary excitement among workers in the field have a tendency to evaporate under closer experimental and/or theoretical scrutiny. Yet the standard model leaves many questions unanswered, and the absence of significant discrepancies leaves us with few signposts that might indicate how to look for their answers. Very generally, one anticipates that measurements resolving increasingly shorter distances will reveal increasingly deeper levels of the laws of nature. The issue is then how to best probe arbitrarily short distances. The most straightforward approach, that has traditionally defined the field of "high energy" physics, is to achieve increasingly high energy for colliding particles, but this route must have some inherent limitation. Alternatives include searches for very rare low energy phenomena, observation of natural high energy collisions provided by cosmic rays, and the more indirect inferrence based on extrapolation of observational cosmology and astrophysics to the physics of the highest energy laboratory – the very early universe. In this article,[1] I will not present any answers but rather I wish to raise some of the questions that will eventually have to be addressed in the planning of future facilities for the study of elementary particle interactions.

In Table 1 are listed some prototype "non high energy accelerator" experiments that can, at least in some indirect way, explore the physics of very short distances, or, equivalently, very high energies, where the energies entered correspond to the equivalent center–of–mass energy that would be required for hard elementary particle (quark or lepton) collisions to study directly the phenomenon being probed. I shall start by elaborating on the entries in Table 1.

1. <u>Proton decay</u>. As is well known,[2] extrapolation of the measured, scale–dependent, coupling constants of the strong, weak and electromagnetic interactions shows that they become equal, within the context of the standard model, at an energy scale of about 10^{15} GeV. This is interpreted as the scale at which the theories describing these interactions become embedded in a unified gauge theory. The disparity among the coupling constants as measured at low energies is attributed to spontaneous symmetry breaking, characterized by the vacuum expectation value (vev) of some scalar field: $v_{GUT} \sim 10^{15}$ GeV. The symmetry breaking induces masses M_{GUT} of the same order for vector bosons that mediate processes violating baryon and/or lepton number conservation. At low energy, $E << M_{GUT}$, such processes are suppressed in rate by a factor

Table 1: Low energy probes of high energy phenomena:

Experiment	Energy Scale
Proton decay	$10^{15} - 10^{16}$ GeV
Axion searches	$10^8 - 10^{12}$ GeV
Monopole searches	$\geq 10^{15}$ GeV
Cosmic rays	≤ 40 TeV
ν-masses, oscillations	
reactor experiments	$\leq 10^{15}$ GeV
solar neutrinos	$10^{11} - 10^{19}$ GeV
Rare decays	
$\mu \to 3e, e\gamma$	$\leq$ (30 - 300) TeV
$K \to \mu e, \pi\mu e$	≤ 400 TeV
$K, \mu \to$ familon + X	$\leq 10^{12}$ GeV
$B - decay$	?
Neutron dipole moment	10^{15} GeV?
Observational astrophysics/cosmology	10^{19}GeV?

$$(E/M_{GUT})^4 \sim (E/GeV)^4 \times 10^{-60}. \tag{1}$$

However, an enrichment of the particle content of the relatively low energy mass spectrum, to include states not presently observed, could change the value of the unification scale. Pushing this scale downwards is ruled out by the non-observation of proton decay to date; in fact present limits on the proton lifetime appear to exclude the simplest model for unification. Very roughly, the proton lifetime is [2]

$$\tau_p \sim M_{GUT}^4 / |\psi(0)|^2 \, m_p^2 \sim (10^{31}yr) \left(\frac{M_{GUT}}{10^{15}GeV}\right)^4, \tag{2}$$

where m_p is the proton mass and $|\psi(0)|^2 \sim m_\pi^3$ is the probability density for two quarks in the proton to be at the same point. It appears unlikely at present that proton decays will be accessible to experimental detection if the proton lifetime is longer than 10^{34} seconds, in which case scales above 10^{16}GeV cannot be probed in this way.

2.Primordial axion searches.[3] "Axion" is the name given to the pseudo–Goldstone boson of a spontaneously broken discreet global symmetry introduced by Peccei and Quinn to assure the suppression of CP–violating effects that could otherwise arise from non–trivial gluon field configurations in the QCD vacuum. Axion phenomenology is determined by the value of the scalar vev v_{PQ} that breaks the PQ symmetry. In particular the axion mass

$$m_a \sim \frac{f_\pi m_\pi}{v_{PQ}} \sim \left(\frac{GeV}{v_{PQ}}\right) \times 13MeV \tag{3}$$

and axion couplings to ordinary matter are inversely proportional to v_{PQ}. Detection of the axion would fix the value of this vacuum expectation value, which would in turn pinpoint a scale of new physics.

Axion searches at accelerator laboratories and considerations of the effects axions would have on stellar evolution rule out values of v_{PQ} up to 10^8 GeV. On the other hand values $v_{PQ} \geq 10^{12}$ GeV would result in an axion energy density in the present universe that is unacceptably large. This leaves a window of allowed values in a range such that the existence of axions could have interesting implications for dark matter in galactic halos and for galaxy formation.[3]

3.Monopole searches[4] Standard grand unified theories entail the existence of very massive Dirac monopoles; typically their masses are somewhat larger than the vacuum expectation value v_{GUT}. Since these entities are stable, some might have survived in the present universe as relics of the Big Bang. Indeed, standard Big Bang cosmology in conjunction with standard grand unified models predicts a monopole density in the present universe that is much too high to be compatible with observation, and an inflationary epoch in the early universe must be invoked to dilute the monopole density. The observation of magnetically charged stable particles would be an exciting signal of new physics; a measurement of their masses would directly reveal its associated energy scale.

4.Cosmic rays. Extensive air shower arrays, such as the Fly's Eye in Utah, have detected showers with primary energies up to about 10^8 TeV, corresponding to a collison center–of–mass energy of about 400 TeV, or an equivalent hard collison center–of–mass energy for quarks and gluons of about 40 TeV, ten times what will be achieved at the proposed Superconducting Supercollider (SSC). Data from even much less energetic cosmic rays may serve to probe very high energy physics because of possible connections to relics of the Big Bang. An example is evidence for cosmic ray anti-protons at energies below the threshold of about 2 GeV, below which an interpretation in terms of secondary $\bar{p}$ production appears difficult. This could be evidence for a variant on the standard GUT scenario,[2] according to which CP is spontaneously broken and different domains of the universe differ in the sign of the baryon asymmetry. An alternative explanation in terms of annihilation of photinos (fermionic supersymmetric partners of photons) in galactic halos has recently been proposed,[6] which, if construed as evidence for supersymmetry, has profound implications for particle theory at very high energies.

5.Neutrino masses and oscillations.[7] The standard model provides no explanation as to why the neutrino is massless. On the other hand, it suffices to assume that the elementary particle spectrum contains no right–handed neutrinos to assure, within the context of the standard model, that the neutrinos are strictly massless. Many theorists find this assumption unattractive. However, if right handed neutrinos exist, one would expect not only the appearance of neutrino masses comparable to other lepton masses, but also lepton flavor violation in charged current interactions similar to the observed quark flavor violation. One is therefore led to introduce new physics that accounts for the observed strong suppression of neutrino masses and mixing. This can be achieved if a scalar that is a singlet with respect to the observed low energy $SU(3)_{color} \times SU(2)_{Left} \times U(1)_{weak}$ gauge group acquires a large vacuum expectation value v_M

that allows right handed neutrinos, also singlets with respect to the observed gauge group, to acquire a Majorana mass. In typical grand unified theories[2] incorporating this mechanism, one obtains relations of the type

$$m_\nu \sim m_u^2/v_M \tag{4}$$

where ν is the neutrino and u the charge $-2/3$ quark for a given fermion generation, and mixing angles for leptons are related to those for quarks. For processes involving predominantly electron neurtrinos, if the dominant mixing is $\nu_e \leftrightarrow \nu_\mu$, one would probe new physics characterized by the scale

$$v_M \sim m_c^2/\Delta m_\nu \tag{5}$$

where Δm_ν is the mass difference to which a particular experiment is sensitive. In reactor experiments, one may be able[7,8] to reach sensitivities on the order of $\Delta m_\nu \sim 0.01 eV$ or, from (4)

$$v_M \sim 10^{11} GeV. \tag{6}$$

On the other hand, the extablishment of oscillations for neutrinos emitted by the sun would be sensitive[2,7,8] to neutrino mass differences in the range $10^{-6} eV \leq \Delta m_\nu \leq 10^{-2} eV$, probing a scale of new physics.

$$10^{11} GeV \leq v_M \leq 10^{15} GeV. \tag{7}$$

Even if there are no right–handed neutrinos, it seems unlikely that neutrino masses (and mass mixing) are absolutely zero, since their vanishing does not correspond to a dynamical symmetry principle like the gauge invariance of quantum electrodynamics that assures a vanishing photon mass. While Majorana masses for the weakly coupled left–handed neutrinos of the standard theory cannot arise within the context of that model, one would expect that they should appear at some level. This is because, using the known (or at least necessary) content of the standard model one can construct an $SU(3)_c \times SU(2)_L \times U(1)$ invariant effective coupling:

$$\mathcal{L}_{eff} = \frac{1}{M} \nu_L^c \nu_L \phi_0^2 + h.c. \tag{8}$$

where ν_L^c is the (right handed) charge conjugate field of the left handed neutrino ν_L, and ϕ_0 is the neutral component of the Higgs field that acquires a vev $v_{WS} \sim 250 GeV$ in the standard electroweak model. In Eq. (8) M is a mass parameter introduced so that $\mathcal{L}_{eff}$ has the correct dimension. It could represent the mass of a heavy fermion exchanged in quantum corrections, arising in the context of a more complete theory, to the effective Lagrangian valid at low energy. At the very least, when gravity is incorporated into the description of elementary particle interactions, one expects[2] that non-renormalizable interactions of the type in Eq. (8) should occur with a scale parameter $M \simeq M_{Planck} \sim 10^{19} GeV$. Detection of solar neutrino oscillations sensitive to mass differences as small as $10^{-6} eV$ has been claimed[8] to be achievable. In the present context this would allow exploration of new physics scales up to the Planck scale:

$$M \sim (250 GeV)^2/m_\nu \sim 10^{19} GeV \left(\frac{10^{-6} eV}{\Delta m_\nu} \right), \tag{9}$$

while reactor experiments sensitive to $\Delta m_\nu \lesssim 0.01$ could probe scales M up to about 10^{15} GeV.

6.Rare decays of metastable particles. Some of these processes may, in fact, be most efficiently measured at what we now consider as "high energy" accelerator laboratories. The point is that rare decays of relatively long–lived particles can be used to probe energy scales far greater than the characteristic energy release in the decay. The prototype is neutral kaon decay that played an important role in the construction of a viable renormalizable theory of weak interactions: the observed strong suppression of decays like $K_L \rightarrow \mu^+\mu^-$ suggested new physics at an energy scale of a few GeV, physics that we now recognize as charmed particles.

Consider first the possibility of measuring muon decay branching ratios to a level of 10^{-12}. I will consider two lepton–number violating modes: $\mu \rightarrow 3e$ and $\mu \rightarrow e\gamma$. The sensitivity of a particular experiment can best be displayed by expressing the branching ratio, within some theoretical context, in terms of an effective mass and the attainable level in branching ratio that I assume here to be 10^{-12}. Thus, for example, we may write

$$\frac{\Gamma(\mu \rightarrow 3e)}{\Gamma(\mu \rightarrow e\nu\bar{\nu})} \simeq \left(\frac{G_{eff}}{G_F}\right)^2 \simeq 10^{-12} \left(\frac{300 \; TeV}{m_V}\right)^4 g^4, \tag{10}$$

where $G_{eff}/\sqrt{2} \equiv g^2/m_V^2$ is the effective Fermi coupling constant for a conjectured effective four-fermion coupling:

$$\mathcal{H}_{eff} = \frac{1}{\sqrt{2}} G_{eff}(\bar{e}\mu)(\bar{e}e) + h.c. \tag{11}$$

The lepton flavor violating process $\mu \rightarrow e\gamma$ could also occur in the presence of couplings like (11). This decay is suppressed by a factor α/π, but this could be compensated for by a large lepton mass ratio if the process is mediated by a heavier charged lepton L. For example, with

$$\mathcal{H}_{eff} = \frac{G_{eff}}{\sqrt{2}}(\bar{e}L)(\bar{L}\mu) + h.c., \tag{12}$$

and provided the couplings in (12) are not pure V±A, one would estimate:

$$\frac{\Gamma(\mu \rightarrow e\gamma)}{\Gamma(\mu \rightarrow e\nu\bar{\nu})} \simeq \frac{\alpha}{\pi} \left(\frac{m_L}{m_\mu}\right)^2 \left(\frac{G_{eff}}{G_F}\right)^2 \simeq 10^{-12} \left(\frac{m_L}{m_\mu}\right)^2 \left(\frac{60 TeV}{m_V}\right)^4 g^4$$

$$\simeq 10^{-12} \left(\frac{m_L}{m_\tau}\right)^2 \left(\frac{300 \; TeV}{m_V}\right)^4 g^4. \tag{13}$$

What might be the origin of effective couplings like those in (11) and (12)? One possibility is that quarks and leptons are composite. Fermions with common constituents are expected[9] to develop contact interactions of this type with $G_{eff} \simeq \Lambda^{-2}$, where Λ is the inverse size of the composite particles. Compositeness scales up to hundreds of TeV can therefore be probed in this way if charged leptons of different flavor have common constituents.

Another way lepton flavor changing currents could be induced is through a spontaneously broken, gauged "horizontal" or "family" symmetry[10] that entails flavor changing couplings to the massive neutral vector bosons of this gauge symmetry. If these couplings are of comparable strength to those of the standard electroweak couplings: $g_{weak}^2 = e^2/8\sin^2\theta_w \simeq 1/17$, the above

127

measurements would probe masses of the relevant vector bosons up to:

$$m_V \sim g_{weak} \times 300 \; TeV \sim 70 \; TeV. \tag{14}$$

However, it is more likely that Fermi couplings of the type in (11), with one flavor changing and one flavor conserving lepton coupling, would suffer a Cabibbo–like suppression factor, giving from (10):

$$m_V \sim \sqrt{\theta_c} \times 70 \; TeV \sim 30 \; TeV. \tag{15}$$

For any given scenario one or the other of (10) and (13) could be forbidden or highly suppressed, but the above discussion illustrates the scales to which such measurements are potentially sensitive.

Similar reasoning can be used to indentify scales probed by studies of rare kaon decays. For example an effective coupling

$$\mathcal{H}_{eff} = \frac{G_{eff}}{\sqrt{2}} (\bar{d}s)(\bar{\mu}e) + h.c. \tag{16}$$

would induce the decays $K_L \to \mu e$ or $\pi^0 \mu e$ with branching ratios

$$B(K_L \to \mu e) \simeq 2.7 \left(\frac{G_{eff}}{G_F} \right)^2 \simeq 10^{-12} g^4 \left(\frac{400 \; TeV}{m_V} \right)^4, \tag{17}$$

$$B(K_L \to \pi^0 \mu e) \simeq 0.27 \left(\frac{G_{eff}}{G_F} \right)^2 \simeq 10^{-12} g^4 \left(\frac{200 \; TeV}{m_V} \right)^4. \tag{18}$$

Couplings of the type (16) are almost certain to be unsuppressed in theories with horizontal gauge symmetries, so measurements of K_L branching ratios down to the 10^{-12} level probe mass scales in this context up to hundreds of TeV.

There are many other variants on the standard model that predict induced flavor changing transitions of the above type. Indeed, present experimental limits provide severe constraints on model building, and have dealt fatal blows to many attempts to extend the standard model. Many of these extensions, like the introduction of a "horizontal" gauge symmetry, are intended to address the question as to why the fermion spectrum contains identical replicas of fermion families, or "generations". An alternative to a horizontal gauge symmetry is a similar ungauged, or global, symmetry that when spontaneously broken would entail[11] the existence of massless Goldstone bosons called "familons" f. One of these could play the role of the axion, discussed above, that acquires a small mass, Eq. (3), induced by QCD instanton effects.[3,11] The others would be strictly massless and could appear as decay products in flavor changing transitions; for example:

$$B(K^+ \to \pi^+ f) = 10^{-10} \left(\frac{5 \times 10^{11} GeV}{F} \right)^2, \tag{19}$$

$$B(\mu \to e f) = 10^{-10} \left(\frac{8 \times 10^{11} GeV}{F} \right)^2, \tag{20}$$

where $F \equiv v_f$ is the scale at which the breaking of the global family symmetry occurs.

Another laboratory for rare phenomena might be provided by copious production of B-mesons. While one would not expect to attain branching ratios anywhere near the levels achievable in K–decay and μ -decay, rare processes could be somewhat enhanced in B–decays because of larger phase space and the relative weakness of b–couplings to ordinary quarks. In any case B–decays potentially provide a probe of CP violation, an issue that is intimately tied in with the fermion mass spectrum and the replication of quark generations. The precision measurements by Steinberger and others of CP violation in the neutral kaon system, that have failed to turn up any deviation from the "superweak" ansatz which confines CP violation to $K^0 - \bar{K}^0$ mixing, are close to establishing a serious embarrassment for the standard model description of CP violation.[12] A closely related issue is the strength of the b-u charged current coupling for which there is as yet no experimental evidence. The absence of this coupling would remove any possibility for understanding CP violation within the framework of the minimal model with three generations of quarks and leptons and a single Higgs particle.

7. The neutron electric dipole moment. As a last example of low energy probes of very high energy physics, consider the electric dipole moment of the neutron. The present limit

$$\frac{d_n}{e} < 4.6 \times 10^{-25} cm = (4 \times 10^{10} GeV)^{-1} \tag{21}$$

suggests that physics in the billions of GeV energy range might be probed by this measurement. However, such an interpretation depends intricately on the details of a model. For example the necessary conjuncture of P and T violating effects are severely suppressed in the standard model which predicts[2] a considerably lower dipole moment than (21), although the energy scale involved is O(100 GeV). However, in another context, the measurement of the neutron dipole moment can be interpreted as a probe of physics at energies far greater than 10^{10}GeV. This is because, in the absence of an axion, some symmetry (for example supersymmetry) is required to assure that non–perturbative properties of the QCD vacuum do not induce CP violating effects, for which the neutron dipole moment is the most sensitive probe. If one sets the dipole moment to zero at tree level, radiative corrections induce a divergent contribution that can be regularized by cutting off the loop integral at a scale Λ for the internal loop momentum; the value of the cut–off parameter that reproduces the experimental result is interpreted as the scale of new physics where the requisite symmetry becomes effective. This is entirely analogous to the cut–off of a few GeV (now understood as the mass scale of charmed particles) needed to suppress strangeness changing neutral currents induced by higher order charged current processes, or the cut–off of a few hundred GeV (now understood as the scale of electroweak symmetry restoration) needed to damp higher order corrections in the Fermi theory of weak interactions. A difference is that, in the case of the neutron dipole moment, the effect depends only logarithmically on the cut–off Λ; this means that it grows very slowly with increasing Λ. In the context of the standard model it has been shown[2] that present data constrain Λ only to lie below a value that is many orders of magnitude larger than the Planck scale, a scale beyond which ordinary perturbative calculations are almost certainly meaningless. However, extensions of the standard model that introduce additional interactions tend to enhance higher order contributions to the neutron dipole moment. It has been argued[2] that extensions of the minimal SU(5) grand unified theory

(GUT) that are necessary to produce a sufficient baryon asymmetry imply, in the absence of an axion, a lower bound

$$d_n \geq 6 \times 10^{-28} \qquad (22)$$

that may be accessible to detection. A measurement at this level could therefore be sensitive to the spectrum of particles with masses $O(M_{GUT})$.

8.<u>Observational cosmology and astrophysics</u>.[13] Attempts to understand the universe as observed today have led to many conjectures that have a direct impact on theories of elementary particles. Examples are the modification of the minimal GUT, mentioned above, that were motivated by the need to account for the observed nucleon density.[2,14] Others include the properties of effective potentials for conjectured scalar fields that acquire vacuum expectation values close to the Planck scale, in order to achieve an acceptable inflationary cosmology,[15] and various candidates for dark matter in galactic halos such as massive neutrinos, axions and the various exotic particles that arise in supersymmetric theories.[13] All of these conjectures are relevant to the physics of extremely high energies.

In the above list I have omitted the class of relatively low energy experiments that address issues related to the non-perturbative properties of QCD such as heavy ion collisions, lattice gauge Monte Carlo calculations (that on the grounds of both budget and technology may appropriately be included with experimental physics!), studies of parton fragmentation, etc. While work in this direction defines an important frontier of the field, the issue of relevance here is how best to explore the physics characteristic of distance scales so small, or equivalently energy scales so large, as to be inaccessible to experimentation at the presently planned complex of accelerator facilities, including the Superconducting Supercollider.

<u>SUPERSTRINGS</u>. As an example, let me cite a specific physical theory that one would like to test. The currently fashionable candidate for the theory that unifies all fields and forces in nature is a supersymmetric string, or superstring, theory in 10 dimensions[16] with an $E_8 \times E_8$ gauge symmetry.[17] Six dimensions are presumed to curl up, or compactify, with a radius of curvature on the order of the Planck length, giving an effective four dimensional field theory[18] that is locally supersymmetric with one E_8 broken to a supersymmetric E_6 GUT embedding the observed gauge interactions, and with the other E_8 describing a "hidden" sector that interacts only gravitationally with observed matter. Just what the doctor ordered: $E_6 \ni SO(10) \ni SU(5)$ has long been recognized[2] as one of the viable candidates for a GUT. However, the present version is able to dispense with the embarrassingly complicated Higgs sector that was found necessary in the straightforward construction of an E_6 GUT. Furthermore, the idea of a hidden sector in the context of local supersymmetry had already been introduced by phenomenologists[19] as a mechanism for obtaining both an acceptable description of the observed low energy world of broken supersymmetry and an acceptable inflationary cosmology.[15] Moreover, the general form obtained[20] for the scalar potential is similar to a class of models[21] that had been advocated as being particularly good for phenomenology. Finally the existance of a hidden sector could have important cosmological consequences,[22] with implications for issues like dark matter and galaxy formation.

130

An essential yardstick for judging a new theory is its ability to predict, not just postdict. While the details of the low energy spectrum depend on just how the extra dimensions curl up, an apparently general feature is the prediction of at least one extra U(1) gauge interaction mediated by an additional massive gauge boson Z' whose mass, at first sight at least, is determined by the general mass scale of electroweak symmetry breaking. In addition, the standard SU(5) spectrum of quarks and leptons must be extended to fill up representations of E_6, including, in particular, right handed neutrinos. Unless a discreet symmetry is invoked to prevent it, neutrinos should acquire Dirac masses m_i comparable to quark and lepton masses. However, when the extra U(1) is broken, the right handed neutrinos can acquire Majorana masses $O(m_{Z'})$. Then when the neutrino mass matrix is diagonalized, the light neutrino mass is suppressed according to

$$m_{\nu_i} \sim m_i^2/m_{Z'}, \tag{22}$$

a mechanism entirely analogous to the one discussed in 5 above, Eqs. (4) and (5). Existing limits

$$m_{\nu_e} \leq 50eV, \; m_{\nu_\mu} \leq 500keV, \; m_{\nu_\tau} \leq 100MeV \tag{23}$$

can be used to parameterize limits on the Z' mass in this context. For example, if we take $m_i \simeq m_{u_i}$ for each generation, the limits (23) imply, respectively

$$m_{Z'} \geq 200GeV, \; 4.5TeV \text{ and } 20TeV. \tag{24}$$

By way of comparison, if the production of 1000 Z' 's in the rapidity region y > 1.5 is required to establish a signal in pp collisions, a collider with a total center–of–mass energy of 100 TeV and an integrated luminosity of 10^{40} cm^{-2} probes[23] Z' masses only up to about 10 TeV, suggesting that improving limits on neutrino masses is the more efficient way to test the string theory! Actually, the limits in (24) were intentionally exaggerated to make a point. It seems[24] that in string theories the usual GUT relations between quark and lepton masses (e.g. Eq. (4)) are lost, so it is probably more appropriate to take $m_i \simeq m_{\ell_i}$ in Eq. (22). Even if lepton masses were related to quark masses, the relations would presumably involve quark masses as measured at the GUT scale, that are smaller than the observed masses by a factor of about three,[2] because of QCD renormalization effects. However, we can obtain far more stringent limits on the Z' mass by turning to cosmology. Cosmological observation limits the energy density of the present universe to less than about twice the critical density, that is, the density required for a closed universe. This limit is exceeded[13] if the mass of any stable neutrino exceeds about 140 eV. Then taking $m_i \simeq m_{\ell_i}$ in Eq. (22) we obtain

$$m_{Z'} \geq \frac{m_\tau^2}{140eV} = 10^4 TeV. \tag{25}$$

The loop–hole is, of course, the stability assumption. The heaviest neutrinos could decay into lighter ones: $\nu_\tau \rightarrow 3\nu$ or $\nu_\tau \rightarrow \nu \gamma$. Although these are flavor changing neutral current transitions, the presence of additional neutral gauge bosons and exotic fermion states in the effective low energy theory could induce such effects. One would expect flavor violating couplings for neutrinos to be similar to those for charged leptons:

$$\tau(\nu_\tau \to 3\nu) \sim \tau(\mu \to 3e)(m_\mu/m_{\nu_\tau})^5 \geq 10^3 \,\text{sec},$$

$$\tau(\nu_\tau \to \nu\gamma) \sim \tau(\mu \to e\gamma)(m_\mu/m_{\nu_\tau})^3 \geq 10^4 \,\text{sec},$$

$$(26)$$

where I have used the present limits on the ν_τ mass, Eq. (23), and on the branching ratios:

$$B(\mu \to 3e) < 2 \times 10^{-9}, \; B(\mu \to e\gamma) < 2 \times 10^{-10}. \tag{27}$$

The bounds in (26) correspond to negligibly short lifetimes on a cosmological scale. However, by expressing the lifetimes in the form:

$$\tau(\nu_\tau \to 3\nu) \sim 10^{12} B(\mu \to 3e)(\tfrac{1}{2}MeV/m_{\nu_\tau})^5 \times 10^{10} yr,$$

$$\tau(\nu_\tau \to \nu\gamma) \sim 10^{12} B(\mu \to e\gamma)(30 keV/m_{\nu_\tau})^3 \times 10^{10} yr,$$

$$(28)$$

one sees that improvements of two or three orders of magnitude in the muon branching ratios and in the τ-neutrino mass could push the ν_τ lifetime to a value comparable to the age of the universe. Thus, by combining precision measurements at low energy with cosmological data, one might be forced to conclude that if models extracted from string theories as presently formulated have any bearing on reality there must be a discreet symmetry that forbids neutrino masses, and indeed it seems that this might be achievable.[24]

The above analysis illustrates the potential of low energy experiments and cosmological observation in constraining models of elementary particle physics, but also their limitations. Should, in fact, non-zero neutrino masses be observed there will be numerous alternative interpretations in terms of the deeper structure of nature as should be clear from the discussion in 5 above. While their non-observation can severely restrict model building, this does not provide much guidance as to what the underlying physics might be.

Inference based on cosmological and astrophysical observation is even more indirect, particularly when it involves the history of the universe before the epoch of nucleosynthesis. Precision measurements of the width of the standard Z particle at $\bar{p}p$ and e^+e^- colliders, that should determine with certainty the total number of neutrinos with standard electroweak couplings, may provide the first direct confirmation of this type of inference, since the description of nucleosynthesis apparently requires[13] that there be no more than the presently observed three neutrino species. However, much of the work on the cosmology/particle interface involves the physics of extremely short distances, high temperatures and early times. One is confronting two unsubstantiated models: a model for the very early universe and a model for the particle physics of very short distances. While dialogue between the two fields is a very useful auxiliary tool for both, it will be difficult to go beyond the realm of speculation without hard data from terrestrial laboratories.

In assessing how best to proceed here on earth, several questions must be addressed.

Will the present generation of rare decay searches reach inherent limits of sensitivity, or are real improvements possible? The scales cited in 6 above, based on $10^{12} K_L$'s or muons and $10^{10} K^+$'s correspond to levels cited as achievable in proposals for studies at facilities such as

Brookhaven, LAMPF, SIN, Fermilab and TRIUMPH. Certainly existing facilities should be exploited as fully as possible, including studies of B–decays at CESR, Fermilab, and, if feasible, the SSC. Until a positive indication for a deviation from the standard model turns up, the theoretical motivation for pushing limits on processes like those discussed in 6 remains strong. The rare decay process with the best defined window of theoretical interest is proton decay. As discussed in 1, extrapolation of low energy data in the context of the minimal model point to a unification scale of about 10^{15} GeV. While extensions of the model can raise this scale considerably, the basis for the successful prediction[2] of a relation between the three low energy coupling parameters, $\alpha_{e.m.}, \alpha_{strong}$ and $\sin^2 \theta_w$, is lost if the unification scale exceeds the Planck scale. Thus, one would ideally want to probe the range between 10^{15} and 10^{19} GeV for signs of a unified theory. This would mean measuring the proton lifetime to a level $\tau_p \sim m_P^4/m_p^2 m_\pi^3 \sim 10^{44}$ years! The point is that there is no lack of theoretical motivation for pushing these limits further than may appear feasible at present; the issue is only a practical one. However, should technological advances make substantial improvements feasible only at a high cost in dollars or person–power, a second question must be addressed.

How well can indirect evidence for physics at large mass scales compete with the direct exploration of such scales in helping to unravel the underlying physical theory? An example of direct evidence would be production of a very massive Z′; indirect evidence for its existence could take the form of non–vanishing neutrino masses or lepton flavor changing transitions.

It is a historical fact that the standard electroweak model, including the structure of neutral currents, the prediction of charmed particles, the precise form of charmed quark couplings, and even the approximate values of their masses, was constructed entirely from data on the weak decays of mesons, nucleons and hyperons. Does this mean that we could have dispensed with the high energy collisions that allowed the study of neutral currents, the production and identification of charmed particles and finally of the W and Z particles? It seems unlikely that any physicist would argue in the affirmative. For one thing neutral current data was needed to pin down the masses of the W and Z, a reflection of the fact that the theory was (and is) incomplete. More generally, and more importantly, without the series of experimental discoveries that ultimately confirmed the theory, it would have remained a set of interesting but speculative ideas, with many competing models considered as viable. Moreover, the study of charmonium states and charmed particle decays have provided new probes of static properties of hadronic matter, related to the whole issue of soft QCD phenomena and confinement: high energy data can contribute to the understanding of long distance phenomena as well as the converse. Finally, it is unlikely that the existence of the third generation of quarks and leptons could have been convincingly inferred from low energy data – although it was indeed postulated [2,12] as one of several possible mechanisms for incorporating the observed CP–violation. In any case, on–going measurements of the weak charged current couplings of the b–quark remain an essential ingredient in attempts to piece together the outstanding puzzles of fermion replication and CP violation.

Just as the observed strong suppression of neutral strangeness–changing effects in kaon decay played a crucial role in the construction of the standard model, limits on these transitions, as well as on lepton flavor violations and neutrino masses, continue to provide important constraints on extensions of that model such as technicolor, supersymmetry, string models, etc. The observation of any discrepancy with the standard model would have a profound impact. However, as is clear from the examples cited above, the observation of an isolated phenomenon of this type would have many possible interpretations. It is far from obvious that sufficient circumstantial evidence could be collected from a variety of measurements of this sort to indicate a unique interpretation, or even to make a case a strongly as could be made, for example, for charm.

A corollary to the above question is: *How well can we exploit the direct evidence from very high energy cosmic rays?* What are the inherent limits in attainable energy as determined by a rapidly falling spectrum? There are also limitations in the intepretation[5] of cosmic ray data at very high energies due to the fact that the primary particles are not observed directly in extensive air showers arrays that permit compensating for low flux with a very large area. In the past cosmic ray physics has played a discovery role due to the fact that the lifetime of a fast particle is dilated. Muons, pions, and kaons were first discovered in cosmic rays, as was the first charmed particle. Consider however the prospects for discovering a quasi–stable particle with a mass of, say, $m \simeq 40 \ TeV$. If the minimum decay length required for detection is[5] $\Delta \ell \geq 10^{-3} cm$, and the energy of the particle is less than 10^8 TeV, one would be sensitive only to lifetimes

$$\tau \geq (5 \times 10^{-7} GeV)^{-1} \sim (10^{-9} m)^{-1}$$

which are rather unlikely in most theoretical contexts. To put it another way, for a particle with decay width $\Gamma \sim \alpha m \sim 10^{-2} m$, the decay path would be sufficiently long for detection only for masses $m \leq 14$ GeV. For example, fermions heavier than the $W^\pm$ with standard electroweak couplings will decay semi-weakly, although some would have Cabibbo–suppressed rates, say, $\Gamma \sim 10^{-4} m$, allowing detection for masses up to $m \leq 140 GeV$. On the other hand, if the top quark has a mass of about 45 GeV, the ground state hadron lifetime would be about 10^{-20} sec , allowing detection for $E > 10^5$ TeV if decay lengths as small as $10^{-3} cm$ are observable.

To summarize, there is little doubt that the direct study of hard, elementary collisions that liberate a given center–of–mass energy provides the most efficient means for probing the physics at that energy scale. The issue then becomes that of technological, financial and societal limits on successive steps to ever more energetic collider facilities, balanced against their scientific justification.

The currently planned collider facilities have target energies with well defined physics goals. The SLC and LEPI are tuned to study in detail the properties of the Z's. The Tevatron at FermiLab will be a copious source of W's as well as Z's. LEPII will exceed the threshold energy for $W^+ W^-$ production, allowing probably the first quantitative check of the Yang Mills structure of non–abelian gauge theories.

The target scale for the SSC is the TeV region of hard collision center–of–mass energies, associated with the scale of electroweak symmetry breaking. At present we have very little insight as to the origin of symmetry breaking, nor to the coupled question as to why this scale is so much smaller than other scales in nature like the "unification" scale of $\geq 10^{15}$ GeV, where strong, weak and electromagnetic coupling strengths become equal, or the Planck scale (10^{19} GeV) where effects of quantum gravity should become important. Attempts to address the latter issue appeal to concepts like compositeness, technicolor and supersymmetry, but there is as yet no evidence to support such ideas. If one of them does play a role in stabilizing the electroweak scale at its (relatively!) small value, some evidence must appear at the TeV scale or below. On very general grounds we know that some new phenomenon associated with electroweak symmetry breaking must occur at or below this scale. This follows from a unitarity argument that is analogous to an earlier argument, based on extrapolation of weak amplitudes in the Fermi theory, that pointed to a scale of a few hundred GeV where the theory had to be modified, a modification that we now recognize as the W and Z particles. A similar extrapolation[2,25] of WW scattering amplitudes in the context of the standard electroweak model points to the TeV scale as the limiting energy beyond which a perturbative treatment of the standard model ceases to be valid without some new physics. This could take the form of a single scalar (Higgs) particle that could have a mass as low as 10 GeV. However, if a Higgs particle much less massive than a TeV does not exist, the vector bosons of the theory will become strongly interacting in the TeV energy region. Whatever the form of the new physics associated with the phenomenon of electroweak symmetry breaking turns out to be, its uncovering will have profound implications both for theoretical speculations on the deeper structure of matter, and for the role of phase transitions in the evolution of the early universe.

There is therefore a clear scientific case for studying hard collisons in the TeV energy region as will be achieved[23] at the SSC. Beyond that we have as yet no clear–cut signposts. Detailed studies of the properties of W's and Z's at CERN, Fermilab, the SLC and LEP may turn up small discrepancies with the standard model that will indicate a scale of new physics, just as studies of the K–meson complex pointed to the few GeV region for charmed particles. There may be hints in hadron colliders of odd phenomena, like events with a large missing energy, that can be confirmed or studied in detail only with the higher yield that is obtained by increasing the hadron collison energy, providing higher parton luminosities for a fixed parton center of mass energy.[23] Alternatively, data from the SSC might point to a multi–TeV e^+e^- collider facility as a cleaner environment for the study of particular events in the TeV mass region. The observation of scaling violations could signal the presence of four–fermion couplings characterized by an effective Fermi constant $G << s^{-1}$. These could arise from the exchange of new, very massive vector bosons of mass $\sim G^{-\frac{1}{2}}$ or could be induced if quarks and leptons are composite with a confinement radius of about $G^{\frac{1}{2}}$. Determining the source of such effects would require probing energy scales of order $G^{-\frac{1}{2}}$.

As an example of a specific scenario, assume that supersymmetry does indeed play a role in determining the scale of electroweak symmetry breaking, and in particular in stabilizing the mass of the Higgs particle at a value below a TeV, so that perturbation theory remains valid at high energies in the bosonic sector of the electroweak theory. This means that the scale of supersymmetry (SUSY) breaking, at least as it appears in the "observable" (as opposed to "hidden") sector of $SU(3)_c \times SU(2)_L \times U(1)$ gauge interactions, cannot be arbitrarily large. The simplest and most popular assumption[19] is that superpartners of the observed quarks, leptons, and gauge particles, all have masses below a TeV. On the other hand there are many models in which some or all of these "sparticles" acquire SUSY-breaking masses only through radiative corrections, and different sparticle masses may be generated at different orders of perturbation theory, therefore differing in their values by an order of magnitude or more. It could be that only the low end of this "sparticle" spectrum falls within energies accessible at the Tevatron or even the SSC, and that their discovery at the presently proposed facilities would indicate the need for higher energies in order to confirm the supersymmetric interpretation.

As another example, consider the case where there is no Higgs particle with a mass below a TeV, and, as discussed above, W's and Z's develop strong interactions in the TeV hard collision energy regime. Explicit calculations[25] within this framework indicate that evidence for a strongly interacting W, Z sector should be discernable at the SSC with the proposed parameters of 20 TeV per beam and a maximum luminosity of $10^{33} cm^{-2} sec^{-1}$. On the other hand, detailed studies of the dynamics of such a system and the spectrum of resonant states will probably require somewhat higher collision energies. A more perverse scenario would be the case where no Higgs particle is detected at any of the currently planned facilities, and data from the SSC shows no indication of strong W, Z interactions. The absence of strong W, Z interactions would imply that there must be a Higgs particle considerably lighter that a TeV. If it is heavy enough to decay into a pair of W's or Z's, these will be the dominant decay modes, and detection of the Higgs particle should be feasible at a hadron collider.[23] A lighter Higgs particle could be identified[2] in e^+e^- collisions via, depending on the Higgs mass, toponium decay, $(t\bar{t}) \rightarrow \gamma + H$, Z^0 − decay, $Z^0 \rightarrow H + \ell^+\ell^-$, or associated production with a Z^0: $e^+e^- \rightarrow H + (Z^0 \rightarrow \ell^+\ell^-)$. In all these reactions the signal for a Higgs particle would be a peak in the invariant mass recoiling against the identified photon or lepton pair. These techniques will allow probing Higgs masses up to

$$m_H \leq 2E_{max} - m_Z \tag{29}$$

where E_{max} is the maximum beam energy that will be reached at LEP, which for LEPII will slightly exceed the charged W mass m_W.

If no Higgs particles are found in the presently proposed e^+e^- colliders, it will be possible to conclude that $m_H \geq 2m_W - m_Z$. On the other hand, absence of the WW and ZZ signals expected either from Higgs decay[23] or from the strong rescattering[25] of W's and Z's would imply $m_H \leq 2m_W$, indicating the need for an e^+e^- collider with beam energies E in the range

$$m_W \leq E \leq m_W + m_Z/2. \tag{30}$$

It is of course more likely that data obtained from presently planned and proposed collider facilities will point to higher, rather than lower energies. In fact, it is frequently said that the most important discoveries at a given facility have been far different from what it was intended for. On the other hand, there have been examples of facilities that did not make a fundamental impact on our understanding of nature. Although good data from all sources ultimately makes a contribution, it is unlikely that we will be supported for increasingly expensive shots in the dark. The problem is that from today's perspective, the only mass scales we can point to with certainty are the TeV scale, to be explored at the SSC, and the Planck scale, which will certainly never be reached in direct collisions.

Then the real question we have to address is: *at what stage will we be forced to abandon successive small steps toward the unreachable Planck scale,* that defines the present goal of theoretical understanding? Hopefully the answer lies in a distant future, but it is clear that one day progress in the field will depend on the interpretation of indirect evidence at energies far below those scales one is trying to elucidate, and on data from the rather uncontrolled single high energy experiment that was the primordial explosion of the Big Bang.

*Work supported in part by the National Science foundation under grant PHY-84-06608 and in part by the Director, Office of Energy Research, Office of High Energy and Nuclear Physics, Division of High Energy Physics of the U.S. Department of Energy under contract DE-AC03-76SF00098.

References

1. This contribution is an elaboration on reflections presented in a panel discussion "Physics beyond the SSC" at the LBL-UCB Conference on Collider Physics, held on the occasion of the 60th birthday of J. D. Jackson.

   Except for very recent work, the following are review articles with references to the original literature:

2. *The standard electroweak model, extensions thereof, grand unification and baryogenisis:* J. Ellis in Gauge Theories and High Energy Physics, Vol. I, p. 157, Eds. M. K. Gaillard and R. Stora (North Holland, Amsterdam, 1983); P. Langacker, Phys. Reports 72C, 185 (1981).

3. *Axions* (see also ref. 11 below): P. Sikivie, Proceedings of the Gif Summer School (1982).

4. *Monopoles:* J. Preskill, Ann. Rev. Nucl. and Part. Phys. 34, 461 (1984).

5. *Cosmic rays:* T. K. Gaisser and G. B. Yodh, Ann. Rev. Nucl. and Part. Phys. 30, 475 (1980).

6. J. Silk and M. Srednicki, Phys. Lett. 53, 624 (1984).

7. *Neutrino masses and oscillations:* F. Boehm and P. Vogel, Ann Rev. Nucl. and Part Sci. 34, 125 (1984).

8. B. Cabrera, L. M. Krauss and F. Wilczek, Phys. Rev. Lett. 55, 25 (1985). L. Krauss and F. Wilczek, Santa Barbara - Harvard preprint HUTP-84/A087, NSF-ITP-85-02.

9. *Compositeness:* M. Abolins et al. in Proc. 1982 DPF Summer Study on Elementary Particle Physics and Future Facilities, Eds R. Donalson, R. Gufstafson and F. Paige p. 274.

10. *Horizontal symmetries:* R. N. Cahn and H. Harari. Nucl. Phys. B176, 135(1980).

11. *Axions and familons:* F. Wilczek, Santa Barbara preprint NSF-ITP-84-14 (Erice lecture 1983).

12. *Quark flavor mixing and CP violation:* L. -L. Chau, Phys. Rep. 95, 1 (1983).

13. *The particle/cosmology interface:* D. N. Schramm in Gauge Theories in High Energy Physics, Vol. I, p. 365 Eds. M. K. Gaillard and R. Stora (North Holland, Amsterdam 1983); K. Oliver and D. N. Schramm, Fermilab preprint FERMILAB-PUB-85/113-A(1985).

14. *Baryogenisis:* M. S. Turner, in Gauge Theories in High Energy Physics, Vol. I. p. 387 Eds. M. K. Gaillard and R. Stora (North Holland, Amsterdam, 1983).

Meson Masses in a Quark Model with Self-Consistent Pion Couplings

E.M. Henley, R. Su*, and M. Li

Institute for Nuclear Theory, Department of Physics, FM-15,
University of Washington, Seattle, WA 98195, USA

S. Kuyucak†

Institute for Theoretical Physics, University of Tübingen,
D-7400 Tübingen, Fed. Rep. of Germany

We investigate the masses of the light (S-state) mesons in a nonrelativistic quark model. The pion is included in the model, and couples to quarks as in chiral bag models. With inclusion of this coupling, we find reasonable fits to all of the masses of the S-state mesons in a self-consistent manner. That is, the mass of the pion which couples to quarks is the same as that of the quark-antiquark pion. The masses are determined by minimizing the energy as a function of confinement radius. The radii of the various mesons are not identical.

A. INTRODUCTION

There have been numerous attempts in the past several years to build a consistent quark model of the baryons and mesons. In most of these models, the baryon masses can be obtained, but the pion mass invariably turns out to be of the order of 400 MeV.[1] Part of the reason is that in these models, the mass splitting between the ρ meson and the pion arises from a spin-spin coupling which cannot be made sufficiently large. A number of solutions to this puzzle have been presented: (1) Since the pion is a very light pseudoscalar meson, it is conjectured that it may be a Goldstone boson, or partially so, so that its mass cannot be obtained in a quark model.[2,3] (2) Another explanation is that a fully covariant treatment is required for the light mesons and especially for the pion, so that the models are not sensible.[3] (3) Most of the models include only single gluon exchanges, and it is conjectured that multi-gluon exchanges are important. Indeed, recently, a semi-phenomenological relativistic model incorporating such effects has been used by Godfrey and Isgur[4] to fit the masses of the baryons and mesons. (4) Recoil or center-of-mass effects are especially important for the light mesons and this has been proposed as a possible remedy for the lack of fit to the pion mass.[5] Other solutions have also been proposed.

In this paper we take a more standard approach to see whether a consistent treatment cannot remove the problem. We find that it may do so. We consider a nonrelativistic model for simplicity, even though we are aware that relativistic effects may be sizeable. We take constituent quarks with masses of 330 MeV for the up and down quarks and 450 MeV for the strange quark. They are confined by a harmonic oscillator potential. Such models have been found to be successful for the masses of the baryons[6] and for nucleon- nucleon scattering.[7] An

---

* Permanent address: Department of Physics, Fudan University, Shanghai, The People's Republic of China

† Present address: School of Physics, University of Melbourne, Parkville, Victoria 3052, Australia

important feature of these models is that they include couplings of quarks to a finite-sized pion.[7] This coupling is often omitted in quark models of the mesons, even though it is included for the baryons. It is our belief that, whether the pion is a Goldstone boson or a quark-antiquark pair, its self-energy contributes to the masses of the mesons as well as baryons, and that it needs to be included in computing these masses. Indeed, the contributions of the pion to the masses of the light mesons has been considered by Detar[9] and others, but in MIT or chiral bag models, where there are center-of-mass problems or/and not in a self-consistent manner. We will elaborate on this last point below.

B. THEORY

Many of the quark models which are used to compute meson and baryon masses use the MIT bag.[10] In such a model, the masses of the baryons and mesons are made up of the following terms of the Hamiltonian

$$H = K_q + V_{conf} + V_{g,el} + V_{g,magn} + \sum_i m_i \,. \tag{1}$$

Here K_q is the kinetic energy of the quarks in the bag, V_{conf} is the confinement potential, $V_{g,el}$ is the gluon exchange electric potential, $V_{g,magn}$ is the magnetic one and m_i is the quark mass. These terms are[10]

$$K_q = \sum_i p_i^2/2m_i \,,$$

$$V_{conf} = -\sum_{i,j} \frac{\lambda_i \lambda_j}{4} \, ar^2 + c \,,$$

$$V_{g,el} = \sum_{i<j} \frac{\lambda_i \lambda_j}{4} \frac{\alpha_s}{r} \,,$$

$$V_{g,magn} = \sum_{i<j} \frac{\pi \alpha_s}{8} \lambda_i \lambda_j \left(\frac{1}{m_i^2} + \frac{1}{m_j^2} + \frac{4}{3} \frac{\vec{\sigma}_i \cdot \vec{\sigma}_j}{m_i m_j} \right) \delta^3(\vec{r}) \,. \tag{2}$$

For the mesons, it is only the last term which gives rise to a mass splitting of the non-strange mesons. Furthermore, in this model, the masses of the pion and "η" are identical, as are those of the ρ and ω mesons. The η and η' are mixtures of the SU(3) singlet and octet mesons,[11] so that the mass of the "η" is not known well[11] and is likely to lie between those of the η (549) and η' (958). Clearly, the masses of the pion and "η" cannot both be fitted in this model. The inclusion of the pion self-energy contribution may ameliorate this situation, but its proper treatment is somewhat more complex. In most treatments severe approximations have been made, which we believe may be justified for the baryons, but not for the light mesons.[1,9] For instance, meson mass differences are neglected, or/and the pion mass is taken to be zero, or/and center-of-mass effects are not treated. In our case, we take a harmonic oscillator confinement potential so that the center-of-mass effect can be taken into account.

In a meson (non-quark) treatment of masses, the self-energy diagrams which contribute[9] are shown in Fig. 1. In the case of a quark model for mesons (where the pion may be partially a Goldstone boson), there are similar diagrams, as illustrated in Figs. 2(a) and 2(b). In addition,

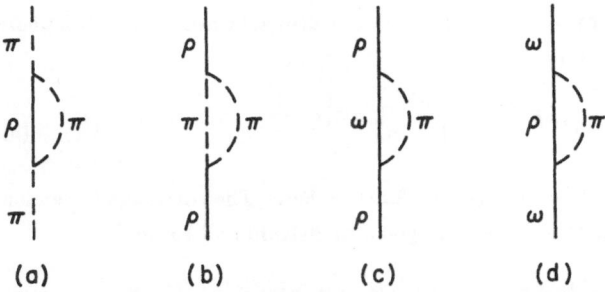

1. Pionic self-energy contribution to meson masses of (a) π, (b) and (c) ρ, and (d) ω

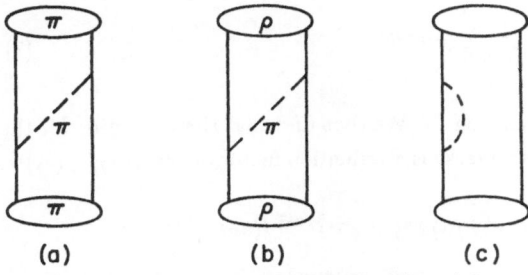

2. Pionic contribution to masses of (a) π, (b) ρ, and (c) quark in a quark model

there are quark self-energy diagrams, as shown in Fig. 2(c), which we assume to contribute to the constituent quark masses, but not to the energy between these constituent quarks which provide the masses of the mesons in a non-relativistic quark model.

The intermediate states in Figs. 2(a) and (b) are similar to those in Figs. 1(a) and (b). If we restrict the intermediate states to S-wave mesons, e.g. π, "η", K, ρ, ω, K^*, ϕ, then the pion contribution to the self-energy of the mesons vanishes for the "η", K, and K^*. We make this assumption.

(1) Pion Coupling Proportional to r^2:

We will show that if the $\pi - q$ coupling is taken to be similar to that used in reference 7 for the description of nucleons[7], then the contribution of the pion self-energy to the masses of the mesons is negligible. On the other hand, if it is taken to be constant, then a reasonable fit to the masses can be obtained. In reference 7 the coupling is taken to be proportional to the square of the distance from the center of the confinement region. There are two reasons for this choice. The first is chiral invariance[12], and the second is to accentuate surface coupling[7], as in the chiral bag models.

We assume that the meson wavefunctions are determined primarily by the quadratic confinement potential, i.e. that they are Gaussian. For two independent quarks or a quark-antiquark we obtain

$$\Psi = \psi(\vec{r}_1)\psi(\vec{r}_2) \,\propto\, e^{-(r_1^2+r_2^2)/2b^2} \,\propto\, e^{-R^2/b^2}\, e^{-r^2/4b^2} \,, \tag{3}$$

where

$$\vec{R} \equiv \frac{\vec{r}_1 + \vec{r}_2}{2}\,, \qquad \vec{r} = \vec{r}_1 - \vec{r}_2\,, \tag{4}$$

for the particular mesons represented by Ψ. However, the center-of-mass wavefunction is a plane wave and thus a more correct wavefunction is

$$\Psi(R,r) = e^{i\vec{p}\cdot\vec{R}} \left(\frac{1}{2\pi b^2}\right)^{3/4} e^{-r^2/4b^2} , \tag{5}$$

where $\vec{p}$ is the momentum of the c.m. We use this wavefunction here. The coupling of the pion to the quarks is taken to be non-local and of the form given by deKam and Pirner[8]

$$\mathcal{L} = F(r) \int \psi^*(\vec{r}+\vec{\eta})\, i\,\vec{\sigma}\cdot\vec{k}\,\vec{r}\cdot\vec{\phi}(\vec{r})\,\psi(\vec{r}-\vec{\eta})\, P(\eta)d^3\eta , \tag{6}$$

with $P(\eta)$ given by

$$P(\eta) = \frac{1}{(\pi b_\pi^2)^{3/2}} e^{-\eta^2/b_\pi^2} , \tag{7}$$

and $F(r) = f_0 \frac{r^2}{b^3}$ as in Ref. 7 or $F(r) = F_0$ as in Ref. 8. We thus find that the correction for the nonlocal coupling of the finite-sized pion to the quarks is a reduction factor of $L(r_1, b_\pi)/|\psi(\vec{r}_1)|^2$, with

$$L(r_1, b_\pi) = \int \psi^*(\vec{r}_1+\vec{\eta})\psi(\vec{r}_1-\vec{\eta})P(\eta)d^3\eta = e^{-r_1^2/2b^2}/(3\pi b_\pi^2)^{3/2} . \tag{8}$$

This reduction is therefore $(2/3)^{3/2}$.

The self-energy contribution of the pion for diagrams such as those of Figs. 2(a) and (b) with only S-state mesons in the intermediate state n is thus

$$\Delta E = -\left(\frac{2}{3}\right)^3 \int \frac{d^3k}{(2\pi)^3} \int d^3r_1\, F\left(\frac{\vec{r}_1}{2}\right) \psi_i^*(\vec{r}_1)i\vec{\sigma}_1\cdot\vec{k}\vec{r}_1\psi_n(\vec{r}_1)\frac{e^{i\vec{k}\cdot\vec{r}_1/2}}{\sqrt{2\omega}}\frac{2}{\gamma-\omega}$$

$$\times \int d^3r_2\, F\left(-\frac{\vec{r}_2}{2}\right) \psi_n^*(\vec{r}_2)(-i)\vec{\sigma}_2\cdot\vec{k}\cdot\vec{r}_2\cdot\psi_i(\vec{r}_2)\frac{e^{i\vec{k}\cdot\vec{r}_2/2}}{\sqrt{2\omega}} , \tag{9}$$

with

$$\gamma = M_i - M_n - k^2/2m_q \approx \delta M .$$

We neglect the recoil, i.e., we keep only static terms. In Eq.(19) i stands for the initial and n for the intermediate meson states. Evaluation of the space integrals finally gives ($\mu = m_\pi$)

$$\Delta E = \frac{f_0^2 \vec{\sigma}_1\cdot\vec{\sigma}_2 \vec{r}_1\cdot\vec{r}_2\, e^{b^2\mu^2/4}}{96\pi^2} \left(\frac{2}{3}\right)^3 \int_\mu^\infty \frac{(\omega^2-\mu^2)^{3/2}}{\omega-\gamma} e^{-b^2\omega^2/4}$$

$$\times \left[9 + \frac{3}{2}b^2\mu^2 + \frac{1}{16}b^4\mu^4 - \left(\frac{3}{2}+\frac{b^2\mu^2}{8}\right)b^2\omega^2 + \frac{b^4\omega^4}{16}\right] d\omega . \tag{10}$$

The value of the integral I, given by

$$I = \int_\mu^\infty \frac{(\omega^2-\mu^2)^{3/2}}{\omega-\gamma} e^{-b^2\omega^2/4}\left[9 + \frac{3}{2}b^2\mu^2 + \frac{1}{16}b^4\mu^4\right.$$

$$\left. - \left(\frac{3}{2}+\frac{b^2\mu^2}{8}\right)b^2\omega^2 + \frac{b^4\omega^4}{16}\right] d\omega , \tag{11}$$

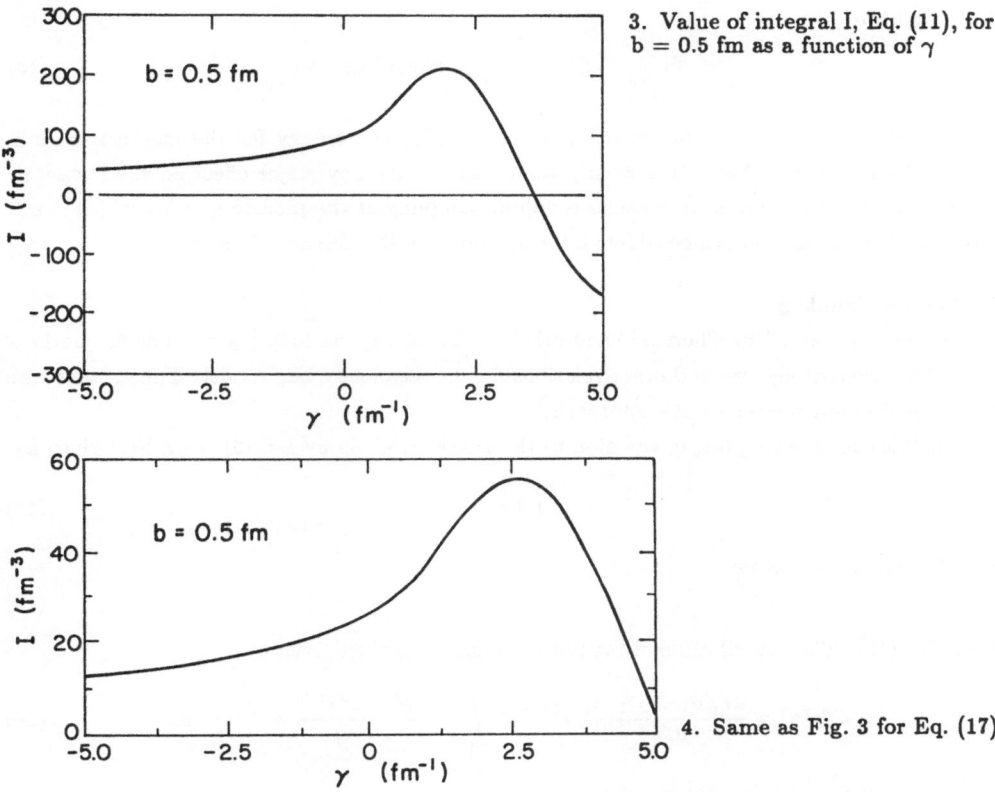

3. Value of integral I, Eq. (11), for $b = 0.5$ fm as a function of γ

4. Same as Fig. 3 for Eq. (17)

is shown in Fig. 3. Its value is always positive in the region $\delta M \lesssim 700$ MeV and grows as a function of *gamma*, reaching a maximum for *gamma* $\cong 400$ MeV. For simplicity, we take $\gamma = \delta M$ (see Table I), and neglect recoil effects. For the pion-quark coupling constant, we take as guidance the fact that the pion-nucleon coupling constant is know to be $g^2/4\pi = 14.2$. We then obtain

$$g < N|i\, \frac{\vec{\sigma} \cdot \vec{k}}{2M_N}\, \vec{\tau}|N> = < N|\sum_i \int L(r,b_\pi)\, F(r)\, i\vec{\sigma}_i \cdot \vec{k}\vec{\tau}_i\, d^3r|N> . \qquad (12)$$

Table I. Relevant mass differences contributing to meson energies, as illustrated in Figs. 1 and 2.

MESON	δM
pion	$M_\pi - M_\rho$
omega	$M_\omega - M_\rho$
rho	$M_\rho - M_\pi$
rho	$M_\rho - M_\omega$

143

Since we have

$$< N| \sum_{i=1}^{3} \vec{\sigma}_i \vec{\tau}_i |N > = \frac{5}{3} < N|\vec{\sigma}_N \vec{\tau}_N|N > ,$$ (13)

it follows that $f_0 = 0.52$. The contribution of the pion self energy for the meson masses is thus of the order of 10 MeV. It is clearly too small to have any major effect on the masses of any meson. We therefore next consider a volume coupling of the pion to quarks, which is not suppressed in the interior, as considered, for instance, by DeKam and Pirner[8].

(2) Volume Coupling:

It has been found by Thomas[13] and others[13] that a volume coupling of pions to quarks is helpful for current algebra and other relationships in the cloudy bag model. Thus it may also be justified in our nonrelativistic approach.

In this case the coupling of the pion to the quarks is given by Eq. (6), with F(r) given by

$$F(r) = F_0 .$$ (14)

For the nucleon, we obtain

$$F_0 = 1.55$$ (15)

from Eq. (12). The contribution of the pion self energy is then given by

$$\Delta E(\gamma) = \frac{4F_0^2 \vec{\sigma}_1 \cdot \vec{\sigma}_2 \vec{\tau}_1 \cdot \vec{\tau}_2}{81\pi^2} e^{b^2 \mu^2/4} \int_{\mu}^{\infty} \frac{(\omega^2 - \mu^2)^{3/2}}{\omega - \gamma} e^{-b^2 \omega^2/4} \, d\omega ,$$ (16)

and the value of the integral I becomes

$$I = \int_{\mu}^{\infty} \frac{(\omega^2 - \mu^2)^{3/2}}{\omega - \gamma} e^{-b^2 \omega^2/4} \, d\omega$$ (17)

with $\gamma = \delta M$ provided by the values in Table I.

One of the differences from previous treatments of meson masses is that we do not neglect the mass differences of Table I, nor the mass of the pion. Equation (16) for the contribution ΔE to the masses of the π, ρ, and ω mesons are then coupled and must be solved in a self-consistent manner. It is this self-consistency which differentiates our treatment from previous ones of the pion self-energy contribution to the masses. That is, we compute the self energies of the pion by iteration and require that the masses obtained correspond to those used in the coupled equations. The masses and confinement radii (i.e., $(5/3)^{1/2}b$) of the various mesons are obtained by minimizing the energy expression, Eq. (18) as a function of b, i.e.

$$H(meson) = C + \Sigma \, m_i + \bar{K}_q(b) + \bar{V}_{conf}(b) + \bar{V}_{g,el}(b) + \bar{V}_{g,magn}(b) + \Delta E(meson) ,$$ (18)

$$\partial H(meson)/\partial b = 0 ,$$ (19a)

$$H_{min} = M(meson) ,$$ (19b)

where $\bar{K}_q(b)$, $\bar{V}_{conf}(b)$, $\bar{V}_{g,el}(b)$, $\bar{V}_{g,mag}(b)$ are the expectation values of K_q, V_{conf}, $V_{g,el}$, $V_{g,mag}$ which are given by Eqs. (2) and (5), and (meson) stands for the particular meson being considered. However, it should be noted that the pion, omega, and rho masses are coupled because $\Delta E(\pi)$, $\Delta E(\omega)$, and $\Delta E(\rho)$ depend on the masses of at least two of these mesons. Thus, the minimization procedure must be carried out in a self-consistent manner, so that the masses found by use of Eqs. (19) are the same as those used in computing $\Delta E(meson)$ from Eq. (16) and Table I. We have carried out this procedure for several choices of C, α_s, and a (see Eq. 2), and show the results in table II.

C. RESULTS AND DISCUSSION

In Table II we display the results for three particular choices of C, α_s, and a. For case 1, the choice of these constants was dictated by fitting the masses of the pion and approximately that of the ω. With this choice, no self-consistent solution is found for the mesons which are coupled by ΔE unless α_s is very small. In that case, the spin-spin (magnetic) contribution to the masses of the mesons is also small, and thus the radii of the vector mesons turn out to be quite large. Also, the splitting of the mesons for which there is no contribution from ΔE ("η", K, K^*) turn out to be quite small, so that all of these mesons have approximately the same mass. The splitting of the K and K^* cannot be fitted with this choice of constants.

Cases 2 and 3 show two examples of fits with larger values of α_s. In these examples, it is possible to approximate the mass splitting of the K and K^*, but at the expense of poorer fits to the π, ρ, and ω masses. Since the η and η' mesons are mixtures of SU(3) octet and singlet[11]

Table II. Table of meson masses for three choices of c, α_s, and a.

Meson	Case 1		Case 2		Case 3		
	b(fm)	M_{th}(MeV)	b(fm)	M_{th}(MeV)	b(fm)	M_{th}(MeV)	M_{exp}(MeV)
π	0.23	139	0.22	123	0.20	105	138
"η"	1.48	692	0.33	557	0.32	458	600–800
ρ	0.96	706	0.52	924	0.50	1034	770
ω	0.97	716	0.50	805	0.48	924	783
ϕ	1.35	914	0.40	855	0.38	855	1020
K	1.35	795	0.35	596	0.31	497	495
K*	1.45	815	0.42	845	0.40	845	892
	$\alpha_s = 0.203$		$\alpha_s = 0.331$		$\alpha_s = 0.405$		
	$a = 0.015$ fm^{-3}		$a = 3.89$ fm^{-3}		$a = 5.04$ fm^{-3}		
	$c = 0$		$c = -4.42$ fm^{-1}		$c = -5.11$ fm^{-1}		

Table III. Table of the variable contributions to the masses of the mesons for case 2 of Table II. All values are in MeV.

	K_q	V_{conf}	$V_{g,el}$	$V_{g,magn}$	ΔE	E
π	1867	150	− 603	− 318	−751	123
"η"	830	337	− 179	− 212	0	557
ρ	334	836	76	− 135	36	924
ω	361	773	86	− 140	− 52	805
ϕ	414	495	90	− 175	0	855
K	658	379	− 119	− 200	0	596
K^*	457	546	115	− 167	0	845

which involve strange mesons, it is not clear what the experimental mass of the mesons we call "eta" should be. The masses obtained for all three examples are reasonable. In the last column of Table II, we show the experimental values of the masses. The best case is perhaps case 2, where all masses are within 20% of the experimental values, and the splitting of the K and K^* meson masses is reasonable. In Table III we show the magnitude of the variable contributions to the masses of the mesons for this case. In this example the mass of the pion is even smaller than 140 MeV, so that the problem of the $\rho - \pi$ mass splitting is certainly resolved. On the other hand the splitting between the ρ and ω is too large due to the contribution of ΔE, and both masses are too large. As α_s increases further (case 3), the π mass becomes smaller still and the ρ and ω meson masses become larger. Thus, the pion self energy term certainly enhances the splitting between the ρ and π masses, and the sum of this term and the magnetic spin-spin term can be of sufficient size to obtain a small pion mass.

If we consider the second example (case 2) as our best one, then it appears that the model we have described here can "explain" or reproduce the small mass of the pion, with a fit to the other mesons which is similar to that obtained in other models.[1,9] Indeed, since we have used a nonrelativistic description throughout, the fit that we obtain to the experimental masses is gratifying.

One of the authors (EMH) thanks Professor A. Faessler for the hospitality shown him at the Institute for Theoretical Physics of the University of Tübingen where this work was begun. He also acknowledges with gratitude a Senior Humboldt award which made this visit possible. Another one of the authors (RKS) thanks Professors W. Haxton, E.M. Henley, G.A. Miller and L. Wilets for their hospitality while visiting the University of Washington. This work was supported in part by the U.S. Department of Energy.

REFERENCES

1. For example, see P.J. Mulders and A.W. Thomas, J. Phys. G: Nucl. Phys. $\underline{9}$, 1159 (1983).

2. See e.g., V. Bernard, R. Brockmann, M. Schaden, W. Weise and E. Werner, Nucl. Phys. $\underline{A412}$, 349 (1984); $\underline{A440}$, 605 (1985).

3. S. Takakura, T. Iwami, and H. Kanada, Progr. Theor. Phys. $\underline{72}$, 379 (1984); N.H. Fuchs and M.C. Scadron, J. Phys. G: $\underline{11}$, 299 (1985).

4. S. Godfrey and N. Isgur, Phys. rev. D $\underline{32}$, 189 (1985).

5. J.F. Donoghue and K. Johnson, Phys. Rev. D $\underline{21}$, 1980 (1975); M.V. Barnhill III, Phys. Rev. D $\underline{30}$, 1126 (1984).

6. N. Isgur and G. Karl, Phys. Rev. D $\underline{18}$, 4187 (1978); $\underline{19}$, 2653 (1979); $\underline{20}$, 1191 (1979); S. Furui, S.B. Khadkikar and A. Faessler, Nucl. Phys. $\underline{A437}$, 619 (1985).

7. G.L. Strobel, K. Bräuer, A. Faessler and F. Fernandez, Nucl. Phys. $\underline{A437}$, 605 (1985).

8. J. DeKam and H.J. Pirner, Nucl. Phys. $\underline{A389}$, 640 (1982).

9. C.E. Detar, Phys. Rev. D $\underline{24}$, 752 (1981); ibid. 762 (1981).

10. A. Chodos, R.L. Jaffe, K. Johnson, C.B. Thorn and V.F. Weisskopf, Phys. Rev. D $\underline{9}$, 3471 (1974).

11. V.P. Efrosin and D.A. Zarkin, Z. Phys. $\underline{C28}$, 211 (1985); F.E. Close, *An Introduction to Quarks and Partons*, (Academic Press, N.Y., 1979), ch. 4.

12. R. Brokmann, W. Weise and E. Werner, Phys. Lett. $\underline{122B}$, 201 (1982); R. Tegen, M. Schedl and W. Weise, Phys. Lett. $\underline{125B}$, 9 (1983).

13. A.W. Thomas, Adv. Nucl. Phys. $\underline{13}$, 1 (1983); G.A. Miller in *Quarks and Nuclei*, ch. 3, ed. W. Weise (World Scientific, Singapore 1984).

Quark and Lepton Mixing in Weak Interactions

K. Kleinknecht

Institut für Physik, Johannes-Gutenberg-Universität,
D-6500 Mainz, Fed. Rep. of Germany

1.   INTRODUCTION

There are indications that a sixth quark, the top quark [1], complements the existing set of five, and that in the lepton sector, a neutrino with the leptonic quantum number of the τ lepton exists. Transitions between the three generations of quarks can be mediated by the flavour-changing charged current weak interaction. In the lepton sector, similar mixing between neutrinos could manifest itself if the neutrinos are massive. This latter mixing, however, would violate lepton number conservation.

In the following review, both quark and lepton mixing is treated in a similar formalism.

2.   QUARK MIXING

2.1   Formalism

The six-quark mixing scheme proposed by Kobayashi and Maskawa [2] serves as a useful parametrization of the connection between generations of quarks. The elements U_{ik} of the quark mixing matrix (i = u,c,t; k = d,s,b) are parametrized in terms of three angles θ_1, θ_2 and θ_3, and one phase δ, possibly related to CP violation (Table 1a). If CP violation is due to quark mixing, then this phase δ is related to the parameter ϵ, describing the admixture of wrong CP parity in the long- and short-lived neutral K-meson states, measured to be $\epsilon = (2.28 \pm 0.05) \times 10^{-3} \times \exp(i\pi/4)$ [3]. An approximate relation derived by Pakvasa and Sugawara [4] is $|\epsilon| = |(m_t - m_c)/m_c| \sin 2\theta_2 \tan \theta_3 \sin\delta/(2\sqrt{2} \cos\theta_1)$ where m_t and m_c are the top- and charm-quark masses. An alternative parametrization of the matrix U has been given by Maiani [5] in terms of angles θ, γ and β and a phase δ' (see Table 1b).

A more detailed calculation of the CP parameter in the K-meson system gives [6-9]

$$\sqrt{2}\,|\epsilon| = B\,\frac{\sin\beta \sin\gamma \sin\delta'}{\sin\theta}\,\left\{-1 + \frac{\eta_3}{\eta_1}\,\ell n\,\frac{m_t^2}{m_c^2} + \frac{\eta_2}{\eta_1}\,\frac{m_t^2}{m_c^2}\,[\sin^2\gamma - \frac{\sin\beta \sin\gamma}{\sin\theta}\,\cos\delta']\right\} \quad (1)$$

where B is the $K^0 - \bar{K}^0$ transition matrix element, normalized to its value for a specific model ("vacuum insertion value"), m_t and m_c are the top and charm quark

Table 1

Parametrizations of the quark mixing matrix

a) Kobayashi-Maskawa parametrization [1]

$$U = \begin{bmatrix} c_1 & s_1 c_3 & s_1 s_3 \\ -s_1 c_2 & c_1 c_2 c_3 - e^{i\delta} s_2 s_3 & c_1 c_2 s_3 + e^{i\delta} s_2 c_3 \\ -s_1 s_2 & +c_1 s_2 c_3 - e^{i\delta} c_2 s_3 & +c_1 s_2 s_3 - e^{i\delta} c_2 c_3 \end{bmatrix}$$

b) Maiani-parametrization [2]

$$U = \begin{bmatrix} c_\beta c_\theta & c_\beta s_\theta & s_\beta \\ -s_\gamma c_\theta s_\beta e^{i\delta'} - s_\theta c_\gamma & c_\gamma c_\theta - s_\gamma s_\beta s_\theta e^{i\delta'} & s_\gamma c_\beta e^{i\delta'} \\ -s_\beta c_\gamma c_\theta + s_\gamma s_\theta e^{-i\delta'} & -c_\gamma s_\beta s_\theta - s_\gamma c_\theta e^{-i\delta'} & c_\gamma c_\beta \end{bmatrix}$$

---

masses and η_1, η_2, η_3 represent QCD corrections. CP violating amplitudes, in this model, are proportional to the product of three, presumably small, angles.

Experimentally, information on the weak mixing angles comes from measurements of weak decays of light and heavy quarks and from neutrino production of charm quarks as observed in dimuon events, as summarized in previous papers [10-14]. New results on the B-meson lifetime [15,16] and on hyperon semileptonic decays [17] give new stringent constraints. In this paper, the impact of all these constraints on the weak mixing angles is analyzed following the procedure of ref. [18], but using recent data on the B lifetime [16,19] and on B meson semileptonic inclusive decays [20,21]. I first go through constraints on the coupling parameters U_{ik}, and then proceed to derive bounds on the mixing angles.

2.2   Constraints on Matrix Elements

2.2.1   Light-quark couplings

Coupling U_{ud}

This coupling parameter has been determined from a comparison of measured rates of nuclear beta decays with that of muon decay. Two different evaluations of this quantity have been made and their results are U_{ud} = 0.9730 ± 0.0024 [11] and U_{ud} = 0.9737 ± 0.0025 [22]. Combining these two, one obtains

$$U_{ud} = 0.9733 \pm 0.0024 \tag{2}$$

Coupling U_{us}

In a series of experiments, the WA2 Collaboration has studies five different hyperon semilaptonic decays, i.e. the leptonic weak decays $\Sigma^- \rightarrow n e \bar{\nu}$, $\Sigma^- \rightarrow \Lambda e^- \bar{\nu}$, $\Xi^- \rightarrow \Lambda e^- \bar{\nu}$, $\Xi^- \rightarrow \Sigma^0 e^- \bar{\nu}$, and $\Lambda \rightarrow p e^- \bar{\nu}$. Including radiative corrections and using in addition the neutron lifetime [23], this experiment gives [17]

$$U_{us} = 0.231 \pm 0.003 \tag{3}$$

This represents a substantial improvement over former analyses [11,22], although the value for U_{us} is exactly the same as the one obtained ten years ago [24]. Lentwyler and Roos [17] deduce a value of $U_{us} = 0.220 \pm 0.002$ from K_{e3} decays, arguing that SU(3) breaking effects are suppressed for vector currents. For the present analysis, eq.(3) was used.

2.2.2 Charm-quark couplings

Coupling U_{cd}

This coupling can be determined from measurements of single charm production in neutrino and antineutrino reactions.

The differential cross-sections for neutrino charm production on isoscalar targets are:

$$\frac{d\sigma^\nu}{dxdy} = \frac{G^2 M E_\nu x}{\pi} \; [U_{cd}^2 \; (u(x) + d(x)) + |U_{cs}|^2 \; 2s(x)] \tag{4}$$

$$\frac{d\sigma^{\bar{\nu}}}{dxdy} = \frac{G^2 M E_{\bar{\nu}} x}{\pi} \; [U_{cd}^2 \; (\bar{u}(x) + \bar{d}(x)) + |U_{cs}|^2 \; 2\bar{s}(x)] \tag{5}$$

where $u(x)$, $d(x)$ and $s(x)$ are the quark density distributions in the proton, G is the Fermi coupling constant, M the nucleon mass, E_ν the neutrino laboratory energy, and x and y the Bjorken scaling variables.

Experimentally, the observation of charm production has been done mainly by three methods: 1. direct observation of the short-lived decay of charmed hadrons in emulsions, 2. observation of semileptonic charm decay $c \rightarrow s + \mu^+ + \nu_\mu$ in neutrino induced dimuon events, $\nu N \rightarrow \mu^- \mu^+ X$, 3. observation of semileptonic charm decay $c \rightarrow s + e^+ + \nu_e$ in dilepton events, $\nu N \rightarrow \mu^- e^+ X$. By far the largest event samples have been collected using the second method.

In order to obtain the coupling parameter U_{cd}, the contribution of charm production from the strange sea s and $\bar{s}$ quarks has to be eliminated. According to the cross-section given in (4) and (5), this can be done [25] by using the weighted difference of neutrino and antineutrino cross-sections:

$$BU_{cd}^2 = \frac{(\sigma^\nu_{\mu^-\mu^+}/\sigma^\nu_{\mu^-}) - (R\sigma^{\bar{\nu}}_{\mu^+\mu^-}/\sigma^{\bar{\nu}}_{\mu^+})}{1 - R} \; \frac{2}{3} \tag{6}$$

where R is the ratio of antineutrino to neutrino total cross-sections, $R = \sigma^{\bar{\nu}}_{\mu^+}/\sigma^\nu_{\mu^-}$

= 0.48 ± 0.02 [25], $R^\nu = \sigma_{\mu\mu}^\nu- / \sigma_\mu^\nu-$ is the dimuon to singlemuon cross-section ratio in neutrino induced reactions corrected for the threshold effects due to the charm mass (slow rescaling) [27], and B denotes the semileptonic branching ratio of that mixture of charmed particles which is produced in the neutrino reactions.

In an analysis of their large sample of neutrino- and antineutrino-induced di-muon events, the CDHS Collaboration [25] obtains $BU_{cd}^2 = (0.41 ± 0.07)10^{-2}$. With a value of B = (7.1 ± 1.3)% based on e^+e^- results and on the composition of charmed particles in neutrino reactions from emulsion experiments, a value of

$$|U_{cd}| = 0.24 ± 0.03 \tag{7}$$

is obtained.

Coupling U_{cs}
$\overline{}$

In charged current reactions this coupling appears always together with the strange-sea structure function xs(x) or its integral S = ∫xs(x)dx, where we use the follo-wing notation: s(x) is the quark density distribution of strange quarks in the proton in the Bjorken scaling variable x, u(x) and d(x) are the distributions of up and down quarks, with U = ∫x u(x)dx and D = ∫x d(x)dx; and similarly for anti-quarks. The density measured is $|U_{cs}|^2 \cdot 2S$. In the absence of an independent determination of S, only the upper limit for 2S given by SU(3) symmetry, $2S \leq \bar{U} d \bar{D}$, and a corresponding lower limit on $|U_{cs}|$ can be obtained. The product $|U_{cs}|^2 \cdot 2S$ can be extracted in three ways from the neutrino and antineutrino dimuon production data [13]. We use here the results from the x distribution of neutrino dimuons [13, 24]

$$\frac{|U_{cs}|^2}{U_{cd}^2} = (6.26 ± 0.73) \frac{1 + \alpha^*}{\alpha} \tag{8}$$

and the one from the cross-sections of neutrino- and antineutrino-induced dimuon production using the semileptonic branching ratio of D mesons:

$$|U_{cs}|^2 = (0.41 ± 0.09) \frac{1 + \alpha^*}{\alpha} \tag{9}$$

where $\alpha = 2S/(\bar{U} + \bar{D})$ is the ratio of momentum fractions carried by strange and non-strange sea quarks in the nucleon, and $\alpha^* = 2S(U_{us}^2 + U_{cs}^2/r_s)/(\bar{U} + \bar{D})$ is the same ratio modified by the threshold suppression factor r_s for the charm-quark mass, which is $r_s = 1.5$ for the experiment considered [25,26].

2.2.3 Bottom-quark couplings

Ratio $|U_{ub}|/|U_{cb}|$
$\overline{}$

At electron-positron storage rings, the reaction $e^+e^- \to \Upsilon(4S) \to B\bar{B}$ can be used as a B meson source. The most sensitive search for decays of b quarks into u quarks can be done by measuring the inclusive lepton momentum spectrum of B meson decays. Data corresponding to an integrated luminosity of 50 pb^{-1} have been collected by the CLEO [21] and CUSB [20] collaboration at the Cornell Electron Storage Ring. The CLEO group measured the momentum spectra of 3750 electron events and 2115 muon events, while the CUSB group reports about the momentum spectrum of 900 electron events. In principle, the two possible semileptonic decays $B \to \ell \nu X_u$ and $B \to \ell \nu X_c$ can be distinguished by measuring the end point of the lepton momentum spectrum. In practice, the analysis is model-dependent because of the theoretical uncertainty about which X_u state with which mass is populated in the decay. Altarelli et al. [28] have calculated the expected lepton momentum spectrum using as an input the observed electron momentum spectrum in decays of charmed D mesons. Based on this model, the limits on the b → u decay width are

$\Gamma(b \to u)/\Gamma(b \to c) < 0.04$ at 90 % C.L.   (CLEO [21])
$\Gamma(b \to u)/\Gamma(b \to c) < 0.045$ at 90 % C.L.   (CUSB [20])

By taking into account the ratio of phase space available, one obtains

$$|U_{ub} / U_{cb}| < 0.12 \text{ at } 90 \text{ % C.L.} \tag{10}$$

B lifetime

The lifetime τ_B of b flavoured hadrons, apart from phase-space factors, depends on the magnitude of the couplings U_{cb} and U_{ub}. In fact [29]

$$\tau_B = 10^{-14}\text{s}/(3.68 \, |U_{cb}|^2 + 7.8 \, |U_{ub}|^2)$$

The previous upper limit obtained by the JADE Collaboration [30], $\tau_B < 1.4 \times 10^{-12}$s at 95 % C.L., is significantly improved by the recent measurements of the Mark II and MAC Collaborations [16,19]. In these experiments, semileptonic decays of b-flavoured hadrons are tagged by identifying an electron in an electromagnetic calorimeter or a muon from its penetration through a layer of iron and by requiring that this lepton has a high transverse momentum p_T relative to the thrust axis of a jet. The transverse momenta required are $p_T > 1$ GeV/c for the Mark II experiment and $p_T > 1.5$ GeV/c for the MAC experiment. In addition, the lepton total momentum is required to exceed 2 GeV/c. The resulting event samples of about 300 events in each experiment contain 80 % leptonic decays of bottom-flavoured hadrons and 20 % charm decays. The distribution of impact parameters Δ of the lepton relative to the interaction point then shows a small shift of the order of 100 μm (about one-third of the resolution) due to the finite lifetime of bottom-flavoured particles. This average shift Δ can be converted into a lifetime τ by the relation $\Delta = \gamma\beta c\tau \sin\alpha$ where α is the angle between the flight path of the decaying particle and the direction of the identified lepton, $\beta = v/c$ is the normalized velocity and γ the time dilatation factor for the decaying particle. The results are:

$$\tau_B = (12.0 \begin{smallmatrix} + 4.5 \\ - 3.6 \end{smallmatrix} \pm 3.0) \times 10^{-13} \text{ s} \qquad\qquad \text{Mark II Collab. [16]}$$

$$\tau_B = (16 \pm 4 \pm 3) \times 10^{-13} \text{ s} \qquad\qquad \text{MAC Collab.} \qquad [19]$$

(11)

2.2.4 Combined fit

Using the constraints of eqs. (2,3) and (7) to (11), we obtain for a minimum $\chi^2 = 1.1/3$ D.F. the values $\sin\theta_1 = 0.231 \pm 0.003$, $\alpha = 2S/(\bar{U} + \bar{D}) = 0.49 \pm 0.07$, and values of $\sin\theta_2$ and $\sin\theta_3$ with the error contours in the ($\sin\theta_2$, $\sin\theta_3$) plane given in fig.1. These contours vary slightly with the values of the phase δ. From the resulting error contours, we obtain a finite value of $0.025 < \sin\theta_2 < 0.06$, and an upper limit for $\sin\theta_3 < 0.02$.

We conclude from this analysis that the second mixing angle θ_2 is smaller than the first one, θ_1, i.e. $\sin\theta_2/\sin\theta_1 < 0.26$. The third angle, θ_3, is still compatible with zero, with the upper limit $\sin\theta_3 < 0.02$ at the 67 % C.L. This pattern of decreasing mixing angles means that weak transitions between members of different quark families are suppressed more for heavy quarks than for light ones.

Analogously for the Maiani parametrization [5], the error contours in the plane of the parameters $\sin\gamma$ (corresponding approximately to $\sin\theta_2$) and $\sin\beta/\tan\theta$ (corresponding to $\sin\theta_3$) are given in fig.2. Here error contours at the one (1σ) and two

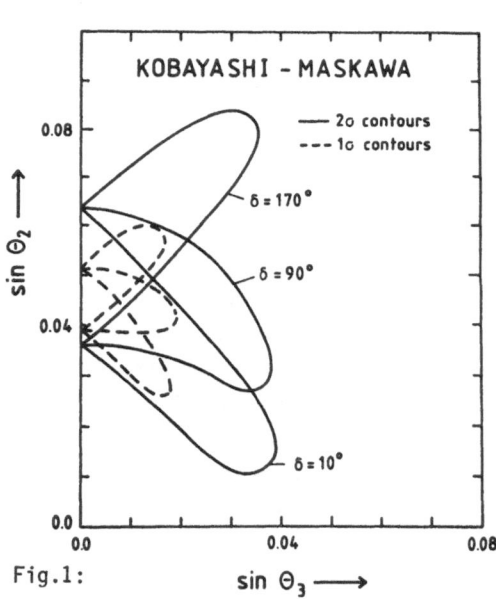

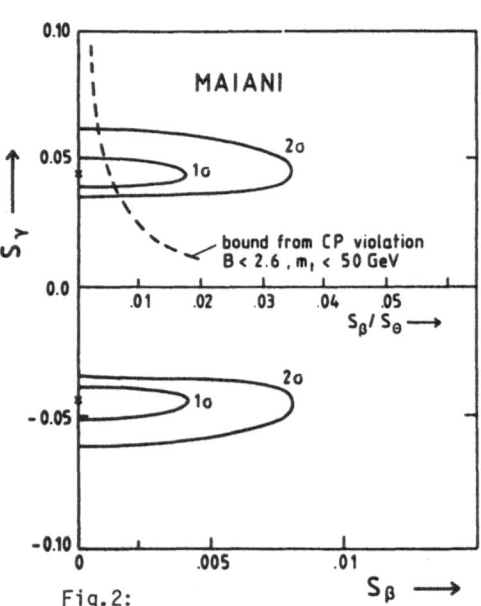

Fig.1: Error contours in the ($\sin\theta_2$, $\sin\theta_3$) plane for three values of the phase δ (10^0, 90^0 and 170^0). One standard deviation contours (dashed line) and two standard deviation contours (solid lines) are shown

Fig.2: Error contours in the plane of mixing parameters $\sin\gamma$ and $\sin\beta/\sin\theta$ in the Maiani parametrization. One standard deviation (1σ) and two standard deviation (2σ) contours are given. Also shown is a limit on the range of angles from the measured value of the CP parameter $|\varepsilon|$

(2σ) standard deviation level are given. These contours are nearly independent of
the phase angle δ' in the Maiani parametrization. The values for the angles are
$\sin\theta = 0.231 \pm 0.003$, $|\sin\gamma| = 0.044 \, {}^{+\,0.007}_{-\,0.005}$ and $\sin\beta < 0.004$.

From the range of values of the mixing angles in either parametrization, the
values of the Kobayashi-Maskawa matrix elements can be obtained. These are given in
table 2. It is evident that the error on these matrix elements from the common fit
is, for most of them, much smaller than the one obtained from individual experimental bounds on one matrix element.

Table 2

Elements of quark mixing matrix $|U_{ik}|$ from fit of experimental constraints
(1 standard deviation range)

	d	s	b
u	0.9723 - 0.9737	0.228 - 0.234	0.000 - 0.004
c	0.228 - 0.234	0.9711 - 0.9727	0.039 - 0.051
t	0.005 - 0.015	0.038 - 0.050	0.9987 - 0.9993

It also appears from the values in table 2 that, apart from the diagonal
elements, and the three off-diagonal elements observed directly up to now, (U_{us},
U_{cs} and U_{cb}), there is only one other non-diagonal element (U_{ts}) whose magnitude is such as to allow direct observation, possibly in the reaction $\nu + s \rightarrow \mu^- + t$
or in the decay $t \rightarrow s + X$. The other two elements, connecting the first and third
families, are of a magnitude which makes their detection very difficult, and therefore represent a challenge to future experimentation. A graphical representation
of the coupling strength between the six quarks is given in fig.3.

This analysis is done in the framework of a six-quark model. If the number of
quark flavours is larger than six, the unitary condition on the elements of the
3 x 3 matrix U_{ik} is replaced by $\Sigma^3_{i=1} \, |U_{ik}|^2 \leq 1$ and $\Sigma^3_{k=1} \, |U_{ik}|^2 \leq 1$. In this
generalized case, the ranges of values for U_{ik} are given in table 3.

The pattern of decreasing mixing angles can be parametrized in still another way
as suggested by Wolfenstein [31]. Realizing that experimentally $|\sin\gamma| \sim (\sin\theta)^2$,

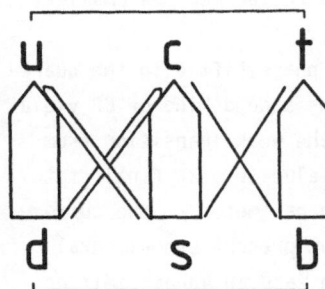

Fig.3:

Graphical representation of
weak coupling strengths between
pairs of quarks

Table 3

Elements of quark mixing matrix U_{ik} from experimental constraints if number of quark flavours is larger than 6

	d	s	b
u	0.9709 - 0.9737	0.228 - 0.234	0.000 - 0.006
c	0.21 - 0.27	0.78 - 0.98	0.039 - 0.051
t	0.00 - 0.12	0.00 - 0.58	0.000 - 0.999

Table 4

Parametrization of the quark mixing matrix according to Wolfenstein [31]

	d	s	b
u	$1 - \lambda^2/2$	λ	$\lambda^3 A(\rho - i\eta)$
c	$-\lambda$	$1 - \lambda^2/2$	$\lambda^2 A$
t	$\lambda^3 A(1-\rho-i\eta)$	$-\lambda^2 A$	1

the parametrization of table 4 is proposed. The experimental limits on these parameters are then: $\lambda = 0.231 \pm 0.003$, $A = 0.82 \begin{smallmatrix} +0.13 \\ -0.10 \end{smallmatrix}$ and $\rho^2 + \eta^2 < 0.2$.

On the theoretical side, there is no satisfactory explanation of these quark mixing angles yet. A few models exist which are able to give values for the Cabibbo angle and the other angles. Typical results are [32]: $\sin\theta \sim |\sqrt{m_d/m_s} + e^{i\alpha}\sqrt{m_u/m_c}|$ or $\tan^2\theta = (-m_d/m_s + m_u/m_c)/(1 - m_u m_d/(m_c m_s))$, and these values agree with the experiment.

If one goes beyond a purely experimental determination of the mixing angles, a description of CP violation in terms of quark mixing requires all three angles to be finite, as shown in eq. (1). This means that $\sin\theta_3$ (or $\sin\beta$), although experimentally compatible with zero, has a lower limit depending on the value of the K^0-$\bar{K}^0$ transition matrix element B. If an upper limit of B < 2.6 from theoretical arguments of Guberina et al. [32] is used, the allowed values for the angles $\sin\beta$ and $\sin\gamma$ in fig.2 have to lie above the hyperbolic line drawn from eq. (1). If the vacuum insertion value B = 1 [6] is assumed, the allowed region becomes smaller, and even more so if the PCAC-value B = 0.33 is taken [8].

The question whether CP violation is indeed due to a phaseshift δ in the quark-mixing matrix can be studied further by searching for the second kind of CP violation not due to the mass matrix but to CP violation in the weak transition from the eigenstate $K_2 = (K^0 - K^0)/\sqrt{2}$ with negative CP eigenvalue to a 2π final state. While the first kind of CP violation is described by the parameter ε, the corresponding amplitude for this second kind is called ε'. From present experimental knowledge the amplitude ε' is still compatible with zero, and an upper limit on the amplitude ratio is $|\varepsilon'/\varepsilon| < 0.01$ at 90 % C.L. [3], consistent with superweak models of CP violation [33]. If, however, CP violation is due to the quark mixing matrix, a finite value of ε' is expected. From the range of mixing angles allowed a lower limit on the ratio of these amplitudes can be derived: $|\varepsilon'/\varepsilon| > 2 \times 10^{-3}$ [34]. Present experiments on the ratio of decay rates of $K_L \rightarrow \pi^0\pi^0$ and $K_L \rightarrow \pi^+\pi^-$ undertaken now at Fermilab, Brookhaven and CERN will show whether this model will emerge as the true picture of CP violation.

3.   Lepton Mixing

In the lepton sector, an analogous formalism can be used. In fact, the weak eigenstates ν_e, ν_μ and ν_τ are related to the mass eigenstates ν_1, ν_2 and ν_3 by a unitary matrix U_{ik} [36]:

$$|\nu_\ell> = \Sigma_k U_{\ell k}|\nu_k> \qquad\qquad (\ell = e,\mu,\tau; \quad k = 1,2,3) \qquad\qquad (12)$$

and this matrix $U_{\ell k}$ can be parametrized in analogy to the quark mixing matrix.

However, experimentalists searching for neutrino oscillations have usually made the ad hoc assumption that only <u>one</u> angle and <u>one</u> neutrino mass difference is different from zero, and have given their results in terms of these two quantities. This procedure ignores the fact that a null result in an oscillation experiment is, in principle, consistent with an oscillation from a flavour k into another flavour ℓ, then to another flavour m and a subsequent oscillation back to flavour k. This means that limits on oscillation parameters done in this way potentially overestimate the significance of the experiments.

Following ref. [37], I give here results on an analysis of available experimental data with five free parameters, i.e. three angles and two neutrino mass differences.

The time evolution of an initially pure ν_ℓ state can be expressed in terms of mass differences and the neutrino energy E. If this energy is large compared to the neutrino mass, the probability to find a $\nu_{\ell'}$ after a distance L(m) from the production point of an initially pure ν_ℓ state of energy E(MeV) is [36]:

$$P_{\ell\ell'} = \sum_{k,k'=1} U_{\ell k} U_{\ell k'} U_{\ell' k} U_{\ell' k'} \cos (2.54 \Delta m^2_{k'k} L/E) \qquad\qquad (13)$$

where $\Delta m^2_{k'k} = |m^2(\nu_{k'}) - m^2(\nu_k)|$ is measured in (eV^2).

As a simplification, CP violation is neglected, i.e. the elements $U_{\ell k}$ are assumed to be real. This leaves us with five free parameters to be determined, because for three neutrino masses only two Δm^2 values are independent.

Of the two parametrizations mentioned above, the one of Maiani is particularly adapted to this problem, because in the limiting case of only one angle being finite, this angle can be directly related to oscillations between two flavours of neutrinos. The Maiani angles θ, β and γ are thus uniquely related to the oscillation channels $\nu_e \leftrightarrow \nu_\mu$, $\nu_e \leftrightarrow \nu_\tau$ and $\nu_\mu \leftrightarrow \nu_\tau$ respectively, and they correspond to the mixing angles used in earlier analyses in the framework of two-neutrino oscillations. We therefore call these angles $\theta = \alpha_{e\mu}$, $\beta = \alpha_{e\tau}$ and $\gamma = \alpha_{\mu\tau}$. We note that this decoupling of angles <u>does not occur</u> in the Kobayashi-Maskawa notation.

The experimental data on neutrino oscillations can be divided into two classes:

1. Disappearance experiments, where neutrinos of one flavour ℓ are observed at two distances from their production point, i.e. at two distant locations along their flight path. The ratio R of neutrino fluxes at the two positions is then:

$$R = P_{\ell\ell}(L_1/E)/P_{\ell\ell}(L_2/E)$$

where L_1 and L_2 are the two distances at which the fluxes are measured.

2. Appearance experiments, where neutrinos of one flavour k travel a distance L up to a detector tuned to observe neutrinos of a different flavour $\ell \neq k$. The measured quantity is then the flux of appearing neutrinos of flavour ℓ compared to the original flux of k-neutrinos, given by $P_{k\ell}(L/E)$.

For the simultaneous fit of oscillation experiments of different kinds all data from reactor and accelerator experiments available at this time are used. They are listed in Table 5 [38]. For each experiment, the original data are used, such that for a given neutrino energy each experiment gives a contribution to the global χ^2 in the fit. For each experiment, making the restrictive assumptions of the authors, the procedure used here gives back the quoted result of the authors in the two-parameter model with only one θ and one Δm^2.

A total chisquare function is constructed from the measured quantities Q^m (ratios of oscillation probabilities at different L/E or absolute probabilities

Table 5

Oscillation experiments included in the fit [38]. The last column is the χ^2 contribution of the experiment for the hypothesis of no oscillation

Experiment	Measured quantity	Δm^2 range [eV^2]	$\chi^2_{no\ osc.}$/NDF
(1) CCFRR	$P_{\mu\mu}$ ratio	30. - 1000.	15.5/14
(2) CDHS	$P_{\mu\mu}$ ratio	0.24 - 90.	15.3/14
(3) CHARM	$P_{\mu\mu}$ ratio	0.60 - 20.	1.8/3
	$P_{e\mu}$ difference	1. - 10.	0/0
(4) GOESGEN	P_{ee} absolute	> 0.01	15.5/15
(5) GOESGEN	P_{ee} ratio	0.03 - 3.	7.4/15
(6) BUGEY	P_{ee} ratio	0.02 - 5.	20.9/8
(7) ν_τ Exp.'s	$P_{\mu\tau}/P_{\mu\mu}$	> 0.2	6.3/5
(8) BNL	$P_{\mu e}$ absolute	> 0.43	2.4/6

from a comparison to calculated initial fluxes) and the corresponding computed values Q^c. The data points of each experiment i are allowed to vary together by a scale factor N_i within the quoted normalization uncertainty $\sigma(N_i)$. The expression

$$\chi^2 = \sum_{i=1} [\sum_j (\frac{N_i Q^m_{ij} - Q^c_{ij}}{\sigma(Q^m_{ij})})^2 + (\frac{N_i - 1}{\sigma(N_i)})^2] \qquad (14)$$

is minimized by fitting the mixing angles and mass differences, where the index i labels the experiments, and the index j gives the bin in L/E for a specific experiment.

There are two best fit solutions with a χ^2_{min} = 78 for 84 D.F. One of them is (solution a) reached for the parameters Δm^2_{21} = 34 eV^2 and Δm^2_{31} = 0.2 eV^2. For this solution, the angles are constrained to be $\sin^2 2\alpha_{e\tau} < 4 \times 10^{-3}$, $\sin^2 2\alpha_{e\tau} < 0.13$ and $\sin^2 2\alpha_{\mu\tau} < 0.02$, and the mixing matrix is given in Table 6a, with 90 % C.L. limits. The values for Δm^2 are different from zero only at the 1.5 standard deviation level, i.e. the finite values are not significant.

Table 6a

90 % C.L. limits on mixing matrix elements for best fit values Δm^2_{12} = 34 eV^2 and Δm^2_{13} = 0.2 eV^2

$$U = \begin{bmatrix} 1.00 - 0.98 & 0. - 0.03 & 0. - 0.18 \\ -0.04 - 0. & 1.00 - 0.99 & 0. - 0.07 \\ -0.18 - 0. & -0.07 - 0. & 1.00 - 0.98 \end{bmatrix}$$

$\sin^2 2\alpha_{e\mu} < 0.004$; $\sin^2 2\alpha_{e\tau} < 0.13$; $\sin^2 2\alpha_{\mu\tau} < 0.002$

A second local minimum of χ^2 is found at Δm^2_{21} = 0.2 eV^2 and Δm^2_{31} = 38 eV^2, i.e. interchanging the two values from solution a. Here also χ^2_{min} = 78, the 90 % C.L. limits on the angles are $\sin^2 2\alpha_{e\mu} < 0.15$, $\sin^2 2\alpha_{e\tau} < 0.09$ and $\sin^2 2\alpha_{\mu\tau} < 0.02$, and the range of matrix elements is given in Table 6b.

Table 6b

90 % C.L limits on mixing matrix elements for fit values Δm^2_{12} = 0.2 eV^2 and Δm^2_{13} = 38 eV^2

$$U = \begin{bmatrix} 1.00 - 0.97 & 0. - 0.19 & 0. - 0.15 \\ -0.21 - 0. & 1.00 - 0.98 & 0. - 0.07 \\ -0.14 - 0. & -0.10 - 0. & 1.00 - 0.98 \end{bmatrix}$$

$\sin^2 2\alpha_{e\mu} < 0.15$; $\sin^2 2\alpha_{e\tau} < 0.09$; $\sin^2 2\alpha_{\mu\tau} < 0.02$

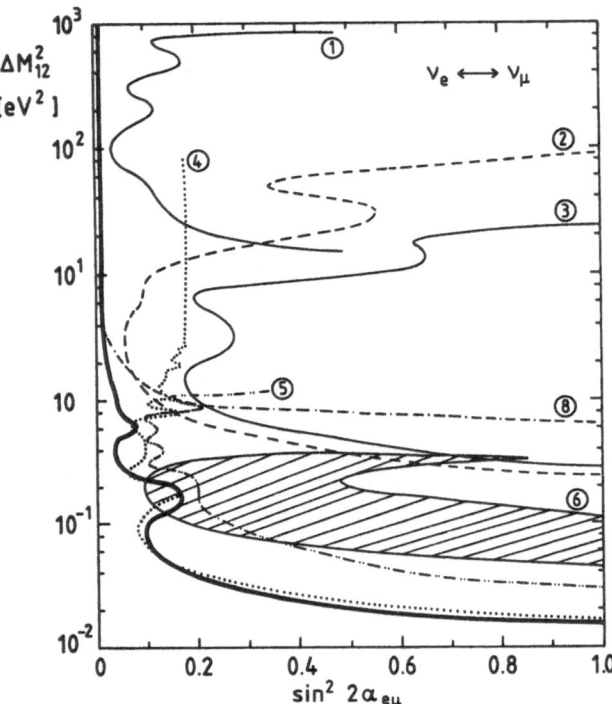

Fig. 4a: Limits on mixing parameter $\sin^2 2\alpha_{e\mu}$ vs. neutrino mass difference ΔM_{12}^2.
The thin lines are 90% C.L. upper limits on $\sin^2 2\alpha_{e\mu}$ from individual
experiments(1)(2)(3)(4)(5) and(7) in ref. [38] <u>assuming</u> $\alpha_{e\tau} = \alpha_{\mu\tau} = 0$,
the shaded area is the allowed range from the Bugey experiment (6) in
ref. [38]). The broad line is the 90% C.L. upper limit on $\sin^2 2\alpha_{e\mu}$ from
the three-flavour oscillation analysis, allowing the other two mixing
angles to vary over the whole range $0 \leq \sin^2 2\alpha_{e\tau} \leq 1$ and
$0 \leq \sin^2 2\alpha_{\mu\tau} \leq 1$

Since there is no way of discriminating between the two solutions, global limits
on angles for mass differences in the range $\Delta m_{12}^2 > 0.06$ eV^2, $\Delta m_{13}^2 > 0.04$ eV^2 and
$\Delta m_{23}^2 > 2$ eV^2 are $\sin^2 2\alpha_{e\mu} < 0.15$, $\sin^2 2\alpha_{e\tau} < 0.13$ and $\sin^2 2\alpha_{\mu\tau} < 0.02$. For any specific mass difference $\Delta m_{k'k}^2$, the 90 % C.L. limit on the corresponding $\sin^2 2\alpha_{\ell\ell'}$
can be derived from (14) in three parameter planes ($\sin^2 2\alpha_{e\mu}$, Δm_{12}^2), ($\sin^2 2\alpha_{e\tau}$,
Δm_{13}^2), and ($\sin^2 2\alpha_{\mu\tau}$, Δm_{23}^2). The curves in Fig.4 give the largest allowed $\sin^2 2\alpha$
values (for 90 % C.L.) as a function of the corresponding Δm^2. Each such point has
been obtained under the condition that in the other two planes all combinations
$(0. < \sin^2 2\alpha < 1., 0.01 < \Delta m^2 < 1000$ eV$^2)$ are possible,

Fig. 4b: Limits om mixing parameter $\sin^2 2\alpha_{e\tau}$ vs. ΔM_{13}^2. Thin lines are 90% C.L.
upper limits on $\sin^2 2\alpha_{e\tau}$ from individual experiments assuming
$\alpha_{e\mu} = \alpha_{\mu\tau} = 0$. Broad line from this three-flavour analysis.

Fig. 4c: Limits on mixing parameter $\sin^2 2\alpha_{\mu\tau}$ vs. ΔM_{23}^2. Thin lines are limits on
$\sin^2 2\alpha_{\mu\tau}$ assuming $\alpha_{e\mu} = \alpha_{e\tau} = 0$. Broad line is from three-flavour analysis.

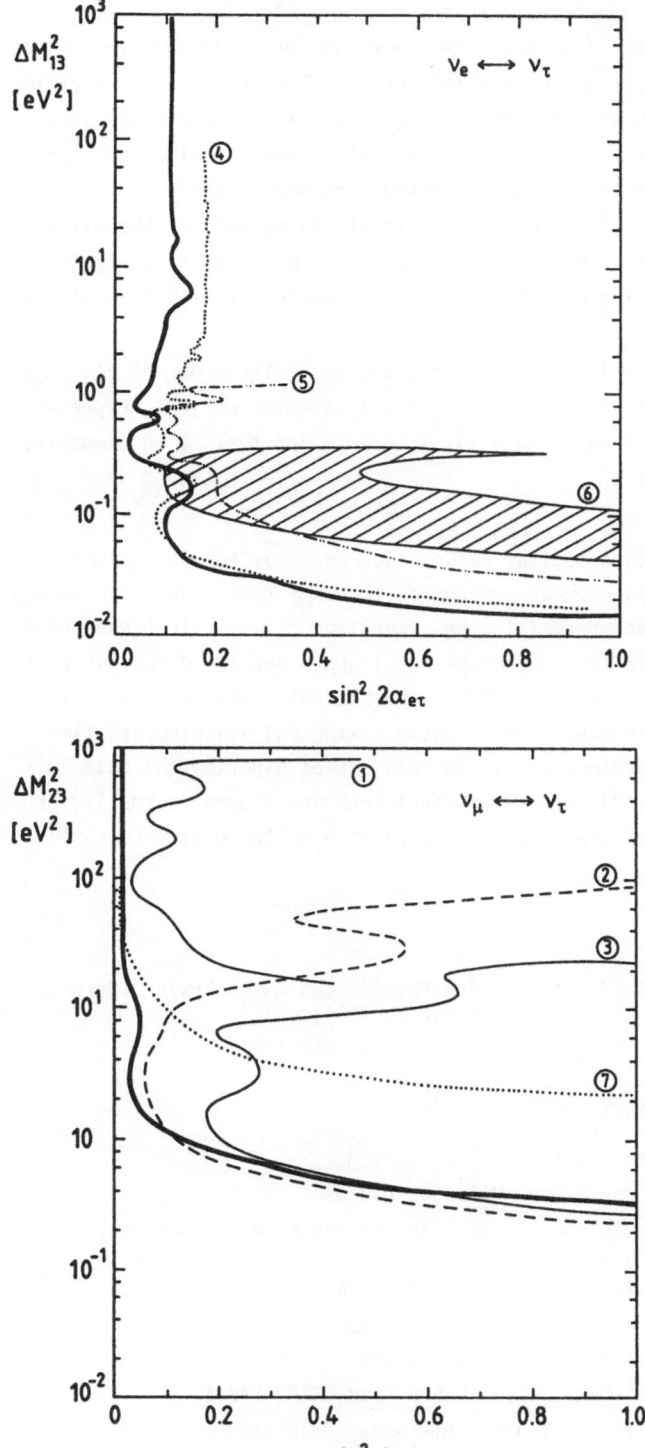

Fig. 4b Caption see
page 160

Fig. 4c Caption see
page 160

For the hypothesis of no oscillation, i.e. all mixing angles fixed to zero, a χ^2 of 85 for 87 D.F. is obtained, i.e. the data are consistent with this hypothesis.

This analysis of existing data on oscillations of three neutrino flavours yields limits on the mixing matrix and on the three mixing angles each corresponding to one channel of oscillations. While this method does not in general allow stringent constraints on the mixing parameters to be extracted from one experiment, the combination of the different experimental data gives limits on angles and the mixing matrix elements which are about as restrictive as earlier two-neutrino analyses in separate oscillation channels. The ensemble of data is consistent with no oscillation occuring.

If neutrino masses are finite [39], then mixing angles of the order of $\sqrt{(m_e/m_\mu)}$ or $\sqrt{(m_\mu/m_\tau)}$ are expected in some models invoking family symmetries [40]. Experiments have nearly reached the level of sensitivity needed for testing such models.

4.   Conclusion

The coupling strength of weak interactions between quarks of different flavour shows a clear pattern: couplings between members of the same family are much stronger than the ones between neighbour families and couplings between first and third family have escaped detection so far. The coupling strength between different families decreases with increasing quark mass, but a theoretical explanation of this pattern is still missing. On the other side, lepton mixing and neutrino oscillations have not been seen in experiments yet. An analysis of experimental data in the framework of simultaneous oscillations of three neutrino fluxes yields limits on the three mixing angles which are about as restrictive as the ones from earlier two-neutrino analyses.

REFERENCES

1   C.Rubbia, Proc.11th Int. Conf. on Neutrino Physics and Astrophysics, Nord-
        kirchen 1984 (World Scientific, Singapore, 1984) p.1

2   M.Kobayashi and K.Maskawa, Progr. Theor. Phys. 49, 652 (1973)

3   C.Geweniger et al., Phys. Lett. 48B, 487 (1974)
    K.Kleinknecht, Ann. Rev. Nucl. Sci. 26, 1 (1976)
    R.H.Bernstein et al., Phys. Rev. Lett. 54, 1631 (1985)
    J.K.Black et al., Phys. Rev. Lett. 54, 1628 (1985)

4   S.Pakvasa and H.Sugawara, Phys. Rev. D14, 305 (1976)

5   L.Maiani, Int. Symp.on Lepton and Photon Interactions at High Energies,
        Hamburg 1977 (DESY, Hamburg, 1977), p.877

6   M.K.Gaillard and B.W.Lee, Phys. Rev. D10, 897 (1974)

7   F.J.Gilman and M.B.Wise, Phys. Rev. D27, 1128 (1983)

8   J.F.Donoghue et al., Phys. Lett. 119B, 412 (1982)

9   E.A.Paschos, B.Stech and U.Türke, Phys. Lett. 128B, 240 (1983)

10 K.Kleinknecht, Proc.10th Int. Neutrino Conference, Balatonfüred, 1982
        (Central Res. Inst. Physics, Budapest, 1982) Vol.1, p.115

11   E.A.Paschos and U.Türke, Phys. Lett. 116B, 360 (1982)

12   S.Pakvasa, Proc. 21st Int. Conf. on High Energy Physics, Paris, July 26-31,
     1982. J. Phys. 43, Supp.12, C3-234 (1982)

13   K.Kleinknecht and B.Renk, Z. Phys. C16, 7 (1982); Z. Phys. C20, 67 (1983)

14   L.L.Chau et al., Phys. Rev. D27, 2145 (1983); L.L.Chau, Phys.Rep. 95, 3 (1983)

15   E.Fernandez et al., MAC Collaboration, Phys. Rev. Lett. 51, 1022 (1983)

16   N.S.Lockyer et al., Mark II Collaboration, Phys. Rev. Lett. 51, 1316 (1983)

17   M.Bourquin et al., WA2 Collaboration, Z.Phys. C21, 27 (1983)
     H.Leutwyler and M.Roos, Z. Phys. C25, 91 (1984)

18   K.Kleinknecht and B.Renk, Phys. Lett. 130B, 459 (1983)

19   J.Yelton, MAC Collaboration, paper given at Rencontre de Moriond, La Plagne,
     Febr.26 - March 4, 1984
     W.T.Ford, paper given at Europhysics Study Conference on Flavour Mixing in
     Weak Interactions, Erice, March 4-10, 1984

20   C.Klopfenstein et al., CUSB Collaboration, Phys. Lett. 130B, 444 (1983)
     J.Lee-Franzini, paper given at Europhysics Study Conference on Flavour Mixing
     in Weak Interactions, Erice, March 4-10, 1984

21   A.Chen et al., CLEO Collaboration, Phys. Rev. Lett. 52, 1084 (1984)
     P.Avery, paper given at Europhysics Study Conference on Flavour Mixing in
     Weak Interactions, Erice, March 4-10, 1984

22   R.E.Shrock and L.L.Wang, Phys. Rev. Lett. 41, 1692 (1978) and 42, 1589 (1979)

23   C.H.Christensen et al., Phys. Rev. D5, 1628 (1972)
     J.Byrne et al., Phys. Lett 92B, 274 (1980)

24   M.Roos, as quoted in K.Kleinknecht, Weak decays and CP violation, Plenary
     report at 17th Int. Conf. on High Energy Physics, London, July 1974,
     ed. by J.R.Smith, Science Research Council London 1974, p.III-23

25   H.Abramowicz et al., Z. Phys. C15, 19 (1982)

26   H.G.J.de Groot et al., Z. Phys. C1, 143 (1979)

27   R.Brock, Phys. Rev. Lett. 44, 1027 (1980)

28   G.Altarelli et al., Nucl. Phys. B208, 365 (1982)

29   M.K.Gaillard and L.Maiani, Proc. Summer Institute on Quarks and Leptons,
     Cargèse, 1979 (Plenum Press, New York, 1980), p.433; the phase space
     values are taken from a more recent analysis of J.Lee-Franzini (Ref.20).

30   W.Bartel et al., Phys. Lett. 114B, 71 (1982)

31   L.Wolfenstein, Phys. Rev. Lett. 51, 1945 (1983)

32   H.Fritzsch, Nucl. Phys. B155, 189 (1979)
     B.Stech, Phys. Lett. 130B, 189 (1983)

33   B.Guberina et al., Phys. Lett. 128B, 269 (1983)

34   L.Wolfenstein, Phys. Rev. Lett. 13, 562 (1964)

35   F.J.Gilman and J.S.Hagelin, SLAC-PUB-3226 (Sept.1983)

36   S.M.Bilenky and B.Pontecorvo, Phys. Rep. 41C, 225 (1978)
     A.deRujula et al., Nucl. Phys. B168, 54 (1980)
     S.P.Rosen, Preprint Los Alamos LA-UR-84-1789 (1984)

37   H.Blümer and K.Kleinknecht, Phys. Lett. 161B, 407 (1985)

38   Experimental results are taken from the following references:
     CCFRR   (1) C.Haber et al., Phys. Rev. Lett. 52, 1384 (1984)
     CDHS   (2) F.Dydak et al., Phys. Lett. 134B, 281 (1984)
     CHARM   (3) F.Bergsma et al., Phys. Lett. 142B, 103 (1984)

Goesgen	(4)	J.Vuilleumier et al., Phys. Lett. 114B, 298 (1982)
Goesgen	(5)	K.Gabathuler et al., Phys. Lett. 138B, 449 (1984)
Bugey	(6)	J.F.Cavaignac et al., Phys. Lett. 148B, 387 (1984)
ν_τ search	(7)	N.J.Baker et al., Phys. Rev. Lett. 47, 1576 (1981)
		N.Armenise et al., Phys. Lett. 100B, 182 (1981)
		O.Errique et al., Phys. Lett. 102B, 73 (1981)
		N.Ushida et al., Phys. Rev. Lett. 47, 1694 (1981)
		G.N.Taylor et al., Phys. Rev. D28, 2705 (1983)
		H.C.Ballagh et al., Phys. Rev. D30, 22/1 (1984)
BNL	(8)	L.A.Ahrens et al., Phys. Rev. D31, 2732 (1985)

39   V.A.Lubimov et al., Phys. Lett 94B, 266 (1980); Proc. EPS Conf. on High
        Energy Physics, Brighton 1984, p.386
      K.E.Bergkvist, Phys. Lett. 154B, 224 (1985)
      J.J.Simpson, Phys. Rev. Lett. 54, 1891 (1985)
      T.Ohi et al., Phys. Lett. 160B, 322 (1985)

40   Z.G.Berezhiani, Phys. Lett. 150B, 177 (1984)
      H.Fritzsch and P.Minkowski, Phys. Rep. 73, 67 (1981)
      L.Wolfenstein, Proc. 11th Int. Conf. on Neutrino Physics and Astrophysics,
        Nordkirchen 1984 (World Scientific, Singapore, 1984) p.730

Physics with the Tevatron

L.M. Lederman

Fermi National Accelerator Laboratory, Batavia, IL 60510, USA

1. INTRODUCTION

Jack Steinberger "initiated" the electron synchrotron at Berkeley in 1950 with an elegant experiment [1] which did much to establish the existence of the π^0. He arrived at Columbia's Nevis Labs in 1951, just as the world's highest energy accelerator, the 385 MeV synchrocyclotron at Nevis, was coming on line. Since then we have together witnessed the successive initiations of the Cosmotron (1952), the AGS (1960), the PS (1961, CERN), the ISR (1972, CERN) and, somewhat separately, Fermilab and the SPS (1972-1976). As Steinberger looks forward to initiating LEP it seems appropriate to use this ceremonial occasion to describe the initiation of the TEVATRON, its origins, its commissioning and the physics expected and, indeed, already glimpsed as of the fall of 1985. (This is the one he missed!)

2. PRE-HISTORY

In 1973, R. R. Wilson, the founding director of Fermilab, started a research and development program, aimed at acquiring the technology of ramped superconducting magnets. In 1975, he was joined by Alvin Tollestrup, on leave from California Institute of Technology and many series of 0.6 m long models were constructed. In 1976, an assembly line was created and full scale (7m) magnets were assembled in a mass-production mode with frequent changes in the tooling, materials, assembly technique. These changes were based on the results of performance of groups of assembled magnets. The objectives were to construct a new ring of superconducting magnets in the tunnel which carried the 400 GeV main ring. The purposes of the machine were to double the energy (400 to 800 GeV, hence the name Energy Doubler) and to save substantial amounts of electrical energy (60 megawatts of magnet pulsed power reduced to 15 megawatts of refrigeration power - hence the name Energy Saver).

The U.S. DOE authorized the project to proceed in July of 1979 and the installation of 776 dipoles each 6.2 m long, 226 quadrupoles, each 4m long and 226 complex spool pieces containing beam diagnostics, correction windings, safety leads etc., was complete by April of 1983. At the same time, a mammoth cryogenic system was created to insure an ample supply of liquid helium to keep the 6.28 km of superconducting magnets at 4.5°K. During this interval, the physics goals of the TEVATRON project were refined and two additional projects started, whose function was to realize the physics potential of the accelerator. The scientific input came from numerous user's workshops, the Fermilab Physics Advisory Committee, the Science Committee of the trustees of the management, the Universities Research Association and the HEPAP group of DOE. It was modulated by the trends in physics emerging from the e^+e^- colliders and from the increasingly powerful CERN fixed target program. The CERN Sp$\bar{p}$S, with the AA ring as a source of antiprotons also played an important role as guide and yardstick.

The new projects were labelled Tevatron I and Tevatron II. Tevatron I was to build an antiproton source and to modify the Fermilab accelerators so as to provide proton-antiproton collisions at maximum energy. The Tevatron II program was charged with the upgrading of extraction, switchyard and beam-line components to operate a fixed target program at energies up to 1000 GeV. In the specifications, two new beam-lines and many modified beam lines were included.

3.   FIXED TARGET PHYSICS

3.1 General Introduction

The program of Tevatron physics that emerged out of the embers of the 400 GeV research and out of the program committee's recommendations contained proposals submitted in the period 1979-1983. One ingredient that is immediately noted is the increase in complexity, ambition, cost and size of group involved in this research. This is consistent with the trend of the field and understandable in a subject which is mature and largely programmatic. The broad range of topics can be classified as: 1) heavy quarks, ii) lepton-induced processes, iii) hard collisions and tests of QCD, iv) weak decays, CP violation and magnetic moments and v) studies using polarized beams of protons and antiprotons. The program of some two dozen experiments scheduled over the period 1983-1988(!) makes use of beams of protons, neutrons, pions, kaons, sigma hyperons, neutrinos, muons, photons tagged and photons untagged, antiprotons and polarized lambdas to make polarized protons and antiprotons.

It is appropriate to survey the status of our understanding. The base of a continuing synthesis and the essential armament we take with us into the frontier is the Standard Model, the notion that the world is made of fermions, six quarks and six leptons and controlled by strong and partially unified electroweak forces. These forces are defined by 12 spin one gauge bosons. In detail, there are weaknesses in this theoretical structure which are very profound, i.e., the spontaneous symmetry breaking, the origin of CP violation, the chiral nature of the known couplings, the overall aesthetics. Other weaknesses are partly a matter of gaps in our information: how do we continue to test and guide the development of QCD so that we

166

fashion a working theory of strong interactions? How do we improve the precision and the detail of heavy flavor production and dynamics? Do we have a precise enough record of CP violating parameters, again to guide the theory? What about all the other elements of the K/M mixing matrix? Must this be enlarged to include neutrino masses? Why are hyperons polarized at high energy and can we learn to track constituent spins? What are the wave functions of the hadrons, i.e., how do quarks bind in baryons and mesons as well as in the "onia". What is the evolution with Q^2 of the fundamental angles and coupling constants? Can we understand how constituents cleverly generate the magnetic moments of the baryons?

Enlarging the domain of the standard model somewhat, we have a set of issues that arise from the coexistence, in a crowded place, of many nucleons, i.e., many packages of assorted quarks and gluons. The issues here have to do with the permeability of the walls separating the packages (bags) or, if indeed these walls do exist when observed by high energy probes. There are a variety of measurements, "EMC effect", "A^α effect" which beg for clarification - which may (or may not) be relevant to intrinsic properties of the constituents, enhanced in an intense social environment. There is enough to do here.

3.2 The Tevatron

The decision to invest in a qualitative advance in fixed target facilities will be judged by scientific history. However it is useful to review the improvements that are available to the physicists who select from the problems outlined above.

1. Energy
The passage from 400 GeV to 800-1000 GeV represents a factor of 40-50% increase in CM energy. However, the domination of constituent physics in the research program makes for a much greater effective extension of parameters. At 200-400 GeV, we are just begining to see quark gluon effects in hadronic and lepton-hadron collisions. For example, at 200 GeV, the upsilon (a hard threshold is 50 GeV) just barely shows up. Scaling violations in muon scattering are barely visible in this domain. Dilepton production (Drell-Yan process) becomes informative on quark structure functions when $m = \sqrt{Q^2} > 4-5$ GeV and the 400 GeV data plays out at 12-14 GeV. Hadronic "jets" i.e., outgoing partons, are seen only with imaginative assistance at 400 GeV. In short, the quantitative study of constituent physics has <u>soft</u> thresholds in the 300-400 GeV domain - the availability of of 800-1000 GeV, then greatly extends the range of parameters.

2. Duty factor
The old machine has a spill structure of 1 sec every 10-14 seconds, limited by the cost of electrical energy. The TEVATRON can deliver beam over a 23 sec spill

every 60 seconds, i.e., an improvement in duty factor of more than a factor of four. For experiments that are not proton limited, this results in at least four times the data taking rate.

3. Flux
The flux of secondary hadron beams is increased, per primary proton, by a large factor due to the larger effective acceptance of the new beam lines.

4. Lifetimes
There is an extension in the lifetime of short-lived particles due to the Lorentz factor. This provides an improved handle on charm, tau, beauty and more intense, more useful and purer beams of hyperons.

5. Rapidity
The extra two-thirds of a unit of rapidity which is available in produced phase-space at the higher energies allows better separation of the various fragmentation regions for ordinary processes. In particular there is emergence of the "central plateau" separating target and projectile fragmentation regions. This is important for studies which attempt to sort out production mechanisms and is especially relevant for A-dependence studies. The difference between 200-400 GeV and 700-1000 GeV (secondary beams) is very significant.

6. General
There is a loss of overall intensity due to the slower cycle time but in many cases this is compensated by higher secondary beam efficiencies. In the case of neutrinos, the cross section is increased and the divergence of the neutrino beam is decreased leading to a major compensation.

3.3 The TEVATRON Program

A strong physics proponent, guide and philosopher of fixed target program is J. D. Bjorken. He has classified the research program in terms of major areas, i.e., i) Electroweak, ii) Magnetic Moments, CP violation and Weak Decays, iii) Heavy Quarks, iv) Hard Collisions and v) Others. The currently (Fall'85) approved program is tabulated (Table 1) under these headings. A map of Tevatron II beam lines is given in Figure 1 which can be used to locate the major fixed target experiments.

   To convey some sense of the breadth of the program we give brief comments on some of the major experiments. This breadth comes from the diversity of secondary beams and detection devices that are deployed to address the physics issues.

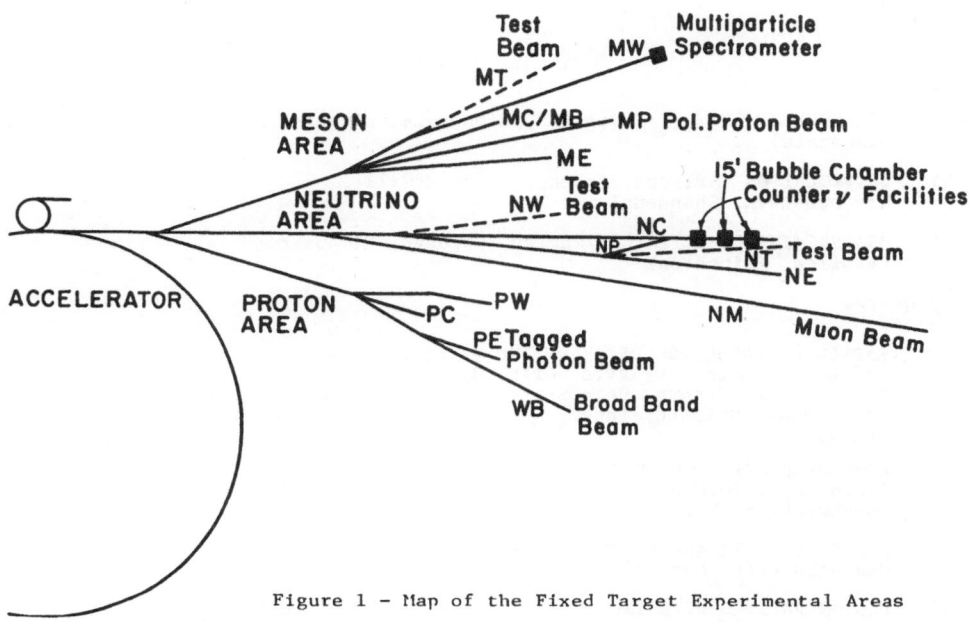

Figure 1 - Map of the Fixed Target Experimental Areas

Table I Glossary of Approved Experiments in the Fermilab Fixed-Target Program

Electroweak

E-632   WIDE BAND NEUTRINOS IN THE 15 FT BUBBLE CHAMBER
(Berkeley, Birmingham, Brussels, CEN/Saclay, CERN,
Fermilab, Hawaii, IIT, Imperial College, MPI/Munich,
Oxford, Rutgers, Rutherford-Appleton, Stevens, Tufts)

E-635   SEARCH FOR AXION-LIKE OBJECTS (Fermilab, VPI)

E-636   STUDY OF BEAM DUMP PRODUCED NEUTRINOS (Beijing, Brown,
Fermilab, Haifa, Indiana, MIT, ORNL, Seton Hall, Tel-
Aviv, Tennessee, Tohoku, Tohoku Gakuin)

E-646   STUDY OF PROMPT NEUTRINO PRODUCTION (Berkeley, Columbia,
Fermilab, Hawaii, Rutgers)

E-649   NUCLEON STRUCTURE FUNCTIONS AT HIGH Q^2 (Fermilab, MIT,
Michigan State)

E-652   NEUTRINO PHYSICS AT THE TEVATRON (Chicago, Columbia,
Fermilab, Rochester)

E-665   MUON SCATTERING WITH HADRON DETECTION (Argonne, Cracow,
CERN, Fermilab, Freiburg, Harvard, Maryland, MIT,
MPI/Munich, San Diego, Washington, Wuppertal, Yale)

E-733   NEUTRINO INTERACTIONS WITH QUAD TRIPLET BEAM (Fermilab,
Florida, MIT, Michigan State)

E-744   NEUTRINO PHYSICS WITH QUAD TRIPLET BEAM (Chicago,
Columbia, Fermilab, Rochester)

E-745   NEUTRINO PHYSICS WITH QUAD TRIPLET BEAM (Beijing, Brown,
Fermilab, Haifa, Indiana, MIT, Nagoya, ORNL, Tel-Aviv,
Tennessee, Tohoku, Tohoku Gakuin)

Table 1 (cont.)

Decays and CP

E-621   MEASUREMENT OF η_{+-0} (Michigan, Minnesota, Rutgers, Wisconsin)

E-721   CP VIOLATION (Arizona, Athens, Duke, McGill, Northwestern, Shandong)

E-731   MEASUREMENT OF ϵ'/ϵ (CEN/Saclay, Chicago, Elmhurst, Fermilab, Princeton)

Heavy Quarks

E-653   HADRONIC PRODUCTION OF CHARM AND B (Aichi, Carnegie-Mellon, Chonnam, UC/Davis, Gifu, Gyeongsang, Jeonbug, Kobe, Korea, Nagoya, Ohio State, Okayama, Oklahoma, Osaka City, Osaka Sci. Ed. Inst., Sookmyong Womans, Toho, Won Kwang)

E-687   PHOTOPRODUCTION OF CHARM AND B (Colorado, Fermilab, Illinois, INFN/Frascati, INFN/Milano, U. Milano, Northwestern, Notre Dame)

E-690   STUDY OF CHARM AND B PRODUCTION (Columbia, Fermilab, Massachusetts, Mexico)

E-691   PHOTON PHYSICS WITH TAGGED PHOTON SPECTROMETER (UC/Santa Barbara, Carleton, CBPF/Brazil, Colorado, Fermilab, NRC/Canada, Oklahoma, Sao Paulo, Toronto)

E-705   CHARMONIUM AND DIRECT PHOTON PRODUCTION (Arizona, Athens, Duke, Fermilab, McGill, Northwestern, Shandong)

E-743   CHARM PRODUCTION IN PP COLLISIONS (Aachen, Brussels, CERN, Duke, Fermilab, Florida State, Coll. of France, Kansas, LPNHE/France, Michigan, Michigan State, Mons, Notre Dame, Strasbourg, Vanderbilt)

Hard Collisions

E-605   LEPTONS AND HADRONS NEAR THE KINEMATIC LIMIT (CERN, Columbia, Fermilab, KEK, Kyoto, Saclay, SUNY/Stony Brook, Washington)

E-672   HIGH P_T JETS AND HIGH MASS DIMUONS (Arizona, Caltech, Chicago Circle, Fermilab, Florida State, George Mason, Indiana, Maryland, Rutgers, Serpukhov)

E-683   PHOTOPRODUCTION OF HIGH P_T JETS (Arizona, Fermilab, Lehigh, Rice, Vanderbilt, Wisconsin)

E-704   EXPERIMENTS WITH POLARIZED BEAM FACILITY (Argonne, Austin, UC/Berkeley, Fermilab, KEK, Kyoto, LAPP/France, LBL, Northwestern, Rice, Saclay, Serpukhov, Trieste)

E-706   DIRECT PHOTON PRODUCTION (Delhi, Fermilab, Michigan State, Minnesota, Northeastern, Pennsylvania, Pittsburgh, Rochester, Rajasthan)

E-711   CONSTITUENT SCATTERING (UC/Davis, Fermilab, Florida State, Michigan)

Others

E-466   NUCLEAR FRAGMENTS (Argonne, Chicago, Chicago Circle, Purdue)

E-508   EMULSION/MULTIPARTICLE PRODUCTION (Cracow, Louisiana State, Tashkent)

E-524   EMULSION/PROTONS GREATER THAN 500 GEV (Washington)

E-576   EMULSION/500 GEV PROTONS (Belgrade, Fermilab, Lund, Lyon,
         Nancy, Ottawa, Paris VI, Santander, Strasbourg, Valencia)

E-750   EMULSION/MULTIPARTICLE PRODUCTION (Delhi)

E-751   EMULSION/1 TEV PROTONS (SUNY/Buffalo)

E-753   CHANNELING STUDIES (Bell Northern Research, Chalk River,
         Fermilab, New Mexico, SUNY/Albany)

E-754   CHANNELING TESTS (Case Western Reserve, Fermilab, GE R&D
         Center, Sandia, SUNY/Albany)

3.3.1 Electroweak Processes

The dominating activity is neutrino physics which typically alternates between narrow band and wide band modes. In 1986, a new tevatron muon beam will be deployed. Finally, a prompt lepton facility has been in the planning phase at Fermilab for many years. The major instruments are held over from the 400 GeV program but with substantial improvements in performance. The discovery of approximate scaling and the qualitative agreement of QCD with the observed level of scale breaking as well as increasingly precise measurements of structure functions at CERN and Fermilab have already had a major impact on the development of the Standard Model. This is the best way to determine α_s and much must still be done to improve accuracy and Q^2 range. The tevatron should open up the range for both neutrino and muon scattering and extend the determinations of F_2 and xF_3. Gluon structure functions must also be measured better.

The "standard" neutrino detector [E-652/744, Table 1] evolved from E-616 (CCFR collaboration), consists of six modules of thick plate calorimetry and three large toroid sections for determination of muon momenta. The objectives are high statistics measurements of structure functions and a new look at such anomalies as the same-sign dimuon events.

A fine grain electronic detector (Lab C) contains 300 tons of thin plates interspersed with flash-chamber proportional tubes designed to measure the direction of the recoiling hadron shower. Muons are measured behind magnetic toroids. Data (5 x 10^{17}, 800 GeV protons) has been taken on weak neutral currents, inverse muon decay and coherent neutrino scattering.

The 15' bubble chamber has been outfitted with laser holography designed to improve the vertex resolution from $\sim$ 400μ to 100μ [E-632].

A new detector, the Tohoku 36" bubble chamber has high resolution (30μ) holography and detailed downstream particle identification. It was designed for studying "prompt" sources of neutrinos, some 60m downstream of a dump. A combined physics-engineering run (E-745) examines vertices of neutrino induced charm decays.

The new muon beam will go to energies as high as 750 GeV and will have intensities at 600 GeV up to 10^7 muons per pulse for 5 x 10^{12} protons on target. A very general spectrometer (E-665) combines two large magnets: the CERN (EMC) vertex magnet and the Chicago Cyclotron Magnet. The properties of recoiling hadrons furnishes data complimentary to e^+e^- fragmentation studies. The structure functions, A dependence and extended domain notably in the region of small x and high Q^2 will be measured.

3.3.2 Weak Decays, Magnetic Moments, CP

Here E-731 is continuing to increase the precision of CP violation parameters. Data taken in 1985 has a statistical value of five times the previous (E-617) experiment. The objective is an error in ε'/ε of $\pm$ .001 which would severely constrain the models for CP violation. The 800 GeV protons and excellent duty cycle provide useable fluxes of 100 GeV/c K_L's five times higher than ever before.

The 1985 run saw the completion of the measurement of η_{+-0} by measuring the interference between K_S and K_L decaying to $\pi^+\pi^-\pi^0$ near the Kaon production target (E-621).

The magnetic moment program at Fermilab goes back many years - now systematic measurement of polarization and magnetic moments of leading hyperons is nearing completion. See Table II. The missing Ω^- measurement is scheduled (E-756) for data taking in 1987.

Finally, the long-standing puzzle of the sign of the asymmetry in the β-decay of the Sigma hyperon was settled by E-715 where some 80.000 β-decays are being analyzed in detail [2].

Table II   Baryon Magnetic Moments[a]

Baryon	Experimental μ, units $e^+n/2m_pc$	Quark Model Prediction	$\mu - \mu Q$	$g/2-1$
p	2.7928456 (11)	input	-	1.79
n	-1.91304184 (88)	input	-	-
Λ	-0.6138 ± 0.0047	input	-	-
Σ^+	2.357 ± 0.012	2.67	-0.30 ± 0.01	2.00 ± 0.014
$\|\Sigma^0 \to \Lambda\|$	1.82 $^{+2.5}_{-.81}$	-1.63	-0.19 $^{+.28}_{-.18}$	-
Σ^-	-1.151 ± 0.021	-1.09	-0.06 ± 0.021	0.47 ± 0.03
Ξ^0	-1.253 ± 0.014	-1.43	+0.18 ± 0.014	-
Ξ^-	-0.69 ± 0.04	-0.49	-0.20 ± 0.04	-0.03 ± 0.05

a) Data from Rev. Mod. Phys. 52, S1 (1980), except for μ_{Σ^+}, μ_{Σ^-}, μ_{Ξ^0}, and μ_{Ξ^-}.

±(10-15)% agreement with quark model

3.3.3 Heavy Quark Physics

The numbers are spectacular! Charm is produced in 0.1 % of the inelastic collisions. Consider a modest 3×10^6 interactions/sec, i.e., 6×10^7 in the 20 sec, spill, and with a 30 % acceptance, one has 10^9 charm events and several times 10^7 $b\bar{b}$ events in a year containing 1000 good hours. This is the dominant challenge to the instrumentalists. Progress in fast, on-line processing and vertex detection gives some hope that a breakthrough of this order (or a factor 10-100 less!) can be had in the next several years. The payoff is that each channel of production and decay can be studied with very good data. The spectroscopy of baryons with heavy quarks delights the acronym specialist, e.g., ccd, ssc, cub, bcs, ccc, etc. Bjorken points out that production dynamics will be instructive for QCD. The existence of cross-sections with energy dependence, A-dependence, x-behaviour etc. must stimulate QCD calculations. He goes on to discuss spectroscopy and decay properties [3].

Tevatron experiments are in these directions. E-653 makes use of a hybrid emulsion spectrometer and is a collaboration of institutions in U.S. and Japan. The few micron spatial resolution of nuclear emulsion is supported by an elaborate downstream spectrometer designed to measure the decay products of charm and beauty quarks and to trace back to the vertex of the event located in the emulsion. Here Silicon Microstrip Detectors (SMD) locate vertices to ± 10μ transverse and 200μ along the beam direction. Some 30ℓ of emulsions were exposed in the 1985 run and this should yield 10.000 associated charm decays and about 150 b-decays.

E-687 makes use of an intense wide band beam of photons ($\sim 5 \times 10^6$ photons, $E > 200$ GeV per 10^{12} protons) to produce c and b states. Both the new beam and elements of the detector were comissioned in the 1985 run. The spectrometer, evolving from a series of earlier experiments (E-87, E-400) has larger aperture, more e/γ segmentation, improved muon and vertex detection.

E-690 uses incident hadrons and fully reconstructs all events "worth reconstructing", i.e., at the rate of 10^5 events/sec with little deadtime. The hardware for this pipelined digital computation has been exercised at BNL for several years and should be installed in FY 87.

E-691 produces heavy quarks via a tagged photon spectrometer. SMD are added to help resolve heavy quark vertices. Again this is an evolution of a program (E-516) which recorded about 10^8 events in the 1985 run.

E-705 looks at a variety of incident beams: $\bar{p}$, p, π^+, π^- and intends to study direct photons and charmonium production via a very high resolution electromagnetic shower counter. Preliminary data was taken by this experiment in 1985. This spectrometer is an upgrade of the spectrometers built for E-537.

E-743 is the Small European Bubble Chamber (LEBC) married, temporarily, to the Fermilab Multiparticle Spectrometer, supplemented by the CERN TRD. The bubble chamber as a high resolution vertex detector recorded over one million picutres in

1985 and should provide $\sim$ 1000 clear charm events to determine cross section and energy dependence.

3.3.4 Hard Collisions

Tevatron experiments have the virtue of being uniquely capable of triggering on specific hard collisions: quark-gluon, quark-antiquark etc. The information arriving from collider experiments at CERN only makes the thrust and interpretation of fixed-target hard collisions more interesting. Nuclear matter provides another interesting environment for both initial state collisions and for final states, e.g., hadronization. Shadowing, A-dpendence, dense systems of gluons at small x - are all examples. Large issues of QCD phenomenology are relevant. Gluonic structure functions can be directly observed by hadron induced prompt photon experiments.

Jets, barely visible at 400 GeV, (we know they are there) will also play a role in the analysis of hard collisions.

E-605 attempts to employ maximum luminosity in order to study hard collisions near the kinematic limit. These collisions generally fight a $1/M^4$ law where M is a measure of the hardness. Open geometry permits simultaneous study of pairs of hadrons, pairs of leptons and exquisite mixtures e.g., $h^+\mu^-$, $\mu^+ e^-$ etc. A 15m long ring imaging Cherenkov counter, calorimetry, muon tagging all follow a double magnet spectrometer, the target-bearing first magnet having a magnetic kick capability of 9 GeV/c. Data has been collected on all combinations of pairs although ambient radiation from walls, surfaces, frames etc. restricted the open geometry rate to $\sim 10^{10}$ interactions/sec. The acceptance is a few percent of the total solid angle peaked about 90° in the CM and strongly favors high p_t. To do full justice to dilepton physics, a 4' thick wall of lead was necessary to protect the detector planes from soft photons. This permitted the '85 run to take $\sim 10^{11}$ interaction/sec. In 1985, some 20.000 upsilons were recorded and the three states very clearly resolved.

E-672 was designed to study hadronic states in association with high p_t jets and high mass dimuons. This was again an evolution of earlier research that used a high-transverse energy trigger and the MPS facility. Forthcoming runs will see the deployment of improved highly-segmented wide angle em and hadronic calorimetry, an open geometry dimuon detector and very forward hadronic calorimetry and particle I.D. via RICH. The experiment will look at meson induced hard collisions over a complete solid angle coverage. The high p_t and dimuon triggers should serve to tag forward jets generated by quarks, diquarks and color-triplets.

E-704 One of the last understood aspects of hadronic interactions concerns the persistence of spin effects at high energies. Interpretation of polarization of hyperons produced inclusively from an unpolarized initial state has given rise to an extensive literature on the origin of this effect. It is expected that information on spin transfer from initial to final state will enliven the discussion.

174

This experiment creates a beam of 50 % polarized protons at 200 GeV with an intensity of 2 x 10^7 ppp. Upgrading to 400 GeV is a design parameter in creating the beam. The first runs will explore the spin dependence of the interaction by measurement of the total cross section of pp and $\bar{p}p$, comparing states with helicities parallel and antiparallel. The consortium from U.S., France, USSR, Japan und Italy have an ambitious follow-up program including inclusive production of neutral pions at x = 0 and large p_t, of neutral and charged pions at large x and of $\Lambda^0(K^0)$ at large x.

E 706 This experiment concentrates on the gluonic structure in hadrons and on the gluon fragmentation through analyzing the production of direct photons and their accompanying hadrons. The following constituent processes are studied: $gq \rightarrow q\gamma$, $\bar{g}g \rightarrow q\gamma$. Available to the experiment are protons, kaons and pions and of course a variety of nuclear targets. The experimental arrangement is dominated by the Liquid Argon Calorimeter designed to record photons with high precision and very large coverage.

E-711 is a study of the angular and energy dependence of constituent scattering using high p_t single hadrons (roughly back-to-back) as the indicators of recoiling constituents. Emphasis is on attempts to achieve high luminosity in order to follow the cross-section. Initial goals of the experiment were achieved in the 1985 run.

3.3.3 "Others"

It is clear that the major thrust of the fixed target research is programmatic but the history of our subject is filled with incidents of major discoveries cropping up in such agendas. Indicators e.g. the sigma-beta decay problem or the same sign dilepton problem are being addressed. Imaginative and speculative proposals exist and are almost pathetically welcome to PAC's as a lively change of menu.

The direct lepton facility (beam-dump) (E-635, 636, 646) has as its bread-and-butter the establishment of the tau neutrino but clearly the short-lived neutrino search is sensitive to other things much beloved by the speculative literature: axions, neutral leptons and the long-lived neutral objects of supersymmetric theories. We have above alluded to the "ultimate" heavy quark spectrometer. Other forays are possible: there has been a program of crystal channeling which could have applications in providing small septum magnets. Other possibilities include rare decay modes, stored $\bar{p}$'s colliding with gas jets, stored muons to make precision neutrino beams.

4. TEVATRON - THE PROTON-ANTIPROTON COLLIDER

Precursors of colliding beam ideas at Fermilab go back to 1976 when Cline et al. proposed to construct an Antiproton source (Proposal P-492). Other possibilities were also proposed soon after, involving subsidiary rings, pp collisions between

175

the main ring and Energy Saver ring, p$\bar{\text{p}}$ in the main ring (the CERN approach) and finally, $\bar{\text{p}}$p collisions in the SAVER ring. In a fifteen hour "shoot-out" of all contenders on Nov.11, 1978, there developed a sense of consensus that only the latter proposal had the possibility of reaching the highest possible energy and that this would be the ultimate jewel of the TEVATRON program: up to 2 TeV in the CM! In 1978 this seemed like a lot. In 1985, this still seems like a lot. In October 13, 1985, the TeV 1 collider achieved p$\bar{\text{p}}$ collisions (Fig.2) at 1.6 TeV and a feasibility trial of the entire system, including detector, was successfully terminated. In what follows, we say how we got here and describe the detectors and physics objectives, those that are clear as well as those that are dimly perceived. There are three systems to discuss: the $\bar{\text{p}}$ source, the main ring (MR), and TEVATRON accelerators and the detectors.

4.1 TEV 1: The source

The initial design specification is to achieve a peak luminosity of $10^{30} \text{cm}^{-2} \text{sec}^{-1}$. In 1984, a subsidiary plan calls for extension to $6 \times 10^{30} \text{cm}^{-1} \text{sec}^{-1}$. The original design can be achieved with as few as 1.8×10^{11} protons and 1.8×10^{11} antiprotons of appropriate phase space density. To reach this requires several hours in which antiprotons are collected at a 2 sec repetition rate and their phase space density

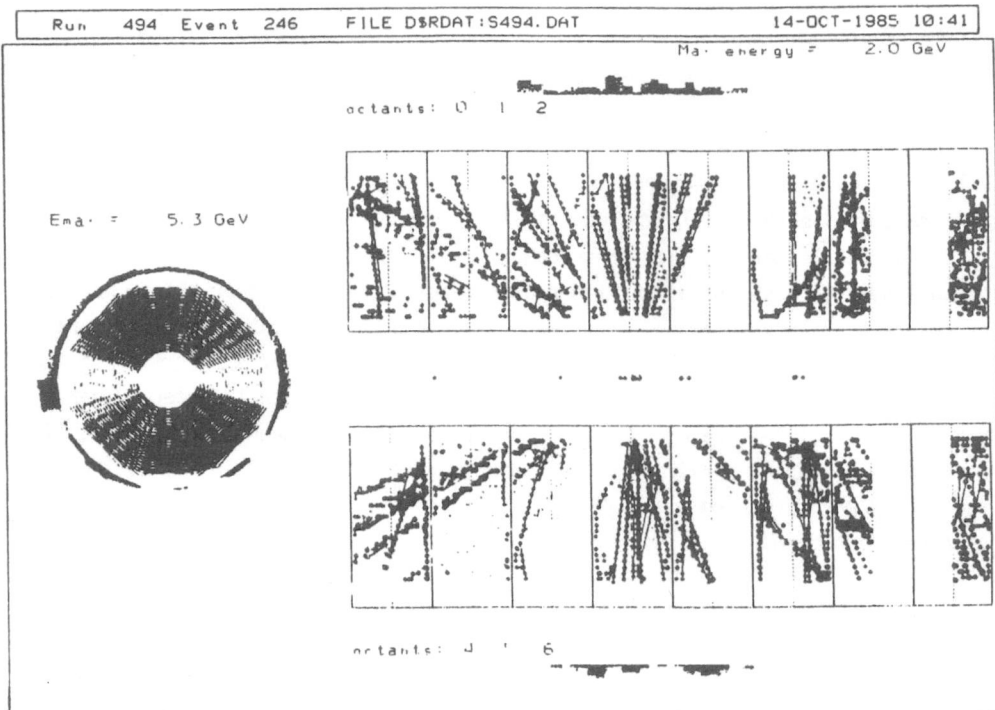

Figure 2 - Proton-antiproton collision at 1.6 TeV in the CM

is effectively increased by six orders of magnitude. To insure an adequate <u>average</u> luminosity, the storage lifetime (> 10 hours) must be long compared to the replenishment time of $\bar{p}$'s (design is $\sim$ 2 hours).

The key idea is of course the method of stochastic cooling invented and implemented at CERN by Simon Van der Meer, Lars Thorndal and co-workers. The work at Fermilab profited enormously from the proof-of-principle and the successful application in the Antiproton Accumulation (AA) ring, and $\bar{p}$p collisions in the CERN SPS. These developments, closely tracked, were both an education and a spur to do better. In fact, a useful variant of the CERN design was invented [4] in 1982. The Fermilab version of the CERN AA ring has two concentric rings, labelled, respectfully, the debuncher ring and the accumulator ring. A key variation is to do rf manipulations on the protons in the Fermilab MR so as to achieve a very narrow time spread (< 1ns) and a concomitant wide energy spread. These protons are extracted at 120 GeV and focussed down to a minimum spot on the $\bar{p}$ production target. The produced $\bar{p}$'s then share the short time spread and small spot size. This maximizes the initial state density of $\bar{p}$'s in phase space and greatly simplifies the subsequent cooling problem.

The short time spread is cashed in for a short energy spread in the debuncher ring by rf phase rotation. This permits the collection of $\bar{p}$'s over a larger energy spread. Here follows the sequence of operations and what has been achieved as of October, 1985 all under the leadership of John Peoples.

4.1.1 Proton Acceleration for $\bar{p}$ Production

Every 2 seconds, one Booster batch (8 GeV injector) containing 2×10^{12} protons in 82 rf bunches is accelerated in the MR to 120 GeV and manipulated by MR rf to produce a 0.7 ns, $\Delta E = \pm$ 185 MeV train of 82 proton pulses. This is extracted and transported to the target. In October, a peak of 1.2×10^{12} protons was extracted every 5 seconds.

4.2.1 $\bar{p}$ Transport

These are collected by an axial focussing magnetic lens, made by a high current discharge through a lithium cylinder. Beam is transported to the debuncher. The momentum spread is 3 % and the beam emittences are 20π mm-rad in each transverse plane. In October, various compromises resulted in a loss of about a factor of 2 from the design.

4.1.3 Debuncher Activities

Here the narrow time structure and large momentum spread are traded into a small momentum spread, ultimately to $\sim$ 0.2 % and a wide time structure. Here also, stochastic cooling of horizontal and vertical emittences takes place to reduce the emittances to 7π mm-mrad. This takes two seconds. The bunches are extracted and

sent to the accumulator ring. The October tests indicated that the debuncher be-
haved as designed, although beam losses were observed.

4.1.4 Accumulator Activities

Successive batches of debuncher $\bar{p}$'s are accumulated by rf stacking at the edge of
the stack. A new batch, density about 7 $\bar{p}$'s/ev is deposited at a radius appropriate
to the stack tail (Fig.3). The fresh batch is moved by the coherent force of the
stochastic cooling system away from the injection channel and toward the center of
the stack. A balance between the diffusion effects and the coherent forces increases the intensity in the core region. Allowing for losses, some 6×10^7 $\bar{p}$'s should
be stacked each 2 sec cycle. In 4 hours, the core should grow to a density of 1 x
10^5 $\bar{p}$'s/ev and $\sim 4 \times 10^{11}$ $\bar{p}$'s. The continuous longitudinal and transverse cooling
will have reduced the emittances to 2π mm-mrad. After this time, a portion of the
core, about 2×10^{11} $\bar{p}$'s, will be extracted towards the TEVATRON, leaving the
source "primed" to begin the 2 hour task of restoring the stack.

In the October run, the sequence of deviation from design performance lead to a
stacking rate of the order of 10^9 $\bar{p}$'s/hour. It should be remembered that source
commissioning had occupied about eight weeks after the mechanical installation was
complete.

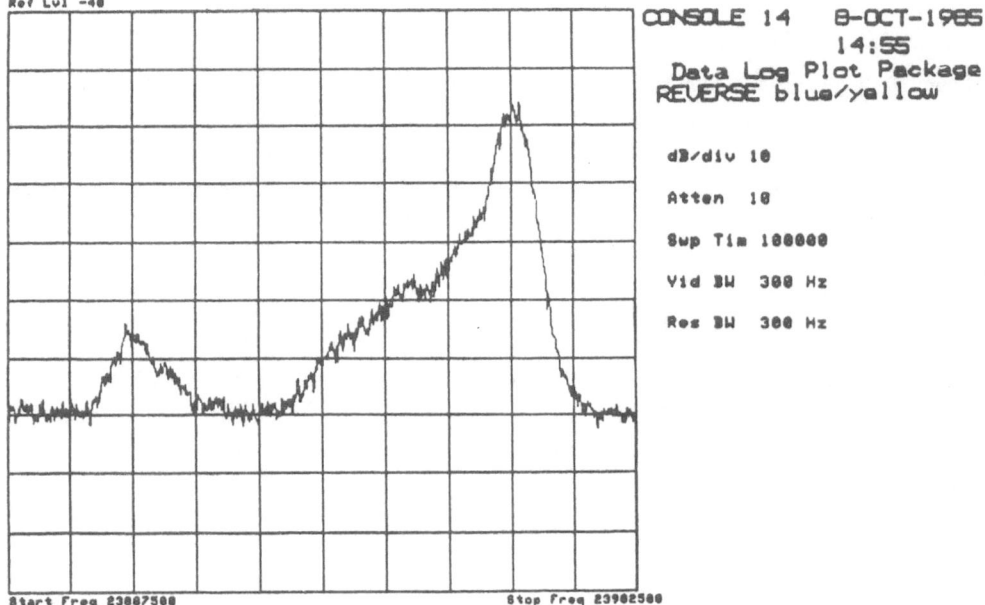

Figure 3 – A frequency pattern which may be read as the intensity vs
radius. The vertical scale is loganthmic. One sees the
fresh batch injected at the left, the stack tail ascending
to the core stack at the right

178

4.1.5 Main Ring Activities

The MR must accept p̄'s rotating in the "opposite" direction, accelerate them to 150 GeV and inject them into the SAVER ring, again in the opposite direction. Here they coast while protons are injected from the Booster to MR to SAVER in the conventional manner. The MR rf is used to coalesce the train of p̄ pulses into one rf bucket. The process is repeated twice so that three equally spaced bunches are accelerated to 150 GeV. The MR must also bypass the major experiments at B-ZERO (CDF) and D-ZERO so that it can resume its function of p̄ production during collisions. The bypasses are a large excursion above the CDF magnet and out-of-plane by 7m at B-ZERO and by 2m at D-ZERO, where a non magnetic detector will sit. The D-ZERO bypass was installed and tested in 1984 and is the first example of the operation of a non-planar synchrotron.

In the October 1985 test, the low p̄ intensity and insufficient detection sensitivity resulted in a loss in these operations of about an order of magnitude. An additional loss of about a factor of five was incurred during p̄ bunch coalescence.

4.1.6 SAVER Activities

The SAVER accepts counter rotating streams of protons and antiprotons and accelerates these to maximum energy, ultimately near 100 GeV. A process of adjusting the interaction points (cogging) takes place so that beam crossings are centered in the detectors. Acceleration to maximum energy then takes place. When coasting full energy beam is achieved, a system of powerful quadrupole magnets is energized to focus the beams down to the maximum density at the interaction point. All of this worked in October although further losses of protons were suffered during the few days of trying all of this out. The accelerator activities were lead by Helen Edwards.

4.2 The Colliding Detector

A map of the accelerator (see Figure 4) can be used to locate the experiments that are briefly described here.

4.2.1 The Colliding Detector at Fermilab (CDF)

The Collider Detector at Fermilab (CDF) is a general purpose detector system designed to explore the physics of 2 TeV proton-antiproton collisions made possible by the Tevatron I Project. It consists of a central magnetic detector that covers the angular range 10^0 to 170^0 with respect to the incident proton direction and two forward/backward detectors that cover the ranges 2^0 to 10^0 and 170^0 to 178^0, respectively. The basic goals of the detector include: 1) the measurement of electromagnetic and hadronic energy flow in fine bins of rapidity and azimuthal angle over the entire angular range of CDF with uniform granularity using systems of shower counters and hadron calorimeters, 2) measurements of the direction of

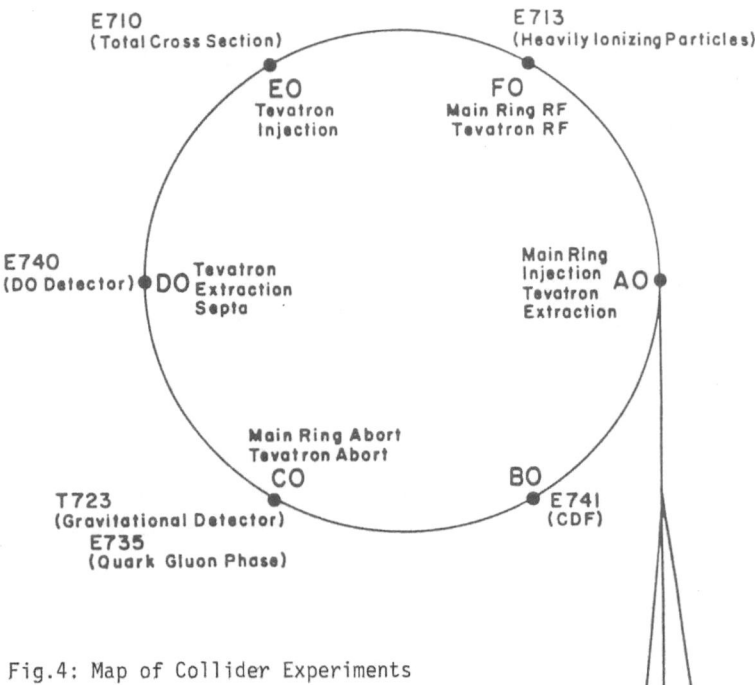

Fig.4: Map of Collider Experiments

charged particles to angles as close to the incident beam directions as technically possible, 3) momentum analysis of charged particles over the angular range 15° to 165°, and 4) identification and momentum analysis of muons over the angular ranges 2° to 20°, 40° to 140°, and 160° to 178°.

The major detector components are:
1. Central detector solenoid magnet with superconducting coil.
2. Charged particle tracking system organized into a central tracking chamber for momentum analysis, a set of vertex time projection chambers to find event topologies, and forward tracking chambers for small angle measurements.
3. Electromagnetic shower counters covering the full angular acceptance of CDF for identifying photons and electrons. There are three subsystems of shower counters, Central, End Plug, and Forward.
4. Hadron calorimeters backing up the shower counters. In addition to the three regions covered by the shower counters, the end wall of the solenoid magnet is instrumented with hadron calorimeters.
5. Muon detectors. The central muon system is behind the central hadron calorimeters; the forward system includes magnetized iron toroids for momentum measurements.
6. Front-end and Data Acquisition Electronics systems and Online Computers for recording data from and monitoring all of the detector systems.

7. Beamline equipment including luminosity monitors and precision vertex detectors.

The detector will be assembled in such a way that it can be staged to match the performance of Tevatron I. In the early stage when the luminosity is relatively low, a subject of the apparatus will be available to study the interesting physics questions that can be addressed at those luminosities. Later, the first stage of the detector will be augmented to make it suitable for the very important high luminosity studies.

4.2.2 The D-ZERO Detector

In contrast to CDF, the D-ZERO is non-magnetic detector except for toroids that are designed to measure muons.

The experiment will study the properties of 2 TeV $\bar{p}p$ collisions with particular emphasis on measurement and identification of leptons (electrons and muons), high transverse momentum jets, and missing energy. Goals of the experiment include high statistics studies of the W and Z bosons enabling precision measurements of their mass, widths and production properties; study of the high p_T multijet and single photon production for testing QCD; and searches for new phenomena beyond the standard model such as new quark generations, heavy leptons, supersymmetric particles, technicolor particles, quark-gluon plasma or quark compositeness.

The proposed detector incorporates three main systems: a central detector, uranium-liquid argon calorimetry over nearly 4π solid angle, and a magnetized iron muon spectrometer. The central detector comprises 24 layers of drift chamber planes, a multicell transition radiation detector for electron identification, and a microvertex detector. There is no central magnetic field. The calorimetry is divided into five angular regions and has a projective tower geometry with 50.000 readout channels. Multiple depth segmentation of the combined EM and hadronic calorimeter is made for enhanced identification of electrons. Energy resolution for hadrons is expected to be 40%/$\sqrt{E}$ with excellent calibration control. The muon system will measure muon momenta to within about 20% up to several hundred GeV/c for angles above 8° with respect to the beams. Three iron toroids provide the field with position and angle measurements given by three concentric sets of proportional drift tubes.

4.2.3 The Elastic Scattering (E-ZERO)

It is proposed to measure $\bar{p}p$ total cross sections, the real parts of forward scattering amplitudes, and forward diffraction elastic scattering over the energy range $\sqrt{s}$ = 300 to 2000 GeV. Data will be normalized to the known coulomb scattering cross section. An alternative method of normalization uses a 4π detector to record all interactions, both elastic and inelastic, in combination with the small angle detectors; such a 4π detector will be built using scintillator counters.

The equipment covers the $|t|$ range from the coulomb region to 0.11 $(GeV/c)^2$ at $\sqrt{s}$ = 300 and 4.8 $(GeV/c)^2$ at $\sqrt{s}$ = 2000. The experiment, which is to be located at the EO intersection region, can be carried out with luminosities of $\sim 10^{26} cm^{-2} sec^{-1}$ or greater.

In order to measure into the coulomb region, it is planned to place detectors in the accelerator lattice, located at the D47 and E14 quadrupoles. This will require $\sim$ 30 cm of room-temperature beam pipe at these locations.

4.2.4 The C-ZERO Quark-Gluon Plasma Detector

A magnetic spectrometer and a central tracking chamber are located at the CO intersection area. This will utilize low luminosity, from 10^{27} to $10^{29} cm^{-2} sec^{-1}$, to measure the transverse momentum distributions, dN/dp_t up to p_t = 1.4 GeV/c for centrally produced $p^{\pm}$, $K^{\pm}$, $\pi^{\pm}$ and γ as a function of the charged particle multiplicity n_c. The temperature T and the size of centrally produced plasma can be extracted from the transverse momentum (p_t) distribution dN/dp_t and from the like pion correlation, respectively.

Evidence for a quark gluon plasma manifests itself by

a. Breaks in $<p_t>$ as a function of dn_c/d_y in the central region $|y|$ 1.
b. Changes in the particle composition as a function of the charged particle multiplicity n_c.
c. Onset of low energy ($E_\gamma \approx$ 200 NeV) direct photon production with increasing charged particle multiplicity n_c.

The central tracking chamber will be a cylindrical drift chamber with endcaps covering almost 4π in solid angle and capable of detecting over 200 tracks. The magnetic spectrometer placed transverse to the cylindrical chamber will cover +0.4 to -1.0 in Y and 21^0 azimuth. Time-of-flight counters at the end of the spectrometer will separate π, K and p from 0.3 to 1.4 GeV/c.

4.2.5 The Search for Highly Ionizing Particles

Thin arrays of plastic track detectors will be used, covering a large solid angle, to search for new particles created in $\bar{p}p$ collisions with ionization greater than that of a relativistic particle with charge 20e. The large center-of-mass energy available for particle production and the special features of plastic track detectors will permit a search for particles with masses much greater than can be produced at other accelerators. The arrays will contain two types of detectors - Lexan and CR-39. The CR-39 has a higher charge resolution than that of any other detector or comparable thickness and a sensitivity adequate to detect magnetic monopoles with β as low as $\sim 10^{-2}$ and charged particles with z/β as low as $\sim$ 10.

4.3 The Physics at Tevatron I

In the face of expected results from CERN by the time of TeV I turn on, 1986-87, it is useful to re-examine the program of physics that can be expected at TeV I. Here we note the more than three-fold increase in energy and luminosity of the TeV I design. The general situation is that the standard model has been enormously successful thus far; no perceivable departures from its predictions are sensed in the present data. On the other hand, much of the theoretical literature of the past several years has focussed on the necessity of augmenting the standard model in some way. Although the popular notions on how to complete the picture may be wrong, it is useful to note that almost all such models postulate observable new phenomena emerging in the mass region $100 \leq m \leq 500$ GeV/c^2 - or in deviations from orthodoxy in W and Z parameters at the level of radiative corrections. Thus the role of experiments at TeV I will be, in our view, to search for evidence of these new ingredients. The combination of good luminosity and high energy should combine to give marked improvement in our ability to find such new phenomena. The fact that the mass scale accessible will be well into the region where new states are predicted to exist will make TeV I a premier search instrument.

In the brief discussion of specific topics below, we shall assume a standard run being four months of running with 50% efficiency, at $L = 10^{30}$ cm^{-2}sec^{-1} ($\int$Ldt = 5 x 10^{36}cm^{-2} = 5 pb^{-1}). Given the standard cross sections, one can then expect about 1500 $Z^0 \rightarrow \mu^+\mu^-$, depending on the scaling violations assumed, and about 15.000 $W^\pm \rightarrow e^\pm\nu$ (and also $W^\pm \rightarrow \mu^\pm\nu$). Thus even for standard studies of the W and Z, the improvement in statistical error could be nearly a factor of three compared with the integrated running of the CERN collider up to 1986. These rates will permit:

1. Precision measurements of the W and Z mass, leading to precise determinations of $\sin\theta_W$ and the ρ parameter. The latter parameter's deviation from the standard-model prediction ($\rho = 1.0$) is sensitive to possible new ingredients in the standard model, e.g., heavy quarks, Higgs, Technicolor Bosons, etc..
2. W & Z width measurements to look for additional questions and to test QCD radiative corrections.
3. Search for narrow states, e.g., $Z^0 \rightarrow e^+e^-\gamma$ as seen in UA1.
4. Precision study of the decay asymmetry in $W^\pm$ production and decay.
5. Search for three boson vertices and gauge-boson coupling via, e.g., $W^\pm\gamma X$ final states. This is sensitive to the W magnetic moment and W, Z compositeness.
6. Detailed study of W and Z production dynamics.

Beyond W, Z physics there is the enhancement in high-p_t QCD studies made possible by the increased energy and luminosity. TeV 1 will extend jets from the CERN project limit of 200 GeV/c to over 500 GeV/c. This probes a new small-distance scale and is a primary handle on the compositeness of quarks.

This is the real gold field for TeV 1. The candidate objects here are far too numerous to list but the richness is suggested by words: technicolored objects, supersymmetry, hypercolor, monopoles, heavy vector bosons, W', Z', heavy leptons, that is, for the generation or new varieties of leptons e.g., neutral heavies, quark-gluon plasma, and more things undreamed of by our theorists.

4.4 Conclusions

We have described a new research facility which should help to advance the subject of high energy physics and carry it to the mid-1990's where, if we are lucky, we can prepare for the final assault on the route to a Grand synthesis. If the apparent summit hides an even more difficult climb, we can only hope that a new generation of "Steinberger-equivalents" emerges although it is not at all clear that such entities exist.

REFERENCES

1 J.Steinberger, W.K.H.Panofsky and J.Steller, Phys. Rev. Lett. 78, 802 (1950)
2 S.Y.Hsueh et al., Phys. Rev. Lett. 54, 2399 (1985)
3 J.D.Bjorken, Proceedings of the Santa Fe Meeting, World Scientific (1984), p. 144
4 A.Ruggiero, $\bar{p}$-NOTES 151 Fermilab (1981)

Quantum Numbers of the Gluon

T.D. Lee

Columbia University, New York, NY 10027, USA

I.

Whenever a new particle is found in physics, a basic task is to determine experimentally its mass, lifetime, decay modes and various quantum numbers, independent of the prevailing theoretical prejudices. Efforts in this direction form the backbone of particle physics. One of the major figures in these endeavours is Jack Steinberger.

I knew Jack first as a fellow graduate student at the University of Chicago in 1946. Beginning with his dissertation, he made the experimental discovery of the continuous e-spectrum in μ-decay, which in turn laid the foundation of the universality of the weak interaction. After Chicago, Jack went to the Institute for Advanced Study in Princeton. There, he made the theoretical calculation of the pion-decay rate, a work that has had major impact on field theory and is extensively referred to even today.

In the early 1950's, through a series of remarkable papers, Steinberger and his associates determined for the first time the spin, parity and other properties of the pion (including the discovery of the Dalitz pair). From the mid-1950's, he pioneered the use of the bubble chamber in particle physics. With these innovative techniques he was able to observe for the first time the existence of Σ^0, to demonstrate parity nonconservation and other properties of hyperon decays. These discoveries established the bubble-chamber as one of the most powerful tools for exploring physics in the sub-atomic region.

In the early 1960's, Jack together with Lederman and Schwartz introduced the spark chamber into high-energy physics; with that they made the discovery of the two neutrinos and launched the field of high-energy neutrino physics. In addition, Jack and his collaboration made the first lifetime measurement of the omega and phi mesons.

The experiments by Jack and his associates on the neutral K-$\bar{K}$ system were executed with a precision and elegance unmatched in high-energy physics. The parameters they determined for the K-$\bar{K}$ complex are still the most accurate. At present, our entire knowledge of CP noninvariance and the violation of time-reversal symmetry exists only in that complex; this uniqueness underscores the importance of neutral kaon physics. Considering that the field has been so intensely studied, it is quite remarkable that more than a decade ago they could have made the analysis that remains definitive.

Since the 1970's, Jack has concentrated on the use of the large multi-wire proportional chamber in high-energy physics. With that he has been able to clear up important questions of quark structure functions, high-y anomaly, di- and tri-muon events, QCD tests, and the determination of the Weinberg angle. It is to be expected that Jack will extend his activities to the multi-TeV region in the years to come.

The progress of physics depends on the close interaction between theory and experiment. Yet it is essential to the good health of the field that these two disciplines should maintain their independence and integrity. At present there are very few experimental high-energy physicists who are not strongly influenced by the current fashion in theory.

Throughout the years Jack Steinberger has always maintained a high intellectual objectivity. Combining that with his superb sense of good physics, he has produced a long series of experiments; each addressed a major problem in physics and produced the decisive measurement that resolved the question. It is in appreciation of that rare and valuable combination of qualities that this paper is written.

II.

From a theoretical point of view, since the gluon is the gauge particle in QCD, its quantum number has been completely assigned; e.g., spin 1, charge 0, SU(3) colour octet, flavour neutral, etc. However, from an experimental point of view, except for its spin, there has not been any direct evidence that these theoretical attributions are valid. The problem is made difficult because the gluon, like the quark, is confined. The quantum numbers of quarks have been measured in various high energy experiments, especially the high-energy neutrino reactions [1]. In this note we want to show that one way to determine the quantum numbers of the gluon is through a study of correlation functions in multi-jet events.

1.   Production of Gluons

The first observational evidence of gluons came from the three-jet events [2,3] in e^+e^- collisions

$$e^+ + e^- \rightarrow \text{jet } (\vec{p}_1) + \text{jet } (\vec{p}_2) + \text{jet } (\vec{p}_3) \tag{1}$$

as shown in Fig.1. Each jet consists of particles whose momenta lie within a narrow cone of half-angle

$$\delta \ll 1 \tag{2}$$

with $\vec{p}_i$ the sum of their momenta (i = 1,2,3). Theoretically, one associates (1) with the reaction

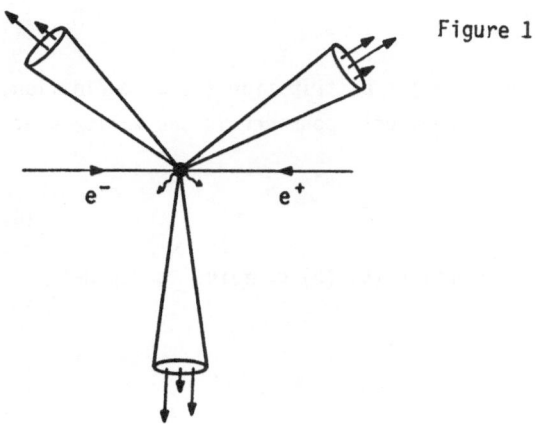

Figure 1

$$e^+ + e^- \to q(\vec{p}) + \bar{q}(\vec{p}') + g(\vec{k}) \tag{3}$$

where $\vec{p}$ is the momentum of the quark q, $\vec{p}'$ that of the anti-quark, and $\vec{k}$ that of the gluon g. To the lowest order of QCD, the distribution of (3) is proportional to [4]

$$\frac{x^2 + y^2}{(1-x)(1-y)} \, dx \, dy \qquad \text{where} \tag{4}$$

$$x = \frac{|\vec{p}|}{E}, \quad y = \frac{|\vec{p}'|}{E} \qquad \text{and}$$

$$2E = |\vec{p}| + |\vec{p}'| + |\vec{k}|$$

is the total energy of reaction (3).

From a single experimental event one cannot tell which jet is generated from a quark and which is from a gluon. Consequently, the momentum $\vec{p}_i$ of the ith jet can be either $\vec{p}$, or $\vec{p}'$, or $\vec{k}$, and the 3-jet distribution can be derived from (4) by symmetrization:

$$P(x_1, x_2, x_3) \, dx_1 \, dx_2 \tag{5}$$

where

$$P(x_1, x_2, x_3) = \frac{x_1^2 + x_2^2}{(1-x_1)(1-x_2)} + \frac{x_2^2 + x_3^2}{(1-x_2)(1-x_3)} + \frac{x_3^2 + x_1^2}{(1-x_3)(1-x_1)}, \tag{6}$$

$$x_1 = \frac{|\vec{p}_1|}{E}, \quad x_2 = \frac{|\vec{p}_2|}{E}, \quad x_3 = \frac{|\vec{p}_3|}{E},$$

and, as before,

$$2E = |\vec{p}_1| + |\vec{p}_2| + |\vec{p}_3|$$

is the total energy. Thus

$$x_1 + x_2 + x_3 = 2, \tag{7}$$

which gives rise to a Dalitz plot. Since the 3-jet distribution is, by definition, symmetric with respect to p_1, p_2 and p_3, we need only consider in the Dalitz plot the domain, say,

$$x_1 \leq x_2 \leq x_3. \tag{8}$$

The experimental verification of the distribution (5)-(6) confirms the spin-1 character of the gluon.

2.   Charge of the Gluon

Consider reaction (1) and differentiate the three jets by arranging their energies in the order (8).

Let $Q(i)$ be the total electric charge, in units of e, of the particles in the ith jet. Due to symmetry under particle-antiparticle conjugation, the average of $Q(i)$ over a large ensemble of jet-events is clearly zero; i.e.

$$<Q(i)> = 0.$$

Next, we consider the two-body correlation product $Q(i)Q(j)$ between two different jets. Because of labeling (8), there are three different such products $Q(1)Q(2)$, $Q(2)Q(3)$ and $Q(3)Q(1)$ in each three-jet reaction. Take, for example, $Q(1)Q(2)$ and average it over an ensemble of 3-jet events. Because the gluon has zero charge, whenever one of the jets (1 or 2) is due to a gluon, the correponding correlation product $Q(1)Q(2)$ would be zero. Therefore it is not difficult to see that the ensemble average is

$$<Q(1)Q(2)> = \lambda \, \frac{x_1^2 + x_2^2}{(1-x_1)(1-x_2) \, P(x_1,x_2,x_3)} \tag{9}$$

where $P(x_1,x_2,x_3)$ is given by (6),

$$\lambda = - \frac{\sum\limits_f (e_q/e)^4}{\sum\limits_f (e_q/e)^2} \tag{10}$$

and e_q is the quark charge. For the total energy 2E below the charm threshold $2m_c$, but above the strange particle production energy,

$$\lambda = - \frac{1}{9} \, \frac{16+1+1}{4+1+1} = - \frac{1}{3} \, ;$$

for $m_c < E < m_b$, the b-quark mass, we have

$$\lambda = \frac{1}{9} \, \frac{16+1+1+16}{4+1+1+4} = - \frac{17}{45}$$

and for $m_b < E < m_t$, the t-quark mass,

$$\lambda = - \frac{1}{9} \ \frac{16+1+1+16+1}{4+1+1+ \ 4+1} = - \frac{35}{99} \ .$$

Through permutations of (1,2,3), the averages of Q(2)Q(3) and Q(3)Q(1) can be similarly derived. Together, they satisfy

$$<Q(1)Q(2) + Q(2)Q(3) + Q(3)Q(1)> = \lambda. \tag{11}$$

Confirmation of (9)-(11) would give strong experimental proof that the gluon has zero electric charge. Similar considerations can be extended to other quantum numbers, such as hypercharge, isospin, ..., of the gluon.

In deriving the above formula, we have neglected higher-order QCD radiative corrections.

REFERENCES

1   H.deGroot et al., Z. Phys. C1, 143 (1979); Phys. Lett. 82B, 292 and 456 (1979)

2   See the proceedings of the International Symposium on Lepton and Photon Inter-
      actions at High Energies, Kyoto, 1985, for detailed references.

3   Other evidence of gluons has also been obtained in high energy neutrino reac-
      tions. See H.Abramowicz et al., Z. Phys. C12, 289 (1982)

4   J.Ellis, M.K.Gaillard and G.Ross, Nucl. Phys. B111, 253 (1976)
      T.A.DeGrand, Y.J.Ng and S.-H.H.Tye, Phys. Rev. D16, 3251 (1977)

The Physics in the Z^0 Energy Regime

U. Nauenberg

University of Colorado, Boulder, CO 80309, USA

I. Introduction

This work covers a range of material aimed at understanding the theoretical ideas and observations that point towards the necessary experiments to test the electroweak theory. These notes are the result of two efforts; (1) my development of a course in elementary particles for fourth or fifth semester graduate students, and (2) my own involvement in experiments, with the SLD collaboration, to test the electroweak theory in the Z° energy domain. By the nature of this contribution I can not cover in detail all of the topics relevant to this study, but I will try to touch on most of them.

I hope that this work expresses my appreciation to Jack, my advisor, for the special learning experience I had under him. His insight and taste in physics left a clear mark in my growth as a high energy physicist. I must add that this growth was aided by the whole Nevis Laboratory (Columbia University) where I worked intensively for at least five years. Very seldom does a graduate student have the opportunity to work in as excellent an intellectual atmosphere as was present at Nevis in the late 50's and early 60's. Among the other colleagues that I recall contributed to my studies at Nevis either by interaction, by deeds, or advise I must mention: Marcel Bardon, Juliet-Lee Franzini, Leon Lederman, Melvin Schwartz, Alan Sachs, Sam Devons, Paolo Franzini, Nicholas Samios, Richard Plano, Carlo Rubbia, Emilio Savattini, Lee Pondrom, Derek Colley, and Jonas Schultz.

II. The Weinberg Salam Electroweak Theory

The Weinberg Salam theory[1] forms the basis of the unification of the weak and electromagnetic interactions. It is this theory, which derives the Z° decay Lagrangian, that will be tested in minute detail when the new colliders LEP at CERN and the SLC at SLAC become operational in the next few years.

The theory is based on the existence of a set of vector U(1) gauge bosons and SU(2) gauge bosons interacting with fermions and scalar bosons as follows.[2]

We assume the fermions appear as left handed doublets and right handed singlets leptons and quarks. This is, in my opionion, rather arbitrary but necessary because the weak interactions that we have observed experimentally are parity violating in a particular way, and there is no natural explanation for this fact. We will assume this state of affairs but later on I will just mention some attempts to correct this.

The leptons and fermions are represented by

$$\begin{pmatrix} \nu_L^e \\ e_L^- \end{pmatrix} , \quad \begin{pmatrix} \nu_L^\mu \\ \mu_L^- \end{pmatrix} , \quad \begin{pmatrix} \nu_L^\tau \\ \tau_L^- \end{pmatrix} , \qquad e_R^-, \; \mu_R^-, \; \tau_R^-$$

$$\begin{pmatrix} u_L \\ d_L' \end{pmatrix} , \quad \begin{pmatrix} c_L \\ s_L' \end{pmatrix} , \quad \begin{pmatrix} t_L \\ b_L' \end{pmatrix} , \qquad u_R, \; d_R', \; c_R, \; s_R', \; t_R, \; b_R'$$

where, for example $\; e_L^- = \dfrac{(1 - \gamma_5)}{2} e^-$, $\quad e_R^- = \dfrac{(1 + \gamma_5)}{2} e^-$.

Let ψ_L refer to the left handed doublets and ψ_R refer to the right handed singlets.

The d_L', s_L', b_L' are mixed states of d_L, s_L, b_L as described by the Kobayashi Maskawa[3] matrix and originally described by Cabibbo.[4] We will neglect this mixing and assume $d_L' = d_L$, $s_L' = s_L$, $b_L' = b_L$. We can do this since, for the study of Z° decays and interactions (our main aim), it is not necessary to include this refinement. It is necessary to include it when you study the charged vector bosons, $W^\pm$, decay and interactions. How these vector bosons appear we will see in the next few pages.

192

We define a doublet scalar field consisting of one charged and one neutral state

$$\phi = \begin{pmatrix} \phi^+ \\ \phi^0 \end{pmatrix}$$

In addition, we define a U(1) gauge field B_μ and an SU(2) gauge field W_μ. The total Lagrangian consists of five parts

$$\mathscr{L}_T = \mathscr{L}_{scalar} + \mathscr{L}_{fermion} + \mathscr{L}_{gauge\ boson} + \mathscr{L}_{lepton\ scalar\ int} + \mathscr{L}_{quark\ scalar\ int}$$

$$\mathscr{L}_{scalar} \equiv \{(\partial_\mu + i\frac{g'}{2}Y_\phi B_\mu + i\frac{g}{2}\tau \cdot W_\mu)\phi\}^\dagger \{(\partial_\mu + i\frac{g'}{2}Y_\phi B_\mu + i\frac{g}{2}\tau \cdot W_\mu)\phi\}$$

$$- \mu^2 \phi^\dagger \phi - \lambda(\phi^\dagger \phi)^2 \tag{1}$$

$$\mathscr{L}_{fermions} \equiv \bar{\psi}_L \gamma^\mu (\partial_\mu + i\frac{g'}{2} Y_L B_\mu + i\frac{g}{2}\tau \cdot W_\mu)\psi_L + \bar{\psi}_R \gamma^\mu (\partial_\mu + i\frac{g'}{2} Y_R B_\mu)\psi_R \tag{2}$$

$$\mathscr{L}_{gauge\ bosons} \equiv -\frac{1}{4}(F_{\mu\nu,i})(F^{\mu\nu}{}_{,i}) - \frac{1}{4}B_{\mu\nu}B^{\mu\nu} \tag{3}$$

$$\mathscr{L}_{lepton\ scalar\ int.} \equiv -G_\ell \{\bar{\psi}_R^\ell \phi^\dagger \psi_L^\ell + \bar{\psi}_L^\ell \phi \psi_R^\ell\} \tag{4}$$

$$\mathscr{L}_{quark\ scalar\ int.} \equiv -\{G_{q,1}\bar{\psi}_L^q (i\tau_2\phi)^\dagger \psi_{R,1}^q + G_{q,2}\bar{\psi}_L^q \phi \psi_{R,2}^q$$

$$+ G_{q,1}\bar{\psi}_{R,1}^q (i\tau_2\phi)\psi_L^q + G_{q,2}\bar{\psi}_{R,2}^q \phi \psi_L^q\} \tag{4'}$$

where Y is the hypercharge $\equiv 2(Q - T_3)$

and $\tau_i \equiv 2T_i$ (i = 1,2,3) where T_i is the isospin, τ_i are the Pauli matrice

$Y\psi_R = Y_R\psi_R = -2\psi_R$ for right handed leptons

$\qquad\qquad = \frac{4}{3}\psi_R$ or $-\frac{2}{3}\psi_R$ for the right handed quarks

$Y\psi_L = Y_L\psi_L = -1\psi_L$ for left handed leptons

$\qquad\qquad = \frac{1}{3}\psi_L$ for the left handed quarks

$Y\phi = Y_\phi \phi = \phi$ for the scalar doublet

$T_3\psi_R = 0$

$T_3\psi_{L,1} = \frac{1}{2}\psi_{L,1}$ $\qquad \psi_{L,1}$ = upper component of ψ_L

$T_3\psi_{L,2} = -\frac{1}{2}\psi_{L,2}$ $\qquad \psi_{L,2}$ = lower component of ψ_L

$\mu^2 \phi^\dagger \phi + (\phi^\dagger \phi)^2$ is the potential associated with the scalar field when $\mu^2 < 0$ which is the case we are considering.

The gauge field terms in the Lagrangian are defined for vector SU(2) and U(1) fields namely

$$F_{\mu\nu,i} = \partial_\mu W_{\nu,i} - \partial_\nu W_{\mu,i} - g\epsilon_{ijk} W_{\mu,j} W_{\nu,k}$$

$$B_{\mu\nu} = \partial_\mu B_\nu - \partial_\nu B_\mu \; .$$

All the Lagrangian terms including the interactions in $\mathcal{L}_{scalar}$ and $\mathcal{L}_{fermion}$ are written so that they are invariant under a gauge transformation.

At this point I would like to mention slightly the arbitrariness of assuming only the existence of left handed doublets for the fermions. This implies that the SU(2) vector gauge bosons only allow interactions between left handed fermions. There may well be SU(2)[5] vector gauge bosons which allow interactions between right handed fermions and, because they may be much more massive than the SU(2) gauge bosons in the present theory, these interactions have not been observed experimentally. The mass difference between these different type of gauge bosons may be due to a phase transition (spontaneous symmetry breaking) during the early phases of the evolution of the universe. This, I feel, is a more reasonable explanation of our present world. Lower limits on the masses of these "right" gauge bosons have been placed in the TeV regime.[6] The search for these possible particles is one of the reasons why we need higher energy colliders.

We continue now with the Weinberg-Salam model. The particles as presented in this structure are all massless because there are no terms in the Lagrangian in the form

$$- m_f \bar{\psi}_L \psi_L \quad \text{or} \quad - m_f \bar{\psi}_R \psi_R \quad \text{or} \quad - \frac{1}{2} m_s^2 \phi^\dagger \phi \quad \text{or} \quad m_W^2 W^{\mu+} W_\mu^- \qquad \text{etc.}$$

One can generate these masses by carrying out a gauge transformation. How this takes place[2] is one of the aspects of the theory, as well as the construc-

194

tion of the Z^0 and W Lagrangians.

Consider the scalar fields which are complex

$$\phi = \begin{pmatrix} \phi^+ \\ \phi^0 \end{pmatrix} = \frac{1}{\sqrt{2}} \begin{pmatrix} \phi_1 + i\phi_2 \\ \phi_3 + i\phi_4 \end{pmatrix} \quad . \tag{5}$$

The scalar potential $\mu^2 \phi^\dagger \phi + \lambda \, (\phi^\dagger \phi)^2$ $(\mu^2 < 0)$ has extrema at

$$\phi^\dagger \phi = 0$$

$$\phi^\dagger \phi = \tfrac{1}{2}(\phi_1^2 + \phi_2^2 + \phi_3^2 + \phi_4^2) = -\mu^2/2\lambda \equiv v^2 \quad . \tag{6}$$

The minimum of the potential $= -\frac{1}{4} \mu^4/\lambda$ occurs for $\phi^\dagger \phi = v^2$.

We can arbitrarily choose this minimum to occur when $\phi_3 = v$ and all other components are 0. This arbitrariness is chosen so that only certain particles acquire mass and we get agreement with experimental observation.

We can rewrite ϕ in the form

$$\phi = e^{i\tau \cdot \xi(x)/2v} \frac{1}{\sqrt{2}} \begin{pmatrix} 0 \\ v + \eta(x) \end{pmatrix}$$

where $\xi(x)$ is arbitrary so that it represent ϕ_1, ϕ_2, ϕ_3, ϕ_4 and so that the vaccum state of ϕ is described by $\phi_3 = v$ all other ϕ's $= 0$. $\eta(x)$ replaces ϕ when we describe the field relative to this vaccum. $\eta(x)$ becomes the observable (Higgs) scalar field. Note that $\eta(x)$ is real.

We can apply a gauge transformation (SU(2) gauge) to this system of Lagrangians. Since these Lagrangians are invariant under gauge transformation, the new Lagrangian describes the same interactions as before but in terms of different parameters.

Let $U(\xi) = e^{-i\tau \cdot \xi(x)/2v}$

To preserve the invariance of the theory under these transformations we must have the various elements of the Lagrangian transform in the following manner

$$\phi' = U\phi = \frac{1}{\sqrt{2}} \begin{pmatrix} 0 \\ v + \eta(x) \end{pmatrix}$$

195

$$\psi_L' = U(\xi)\psi_L$$

$\psi_R' = \psi_R$ the singlet states do not change under an SU(2) transformation

$$\tau \cdot W_\mu' = U(\xi)\tau \cdot W_\mu U^{-1}(\xi) + \frac{i}{g}[\partial_\mu U(\xi)]U^{-1}(\xi) \tag{7}$$

$B_\mu' = B_\mu$ the U(1) gauge boson does not change under an SU(2)
transformation.

The effect of the gauge transformation is to make ϕ_1, ϕ_2, ϕ_4 disappear and, as we will see, selected mass terms appear in the Lagrangian.

To see this we look at the new Lagrangian terms. We consider first the term that describes the lepton and scalar interaction as given by

$$\mathcal{L}_{f.s.int.} = -G_f \{\bar{\psi}_R' \phi'^\dagger \psi_L' + \bar{\psi}_L' \phi' \psi_R' \}$$

$$= -\frac{G_f}{\sqrt{2}} \{\bar{\psi}_R(0, v + \eta(x))U(\xi)\psi_L + \bar{\psi}_L U^{-1}(\xi) \binom{0}{v + \eta(x)}\psi_R\} \quad .$$

It is sufficient, to understand the major terms, to assume ξ_1 small and expand $U(\xi)$ to first order.

$$\mathcal{L} = -\frac{G_f}{\sqrt{2}} \{[0, \bar{\psi}_R(v + \eta(x))][1 - i\frac{\xi \cdot \tau}{2v}]\binom{\psi_{L,1}}{\psi_{L,2}} + (\bar{\psi}_{L,1}, \bar{\psi}_{L,2})[1 + i\frac{\xi \cdot \tau}{2v}]\binom{0}{\psi_R(v+\eta(x))}\}$$

where $\psi_{L,1}$, $\psi_{L,2}$ refer to the top and bottom element of the doublet.

We look at the terms which are independent of ξ_1. These terms are

$$= -\frac{G_f}{\sqrt{2}} \{[v + \eta(x)]\bar{\psi}_R \psi_{L,2} + [v + \eta(x)]\bar{\psi}_{L,2}\psi_R\} \quad .$$

In the case of leptons this gives rise to the term

$$= -\frac{G_\ell}{\sqrt{2}} (v + \eta(x))\bar{\psi}_\ell \psi_\ell \quad . \tag{8}$$

In a fermion Lagrangian a term of the form

$$= -\frac{G_\ell}{\sqrt{2}} v\bar{\psi}_\ell \psi_\ell \equiv -m_\ell \bar{\psi}_\ell \psi_\ell \tag{9}$$

represents a lepton of mass $m_\ell = \frac{G_\ell v}{\sqrt{2}}$.

196

The other term

$$\frac{G_\ell}{\sqrt{2}} \, \eta(x)\bar{\psi}_\ell \psi_\ell \qquad \frac{G_\ell}{\sqrt{2}} = \frac{m_\ell}{v} \tag{10}$$

represents the interaction between the Higgs and the lepton. The coupling is proportional to the lepton mass. The same argument applies to the quarks when using (4'). This is characteristic of the interaction of the Higgs scalar; namely, the strength of the interaction between it and any other particle in the system is proportional to the mass of this other particle. We will determine later (see page 12) that $v \simeq 246$ GeV. Hence the Higgs coupling strength is quite small for $m < \sim 10$ GeV.

To determine the form of the interactions of the gauge bosons we consider next the new, gauge transformed, scalar Lagrangian which we associate with the Higgs particle

$$\mathscr{L}_H = \frac{1}{2}\,(0, v + \eta(x))\left(\partial^\mu - \frac{ig'}{2}\,B'^\mu - \frac{ig}{2}\tau\cdot W'^\mu\right)\left(\partial_\mu + \frac{ig'}{2}B'_\mu + \frac{ig}{2}\tau\cdot W'_\mu\right)\binom{0}{v + \eta(x)}$$

$$- \frac{\mu^2}{2}\,(v + \eta(x))^2 - \frac{\lambda}{4}\,(v + \eta(x))^4 \ .$$

Replacing B'_μ and W'_μ by expressions (7) and disregarding the terms that depend on ξ_i we get

$$\mathscr{L}_H = \frac{1}{2}\,\partial^\mu\eta(x)\partial_\mu\eta(x) + \left(\frac{v + \eta(x)}{8}\right)^2 (0,1)\{g'B^\mu + g\tau\cdot W^\mu\}\{g'B_\mu + g\tau\cdot W_\mu\}\binom{0}{1}$$

$$- \frac{\mu^2}{2}\,(v + \eta(x))^2 - \frac{\lambda}{4}\,(v + \eta(x))^4 \ . \tag{11}$$

If we look, as before, at the terms independent of $\eta(x)$ we get

$$= \frac{v^2}{8}\,\{g^2(W_1^\mu - iW_2^\mu)(W_{\mu,1} + iW_{\mu,2}) + (g'B^\mu - gW^\mu{}_{,3})(g'B_\mu - gW_{\mu,3})\} \ .$$

If we redefine these expressions for the gauge bosons to form two charged and two neutral ones

$$W_\mu^\pm = \frac{(W_{\mu,1} \mp iW_{\mu,2})}{\sqrt{2}} \qquad Z_\mu = -\frac{(g'B_\mu - gW_{\mu,3})}{\sqrt{g^2 + g'^2}} \qquad A_\mu = \frac{gB_\mu + g'W_{\mu,3}}{\sqrt{g^2 + g'^2}} \tag{12}$$

we get

$$= \frac{v^2}{8} \left\{ 2g^2 W^{\mu +} W^-_\mu + \overline{\sqrt{g^2 + g'^2}} \ Z^\mu Z_\mu \right\} .$$

The Lagrangian for massive vector bosons has a mass term $= m^2_W W^{\mu +} W^-_\mu$ while for a neutral one it has a mass term $\frac{1}{2} m^2_Z Z^\mu Z_\mu$.

Hence we end up with three massive vector bosons W^+, W^-, Z°, and one massless one, A_μ, since there is no $A^\mu A_\mu$ term present. The masses are

$$M_{W^\pm} = \frac{vg}{2} \qquad M_Z = \frac{v\sqrt{g^2 + g'^2}}{2} \qquad M_A = 0 . \qquad (13)$$

The vector boson represented by A_μ mediates the electromagnetic interactions since we believe the photon is massless. The other three, which have acquired a mass, mediate the weak interactions.

There are 2 other terms in (11) which describe the interaction of the gauge bosons with the Higgs scalar, namely

$$\left(\frac{v\eta(x)}{4} + \frac{\eta^2}{8} \right) \left\{ g^2 (W^\mu_1 - iW^\mu_2)(W_{\mu,1} + iW_{\mu,2}) + (g'B^\mu - gW^\mu_{,3})(g'B_\mu - gW_{\mu,3}) \right\} .$$

Using (12) and (13) we can rewrite this as

$$\frac{2M^2_W}{v} W^{\mu +} W^-_\mu \eta(x) + \frac{M^2_Z}{v} Z^\mu Z_\mu \eta(x) + \frac{M^2_W}{v^2} W^{\mu +} W^-_\mu \eta^2(x) + \frac{1}{2} \frac{M^2_Z}{v^2} Z^\mu Z_\mu \eta^2(x) . \qquad (14)$$

Relations (10) and (14) define the Feynman rules for the interactions of the Higgs particle with fermions and gauge bosons and are depicted in Fig. 1. We will come back to these when we discuss the experiments to look for the Higgs particle. Notice the first two terms in (14) are very similar to (10).

Finally, the last 2 terms in (11) can be reduced to

$$\mu^2 \eta^2 - (1/4)\eta^4 - \lambda v \eta^3 + (1/4)\mu^4 / \lambda .$$

This gives rise to a Higgs mass term $\mu^2 \eta^2$ and hence $m^2_\eta = -\mu^2 > 0$. Although we can determine a value for v we have no way of estimating μ^2 and hence the mass of the Higgs is not determined by the theory.

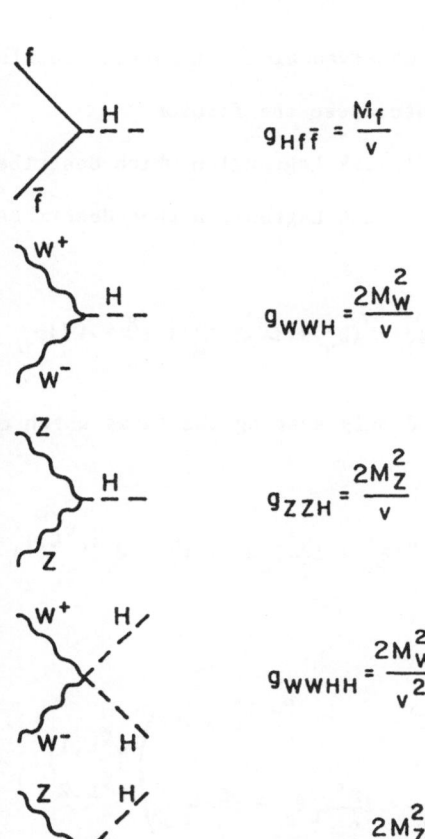

$$g_{Hf\bar{f}} = \frac{M_f}{v}$$

$$g_{WWH} = \frac{2M_W^2}{v}$$

$$g_{ZZH} = \frac{2M_Z^2}{v}$$

$$g_{WWHH} = \frac{2M_W^2}{v^2}$$

$$g_{ZZHH} = \frac{2M_Z^2}{v^2}$$

Fig.1: Couplings of the Higgs to fermions and gauge bosons

At this point, for simplicity, we define a parameter that describes the ratio of the U(1) coupling g' to the SU(2) coupling g. This parameter is known as the Weinberg angle.

$$\tan\theta_\omega \equiv \frac{g'}{g} \qquad (15)$$

Therefore, $\sin\theta_\omega = \dfrac{g'}{\sqrt{g^2 + g'^2}}$ $\qquad \cos\theta_\omega = \dfrac{g}{\sqrt{g^2 + g'^2}}$

$$\frac{M_W}{M_Z} = \cos\theta_\omega \qquad (16)$$

$$Z_\mu = -\sin\theta_\omega B_\mu + \cos\theta_\omega W_{\mu,3}$$

$$A_\mu = \cos\theta_\omega B_\mu + \sin\theta_\omega W_{\mu,3} \;\; . \qquad (17)$$

To connect the couplings g' and g to the observed electromagnetic coupling α, and the weak interaction coupling, G_F, we decompose the fermion Lagrangian. This will also give us the neutral weak Lagrangian which describes the Z° decays and interactions, and the charged weak Lagrangian that describes the $W^{\pm}$ decays and interactions.

$$\mathcal{L}_{\text{fermions}} \equiv \bar{\psi}'_R \gamma^\mu (\partial_\mu + i\frac{g'}{2} Y_R B'_\mu) \psi'_R + \bar{\psi}'_L \gamma^\mu (\partial_\mu + i\frac{g'}{2}Y_L B'_\mu + i\frac{g}{2} \tau \cdot W'_\mu)\psi'_L .$$

Using (7), expanding $U(\xi)$ for small ξ, and only keeping the terms which do not depend on ξ we get

$$= \bar{\psi}_R \gamma^\mu (\partial_\mu + i\frac{g'}{2} Y_R B_\mu)\psi_R + (\bar{\psi}_{L,1} , \bar{\psi}_{L,2})\gamma^\mu(\partial_\mu + i\frac{g'}{2}Y_L B_\mu + i\frac{g}{2} \tau \cdot W_\mu)\binom{\psi_{L,1}}{\psi_{L,2}}$$

$$= \bar{\psi}_R \gamma^\mu (\partial_\mu + i\frac{g'}{2} Y_R B_\mu)\psi_R$$

$$+ (\bar{\psi}_{L,1} , \bar{\psi}_{L,2})\gamma^\mu \begin{pmatrix} \partial_\mu + i\frac{g'}{2}Y_L B_\mu + i\frac{g}{2} W_{\mu,3} & i\frac{g}{\sqrt{2}} W_\mu^+ \\ i\frac{g}{\sqrt{2}} W_\mu^- & \partial_\mu + i\frac{g'}{2}Y_L B_\mu - i\frac{g}{2} W_{\mu,3} \end{pmatrix} \binom{\psi_{L,1}}{\psi_{L,2}}$$

$$\tag{18}$$

where we made use of some of the relations in (12).

To make connection between g and G_F we collect the off diagonal terms which give

$$\mathcal{L}_w = \frac{ig}{\sqrt{2}} \{ \bar{\psi}_{L,1} \gamma^\mu \psi_{L,2} W_\mu^+ + \bar{\psi}_{L,2} \gamma^\mu \psi_{L,1} W_\mu^- \}$$

using $\psi_{L,1} = \frac{1-\gamma_5}{2} \psi_{f,1}$ $\bar{\psi}_{L,1} = \bar{\psi}_{f,1} \frac{(1+\gamma_5)}{2}$ we get

$$\mathcal{L}_w = \frac{ig}{2\sqrt{2}} \{ \bar{\psi}_{f,1} \gamma^\mu (1-\gamma_5) \psi_{f,2} W_\mu^+ + \bar{\psi}_{f,2} \gamma^\mu (1-\gamma_5) \psi_{f,1} W_\mu^- \} .$$

Applying these Lagrangian terms to the $\mu^- \to e^- \nu_\mu \bar{\nu}_e$ decay (Fig. 2) we get the matrix element

$$= (\frac{g}{2\sqrt{2}})^2 \bar{\psi}_{\nu\mu} \gamma^\alpha (1-\gamma_5) \psi_\mu W_\alpha W_\beta \bar{\psi}_{\nu e} \gamma^\beta (1-\gamma_5) \psi_e$$

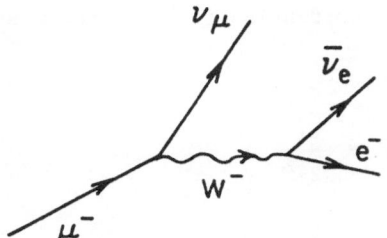

Fig.2: Feynman diagram for μ^- decay

where $W_\alpha W_\beta = -\dfrac{(g_{\alpha\beta} - q_\alpha q_\beta / M_w^2)}{q^2 - M_w^2} = \dfrac{g_{\alpha\beta}}{M_w^2}$ since in μ decay $q \ll M_w$.

The phenomenological matrix element that describes this decay in terms of the Fermi coupling is given by

$$\mathcal{L} = \frac{G_F}{\sqrt{2}} \bar{\psi}_{\nu\mu} \gamma^\alpha (1-\gamma_5) \psi_\mu \, g_{\alpha\beta} \, \bar{\psi}_{\nu e} \gamma^\beta (1-\gamma_6) \psi_e \quad .$$

Comparing these two expressions leads to the connection between the theoretical expression and the measurements

$$\frac{g^2}{8M_w^2} = \frac{G_F}{\sqrt{2}} \quad . \tag{19}$$

The magnitude of G_F is determined from the muon lifetime.[7]

$$\frac{1}{\tau_\mu} = \frac{1}{\hbar} \, G_F^2 \, m_\mu^5 / 192\pi^3 \, \{1 - 8\frac{m_e^2}{m_\mu^2}\}\{1 + \frac{3}{5}\frac{m_\mu^2}{M_w^2}\}$$

$$\times \; \{(1 + \frac{\alpha}{2\pi} (\frac{25}{4} - \pi^2) + \frac{\alpha^2}{3\pi^2} (\frac{25}{4} - \pi^2) \, \ell n \frac{m_\mu}{m_e}\} \tag{20}$$

$$= 1/2.19709 \; \text{sec}^{-1}$$

which gives

$$G_F = 1.16634 \times 10^{-5} \; \text{GeV}^{-2}$$

and $\quad g \approx .668 \quad$ for $\quad M_w \approx 82.3$ GeV .

In section X we discuss the need for expressing the lifetime with some of the radiative corrections included.

An interesting observation can be made using expression (13) for the mass of the W and (19) namely

$$\frac{G_F}{\sqrt{2}} = \frac{g^2}{8M_W^2} = \frac{g^2}{8} \frac{4}{v^2 g^2} = \frac{1}{2v^2} .$$

Therefore

$$v = \left(\frac{1}{\sqrt{2}G_F}\right)^{1/2} = 246 \text{ GeV} .$$

From (6) $v = (-\mu^2/\lambda)^{1/2}$ is a suggestive indicator of the dimensions of the Higgs potential.

In a similar manner we must relate the electromagnetic coupling constant, q $= \sqrt{4\pi\alpha}$, to the values of g and g'. For simplicity we carry out the derivation only for leptons, but the conclusions apply equally well to quarks.

From (17) we have

$$\begin{aligned}
B_\mu &= \cos\theta_\omega A_\mu - \sin\theta_\omega Z_\mu \\
W_{\mu,3} &= \sin\theta_\omega A_\mu + \cos\theta_\omega Z_\mu .
\end{aligned} \tag{21}$$

Also for leptons $Y_R = -2$ and $Y_L = -1$. Considering only the diagonal terms in (18), since the off diagonal terms we have already accounted for, we have

$$\bar{\psi}_R \gamma^\mu \partial_\mu \psi_R + (\bar{\psi}_{L,1}, \bar{\psi}_{L,2}) \begin{pmatrix} \gamma^\mu\partial_\mu & 0 \\ 0 & \gamma^\mu\partial_\mu \end{pmatrix} \begin{pmatrix} \psi_{L,1} \\ \psi_{L,2} \end{pmatrix}$$

$$- ig'\left\{\bar{\psi}_R \gamma^\mu (\cos\theta_\omega A_\mu - \sin\theta_\omega Z_\mu) \psi_R\right\}$$

$$- \frac{ig'}{2} (\bar{\psi}_{L,1}, \bar{\psi}_{L,2}) \gamma^\mu \begin{pmatrix} \cos\theta_\omega A_\mu - \sin\theta_\omega Z_\mu & 0 \\ 0 & \cos\theta_\omega A_\mu - \sin\theta_\omega Z_\mu \end{pmatrix} \begin{pmatrix} \psi_{L,1} \\ \psi_{L,2} \end{pmatrix}$$

$$+ \frac{ig}{2} (\bar{\psi}_{L,1}, \bar{\psi}_{L,2}) \gamma^\mu \begin{pmatrix} \sin\theta_\omega A_\mu + \cos\theta_\omega Z_\mu & 0 \\ 0 & -\sin\theta_\omega A_\mu + \cos\theta_\omega Z_\mu \end{pmatrix} \begin{pmatrix} \psi_{L,1} \\ \psi_{L,2} \end{pmatrix} . \tag{22}$$

For leptons $\psi_{L,1} = \psi_\nu$, $\psi_{L,2} = \psi_\ell$. Since A_μ has remained massless, let us consider it as representing the electromagnetic interactions. Hence we consider only the A_μ terms for the moment.

$$- \frac{ig'}{2} \cos \theta_\omega \{ \overline{\psi}_R \gamma^\mu A_\mu \psi_R + \overline{\psi}_{L,1} \gamma^\mu A_\mu \psi_{L,1} + \overline{\psi}_{L,2} \gamma^\mu A_\mu \psi_{L,2} \}$$

$$+ \frac{ig}{2} \sin \theta_\omega \{ \overline{\psi}_{L,1} \gamma^\mu A_\mu \psi_{L,1} - \overline{\psi}_{L,2} \gamma^\mu A_\mu \psi_{L,2} \} \ .$$

Remembering (15) that $g \sin \theta_\omega = g' \cos \theta_\omega$ and that ψ_R has no partner for $\psi_{L,1}$, we note that the terms with $\psi_{L,1}$ drop out. Hence A_μ does not couple to neutrinos as expected. I should add here that in the case of quarks, since ψ_R has a partner for $\psi_{L,1}$ and Y_R, Y_L are not so simple this does not happen, as hoped, if A_μ represents the electromagnetic field.

Remembering

$$\psi_R = \frac{(1 + \gamma_5)}{2} \psi_\ell \qquad \psi_{L,2} = \frac{(1 - \gamma_5)}{2} \psi_\ell$$

we get

$$= - ig' \cos \theta_\omega \ \overline{\psi}_\ell \gamma^\mu A_\mu \psi_\ell \ . \tag{23}$$

This we can connect with the electromagnetic Lagrangian

$$= - i \ q_\ell \ \overline{\psi}_\ell \gamma^\mu A_\mu \psi_\ell \ .$$

Making connection with experiments requires

$$g' \cos \theta_\omega = q_\ell = \sqrt{4\pi\alpha} \tag{23'}$$

where $\alpha = (137.036)^{-1}$ is the electromagnetic coupling const.

Using (15), (16), (19) and (20) we get

$$M_z = (\frac{\pi \alpha}{\sqrt{2} G_F})^{1/2} \frac{1}{\sin \theta_\omega \cos \theta_\omega} \tag{24}$$

$$M_z = \frac{37.281}{\sin \theta_\omega \cos \theta_\omega} \ \text{GeV} \ . \tag{24'}$$

If we apply radiative corrections to the value of α and G_F[11,14]

$$\alpha^{-1}(M_z) = 128 \quad \text{GeV}$$

$$G_F(M_z) = G_F(0) \qquad (24'')$$

$$M_z = \frac{38.6}{\sin \theta_\omega \cos \theta_\omega} \text{ GeV} .$$

For $\sin^2\theta_\omega = .220$ we get

$$M_z = 93.2 \text{ GeV}$$

$$M_w = 82.3 \text{ GeV}$$

$$g' = .354 .$$

We would arrive at the same conclusion even if we did not restrict the present analysis to leptons only, but included all the quarks.

Our last task is to generate the fermion Z° Lagrangian that describes the Z° decay into and interactions with fermions. Using (18) and (21) and only keeping the terms proportional to Z_μ we get

$$\mathcal{L}_{Z,f} = \frac{ig'}{2} \sin \theta_\omega \{- Y_R \, \overline{\psi}_R \, \gamma^\mu \, \psi_R - Y_L \, (\overline{\psi}_{L,1} \, \gamma^\mu \, \psi_{L,1} + \overline{\psi}_{L,2} \, \gamma^\mu \, \psi_{L,2})\} Z_\mu$$

$$+ \frac{ig}{2} \cos \theta_\omega \{2 T_3^{L,1} \, \overline{\psi}_{L,1} \, \gamma^\mu \, \psi_{L,1} + 2 \, T_3^{L,2} \, \overline{\psi}_{L,2} \, \gamma^\mu \, \psi_{L,2}\} Z_\mu .$$

Using $Y_R = 2Q_R$, $Y_L = 2(Q_L - T_3{}^L)$, and $Q_L = Q_R = Q$ we can write

$$\mathcal{L}_{Z,f} = ig' \sin \theta_\omega \{-2Q(\overline{\psi}_R \, \gamma^\mu \, \psi_R + \overline{\psi}_{L,1} \, \gamma^\mu \, \psi_{L,1} + \overline{\psi}_{L,2} \, \gamma^\mu \, \psi_{L,2})\} Z_\mu$$

$$+ i(g' \sin \theta_\omega + g \cos \theta_\omega)\{T_3^{L,1} \, \overline{\psi}_{L,1} \, \gamma^\mu \, \psi_{L,1} + T_3^{L,2} \, \overline{\psi}_{L,2} \, \gamma^\mu \, \psi_{L,2}\} Z_\mu .$$

Since $Q \psi_\nu = 0$ we can artificially add a right handed neutrino state.

We finally get

$$\mathcal{L}_{Z,f} = \frac{i g}{\cos \theta_\omega} \{T_3^f \, \overline{\psi}_f \, \gamma^\mu \, \frac{(1-\gamma_5)}{2} \, \psi_f - Q^f \sin^2 \theta_\omega \, \overline{\psi}_f \, \gamma^\mu \, \psi_f\} Z_\mu . \qquad (25)$$

Using (16) and (19) this is also written

204

$$\mathcal{L}_{Z,f} = i(\sqrt{2} \, G_F)^{1/2} M_z \{ T_3^{\,f} \, \overline{\psi}_f \, \gamma^\mu \, (1-\gamma_5) \, \psi_f - 2Q^f \sin^2 \theta_\omega \, \overline{\psi}_f \, \gamma^\mu \, \psi_f \} Z_\mu$$

$$= i(\sqrt{2} \, G_F)^{1/2} M_z \{ \overline{\psi}_f \, \gamma^\mu \, (\upsilon_f - a_f \, \gamma_5) \, \psi_f \} \, Z_\mu$$

$$\upsilon_f = T_3^{\,f} - 2Q^f \sin^2 \theta_\omega \tag{26}$$

$$a_f = T_3^{\,f} \, .$$

This is the Lagrangian which we will apply to the study of Z° decays in the next sections. This completes the extraction of the various relevant terms and will now discuss some of the experiments that will test the theory.

III. Z° Decays

Using (26) we can derive the Z° decay rates into various channels.[8]

Using the expression for the partial decay rate in GeV in terms of the matrix element (assuming an unpolarized Z°)[9]

$$\Gamma(Z° \to f\overline{f}) = \frac{|\mathcal{m}|^2}{16\pi M_z} (1 - 4\frac{m_f^2}{M_z^2})^{1/2}$$

where $\mathcal{m} = \langle f\overline{f} | H_{int.} | Z \rangle = - \langle f\overline{f} | L_{Z,f} | Z \rangle.$

$$\Gamma(Z° \to f\overline{f}) = \frac{cM_z^3}{24\pi} \frac{G_F}{\sqrt{2}} (1-4\frac{m_f^2}{M_z^2})^{1/2} \{ |\upsilon_f|^2 + |a_f|^2 + \frac{2m_f^2}{M_z^2} (2|a_f|^2 - |\upsilon_f|^2) \} \tag{27}$$

where the value of c (color) = 1,3 for leptons and quarks respectively.

Using the latest measured values[10] of the parameters

$$G_F = 1.166 \times 10^{-5} \text{ GeV}^{-2}$$

$$\sin^2 \theta_\omega = .220 \pm .008$$

$$M_z = 92.7 \pm 2.0 \text{ GeV}$$

Table I Z° Decay Rates

Channel	υ_f	a_f	Γ(GeV)	Branching Ratio (%)
$Z° \rightarrow \bar{\ell}_i \ell_i$ $i = e,\mu,\tau$	$-1/2[1-4\sin^2\theta_\omega]$	$-1/2$	.0884	3.26
$Z° \rightarrow \bar{\nu}_i \nu_i$ $i = e,\mu,\tau$	$1/2$	$1/2$	.174	6.42
$Z° \rightarrow \bar{q}_i q_i$ $i = u,c$	$1/2[1-(8/3)\sin^2\theta_\omega]$	$1/2$	.306	11.3
$Z° \rightarrow \bar{q}_i q_i$ $i = d,s$	$-1/2[1-(4/3)\sin^2\theta_\omega]$	$-1/2$	.392	14.5
$Z° \rightarrow b\bar{b}^*$	$-1/2[1-(4/3)\sin^2\theta_\omega]$	$-1/2$	.387	14.3
$Z° \rightarrow t\bar{t}^*$	$1/2[1-(8/3)\sin^2\theta_\omega]$	$1/2$	.065	2.39
$\sum_i Z° \rightarrow \bar{q}_i q_i g^*$			.074	2.7
Total			2.71	100.00

* We used M_b = 5GeV, M_t = 40GeV and[9] $Z° \rightarrow q_i q_i g$ = .04 $Z° \rightarrow \bar{q}_i q_i$

The results for the decay rate into each channel and the Z° mass width are
shown in Table 1.

I would like to postpone until a latter section any discussions on the
measurements that will test the electroweak theory.[12,13] In this manner we
will have most of the facts at hand which will allow us to discriminate which
are the best measurements for a given purpose.

IV. $e^+e^- \rightarrow \gamma$ or $Z° \rightarrow f\bar{f}$.

The e^+e^- annihilation into fermions in the energy regime near the Z°
measures the amplitude sum of the electromagnetic and neutral weak interaction
present in the process as shown in fig. 3. We will only consider the first four
graphs. The effect of fig. 3c is to renormalize the value of α^{14} in fig. 3a

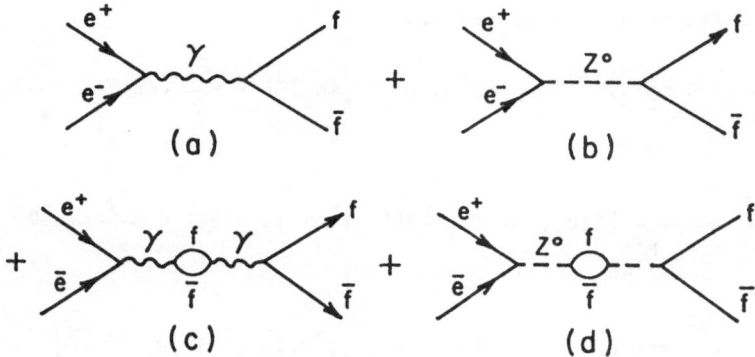

Fig.3: Feynman diagrams that contribute to the reaction $e^+e^- \to f\bar{f}$. Diagrams (c) and (d) can be incorporated into (a) and (b) by renormalizing the electromagnetic and weak coupling constants

$$\frac{1}{\alpha(M_z)} = \frac{1}{\alpha(0)} - \frac{2}{3\pi} \sum_f Q_f^2 \, \ell n(\frac{M_z}{m_f}) + 1/6\pi = \frac{1}{\alpha(0)} - \frac{171}{6\pi} \qquad (28)$$

$$\frac{1}{\alpha(M_z)} \approx 128 \, . \qquad (28')$$

The effects of fig. 3d and other ones are much smaller and do not contribute to this correction.

Given the expression for the differential cross section

$$\frac{d\sigma}{d\Omega_f} = \frac{1}{64\pi^2 s} |\mathcal{M}|^2 \frac{P_f}{P_e}$$

and the matrix element

$$\mathcal{M} = -\frac{4\pi\alpha(M_z)}{s} \, \bar{v}_e \gamma_\mu u_e \, g^{\mu\nu} \, \bar{u}_f \gamma_\nu v_f$$

$$- M_z^2 \sqrt{2}G_F \, \bar{v}_e \gamma_\mu (\upsilon_e - a_e \gamma_5) u_e \, \frac{(g^{\mu\nu} - q^\mu q^\nu / M_z^2)}{s - M_z^2 + i\Gamma_z M_z} \, \bar{u}_f \gamma_\nu (\upsilon_f - a_f \gamma_5) v_f$$

$s = q^2$ is the center of mass energy squared.

$P_f = \frac{\sqrt{s}}{2} (1 - 4 \frac{m_f^2}{s})^{1/2}$ is the fermion momentum we can write out the differential cross section

$$\frac{d\sigma}{d\Omega_f} = \frac{c}{64\pi^2 s}(1-4\,m_f^2/s)^{1/2} \times \left\{16\pi^2\,\alpha_e(M_z)\alpha_f(M_z)(1 + 4\frac{p_f^2}{s}\cos^2\theta + 4\frac{m_f^2}{s})\right.$$

$$+ 2M_z^4\,G_F^2\,\frac{s^2}{(s-M_z^2)^2 + \Gamma_z^2\,M_z^2}\left[(|v_f|^2 + |a_f|^2)(|v_e|^2 + |a_e|^2)(1 + 4\frac{p_f^2}{s}\cos^2\theta)\right.$$

$$+ 8\,v_f\,v_e\,a_f\,a_e\,(\frac{p_f^2}{s}\cos\theta) + 4(|v_f|^2 - |a_f|^2)(|v_e|^2 + |a_e|^2)\,\frac{m_f^2}{s}\Big]$$

$$+ \frac{16\pi(\alpha_e(M_z)\alpha_f(M_z))^{1/2}}{s}\,M_z^2\,\frac{G_F}{\sqrt{2}}\,\frac{s^2}{(s-M_z^2)^2 + \Gamma_z^2\,M_z^2}\,(s - M_z^2)$$

$$\times \left[v_f\,v_e\,(1 + 4\frac{p_f^2}{s}\cos^2\theta + 4\frac{m_f^2}{s}) + 4\,a_f\,a_e\,\frac{p_f}{\sqrt{s}}\cos\theta\right]\right\} \tag{29}$$

where c = color = 1,3 for leptons and quarks respectively, $\alpha_f = (\frac{Q_f}{e})^2\,\alpha_e$ and

where we neglect terms of the order $m_e/\sqrt{s}$.

The values of v_f, v_e, a_f, a_e are given in Table 1.

V. Expected Rates

We calculate the expected number of events in the Z° energy scale. We use the renormalized coupling constants

$$\alpha^{-1}(M_z) = 128.0$$

$$G_F(M_z) = 1.16634 \times 10^{-5}\ GeV^{-2}$$

and[10]

$$\sin^2\theta_\omega = 0.220$$

$$M_z = 92.7\ GeV\ .$$

We have for the electromagnetic cross section

$$\sigma(e^+e^- \to \mu^+\mu^-,\ s = M_z^2) = \frac{4\pi\alpha^2(M_z)}{3s}\,(\hbar c)^2 = 1.16 \times 10^{-35}\ cm^2$$

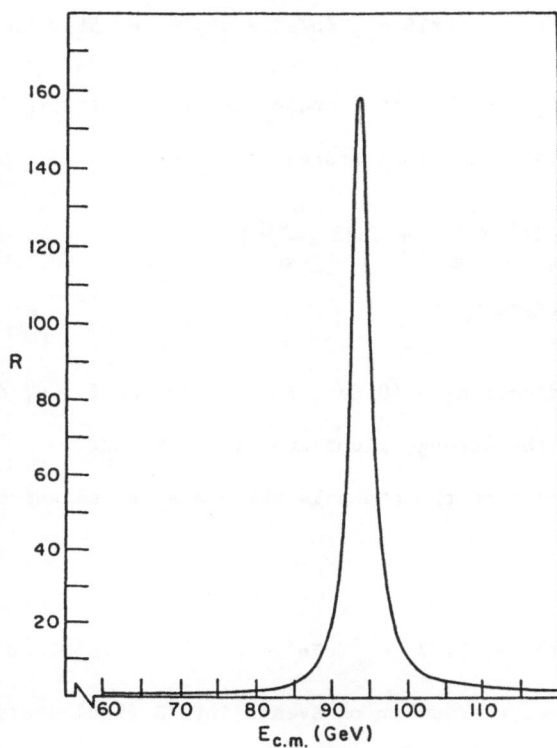

$$R \equiv \frac{\sigma_T(e^+e^- \to \mu^+\mu^-)}{\sigma_{e.m.}(e^+e^- \to \mu^+\mu^-)}$$

Fig.4: Relative cross section for center of mass energy near the Z^0 mass

where we neglect events with emission of photons.

In fig. 4 we show how the muon pair rate changes with energy; we get

$$\frac{R(e^+e^- \to \mu^+\mu^-) \text{ tot}}{R(e^+e^- \to \mu^+\mu^-) \text{ em}} = 158 \text{ at the mass of the } Z \text{ .}$$

Since the weak electromagnetic interference term is zero at $s = M_z^2$, as seen from (29), we can calculate the total rate by separating the contribution from the weak and electromagnetic interactions.

Considering the $\mu^+\mu^-$ final state only, we have

$$\sigma_{weak}(e^+e^- \to \mu^+\mu^-) = \sigma_{tot} - \sigma_{e.m}(e^+e^- \to \mu^+\mu^-)$$
$$= 157 \ \sigma_{em}(e^+e^- \to \mu^+\mu^-) \text{ .}$$

If we use the $Z^\circ \to \mu^+\mu^-$ branching ratio (3.26 %) from table I, we determine the total Z° production cross section

$$\sigma_{weak}(e^+e^- \to Z^0 \to f\bar{f}) = 4816 \; \sigma_{em}(e^+e^- \to \mu^+\mu^-) = 55.9 \text{ nb}.$$

The electromagnetic contribution to the total cross section (neglecting the Moller scattering contribution to the e^+e^- final state) is

$$\sigma_{em}(e^+e^- \to \gamma \to f\bar{f},g) = \{3 + 3[\sum_q (\frac{Q_i}{e})^2 + .693 \, (\frac{Q_t}{e})^2]$$
$$[1 + \frac{\alpha_s(E_{cm} = M_z)}{\pi}]\}\sigma_{em}(e^+e^- \to \mu^+\mu^-)$$

where the top quark contribution assumes $m_t = 40$ GeV, the sum is for $i = u$, d, s, c, b only, and $\alpha_s(M_z) \approx 0.12$ is the strong, gluon coupling constant.

The term $1 + \alpha_s/\pi$ takes into account the channels where a gluon emission takes place.

We finally get

$$\sigma_{em}(e^+e^- \to \gamma + f\bar{f},g) = 7.77 \; \sigma_{em}(e^+e^- \to \mu^+\mu^-) = .0901 \text{ nb}.$$

Using these results we calculate the number of events into a final state. We consider an average luminosity of 10^{30} cm^{-2} sec^{-1} and calculate the number of events in one year of running. The results are in Table 2.

VI. Measurement of $\sin^2\theta_\omega$

a) Measurement of the Z^0 mass

As can be seen in Table II we expect to collect ~ 10^6 events/year for center of mass energies at the Z^0 mass. Hence, in principle, we can measure the Z mass to ~ .1 GeV if the beam energy is stable. This is more likely at LEP than at the SLC, although the expectation at the SLC is that the average energy of the beam will be stable to ± .05%.[15]

Using eq. (24) we can determine how well we can determine $\sin^2\theta_\omega$

$$M_z = (\frac{\pi\alpha(M_z)}{\sqrt{2} \, G_F})^{1/2} \frac{1}{\sin\theta_\omega \cos\theta_\omega}$$

$$\delta(\sin^2\theta_\omega) = \frac{\sin^2\theta_\omega \cos^2\theta_\omega}{1 - 2\sin^2\theta_\omega} \{(\frac{\delta\alpha(M_z)}{\alpha(M_z)})^2 + 4(\frac{\delta M_z}{M_z})^2\}^{1/2}$$

TABLE II No. of Events/year ($E_{c.m.} = M_z$)

Channel	(No. of Events)$_{em.}$	(No. of Events)$_{weak}$	Total
	(10^3)	(10^3)	(10^3)
e^+e^-	.366 *	57.5	57.9
$\mu^+\mu^-$	.366	57.5	57.9
$\tau^+\tau^-$	.366	57.5	57.9
$\sum_i \nu_i\bar{\nu}_i$ (i=e,μ,τ)	0	339.7	339.7
$u\bar{u}$	.487	199.3	199.8
$d\bar{d}$	.121	255.8	255.9
$c\bar{c}$	.487	199.3	199.8
$s\bar{s}$	.121	255.8	255.9
$t\bar{t}$	.337	42.2	42.5
$b\bar{b}$	.121	252.7	252.8
$\sum q_i\bar{q}_i$ g(i=u,d,s,...)	- .067	- 47.8	47.9
All	2.84	1765	1768

*   We do not include the contribution from e^+e^- scattering through t channel
     exchange (Moller Scattering).

and from (28)

$$\frac{\delta\alpha(M_z)}{\alpha(M_z)} = -\frac{2\,\alpha(M_z)}{3\pi}\sum_f Q_f^2 \left(\frac{\delta m_f}{m_f}\right) \qquad \frac{\delta M_z}{M_z} \ll \frac{\delta m_f}{m_f}$$

$$\simeq 1.3 \times 10^{-3}$$

where we assumed $\delta m_f/m_f \simeq$ 30 % for u, d, s; = 10 % for c; 5 % for b; 2 % for
t when measured, and negligible for the leptons.

Using $\delta M_z \simeq 0.1$ GeV, $\sin^2\theta_\omega$ (M_z) $\simeq$ .220 we get

$$\delta(\sin^2\theta_\omega) \simeq .0008$$

where, at this level of accuracy, the error in the radiative corrections begins
to contribute to the over all error.

b) Measurement of the Angular Asymmetry

From eq. (29) we note that there is a forward backward asymmetry in the angular distribution of $e^+e^- \to f\bar{f}$ reactions. If we define

$$A \equiv \frac{\int_0^1 \frac{d\sigma}{d\cos\theta} - \int_{-1}^0 \frac{d\sigma}{d\cos\theta}}{\int_0^1 \frac{d\sigma}{d\cos\theta} + \int_{-1}^0 \frac{d\sigma}{d\cos\theta}} \quad .$$

Then A for the total cross section and, separatedly, for the contribution of the weak interaction is shown, for various final states, in fig. 5 and 6. These do not include the external radiative corrections. Their effect is to reduce significantly the asymmetry for $E_{c.m.} > M_z$.[13]

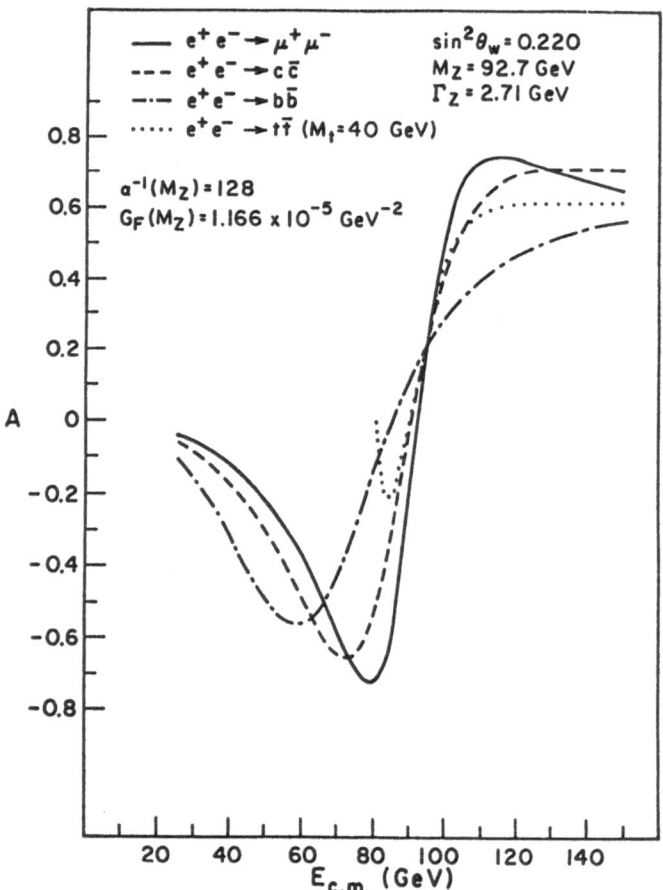

Fig.5: Angular asymmetry due to both the electromagnetic—weak interference and the weak amplitude

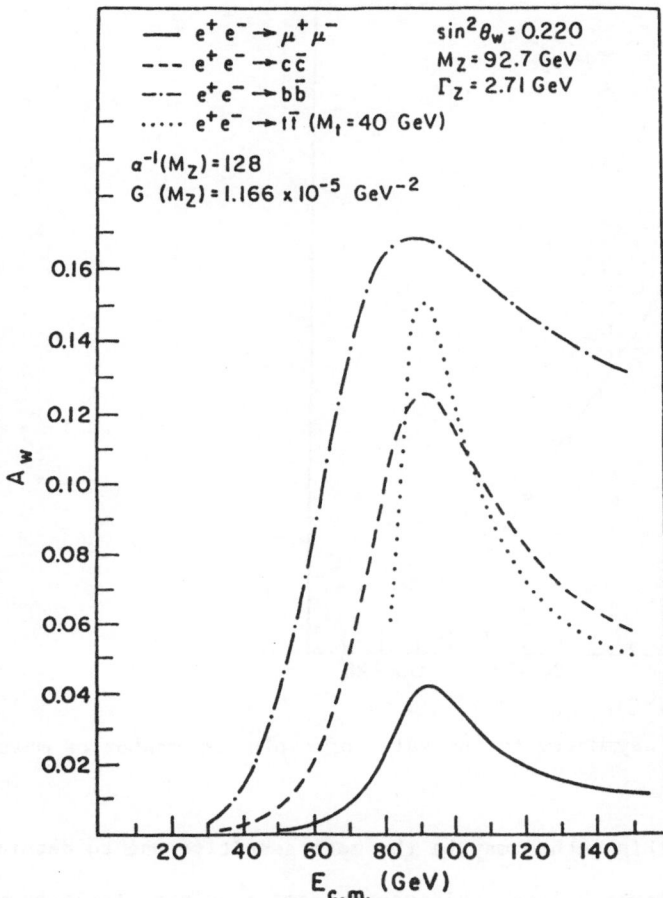

Fig.6: Angular asymmetry due to the weak amplitude alone

If we measure the asymmetry at $E_{c.m.} = M_Z$ then only the weak interaction contributes and we are sensitive, as seen from (29), to

$$A_{weak} \propto v_f v_e a_f a_e = f(\sin^2 \theta_\omega) \; .$$

In fig. 7 we plot this asymmetry (for $e^+ e^- \rightarrow \mu^+ \mu^-$) at $s = M_Z$ for various values of $\sin^2 \theta_\omega$ assuming the dependence shown in table I. Approximately

$$\delta(\sin^2 \theta_\omega) \approx \frac{1}{3} \delta A_w \; .$$

Hence for a 1 % measurement of this asymmetry we get, near $\sin^2 \theta_\omega \approx .220$

$$\delta(\sin^2 \theta_\omega) \approx \frac{1}{3} \times .0004 \approx .0001$$

213

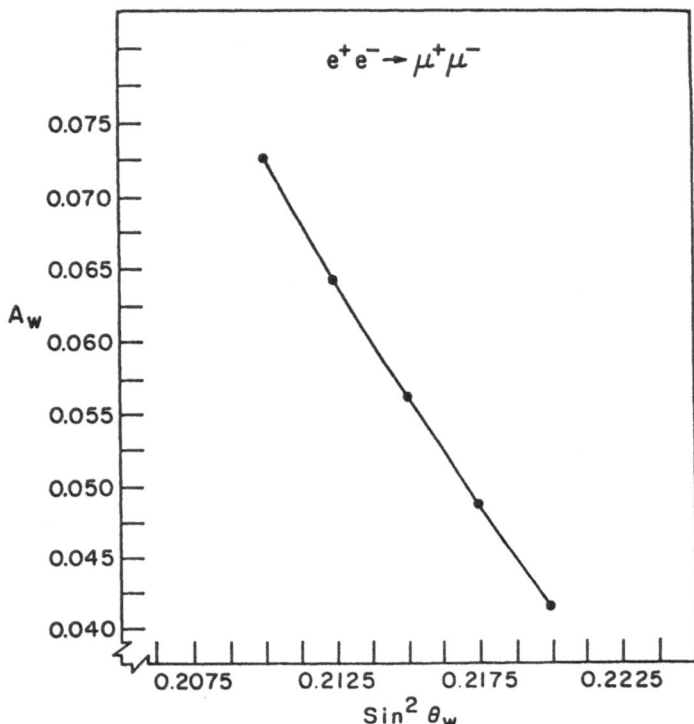

$$e^+ e^- \rightarrow \mu^+ \mu^-$$

Fig.7: Sensitivity of the asymmetry to the value of $\sin^2\theta_\omega$ at center of mass energy equal to M_z

This indicates that this method may be the most sensitive one to determine $\sin^2\theta_\omega$ most accurately. Nevertheless, this measurement requires stable beam conditions because the asymmetry due to the weak electromagnetic interference changes very rapidly in this energy regime and can become a major contributor to the asymmetry if the beam energy is off by ~ .5 %. At SLC, it is expected that the beam energy will be known to .05 %[15] while at LEP it should be known even more accurately.

The measurement of the asymmetry can not measure the relative signs of a_e, a_f, v_e, v_f because they appear as products. The best way to measure the relative sign of these factors is to measure the final state polarization of taus.[12],[13] This can also be accomplished by using polarized electron and positron beams and measured the angular distribution of the final state particles.

214

VI. The Top Quark Mass and the Angular Asymmetry

A curious possibility, if one believes that the electroweak theory predicts the angular asymmetry correctly, is that one may be able to determine the top mass by the measurement of the angular asymmetry in $e^+e^- \rightarrow t\bar{t}$.

For example, in fig. 8 we show the predicted angular asymmetry from eq. (29) versus top mass at $E_{cm} = 85$ GeV and at the Z mass. The sensitivity is evident. Nevertheless, one must caution that these results may have large correction due to the nearby toponium states.[16] This may turn out to be one of the accurate method to determine the top mass. The other methods that have been discussed is the measurement of the toponium ($t\bar{t}$ bound state) spectrum and the excitation of t production near threshold.[17]

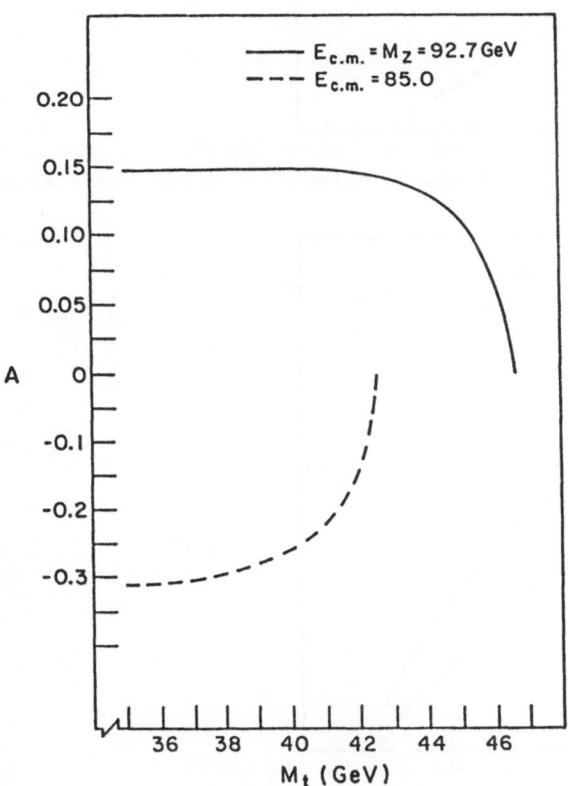

Fig.8: Sensitivity of the angular asymmetry in the reaction $e^+e^- \rightarrow t\bar{t}$ to the value of the top quark mass

VIII. Search for New Families

Although it is unlikely that new families with M < $M_Z/2$ will be present, nevertheless one of the studies will be a deviation of the Z° mass width from the predicted value (2.71 GeV) indicated in Table I. In fig. 9 and 10 we show the partial width that would be contributed by quark or lepton pairs as their mass increases where the Z° decay obeys the electroweak theory.

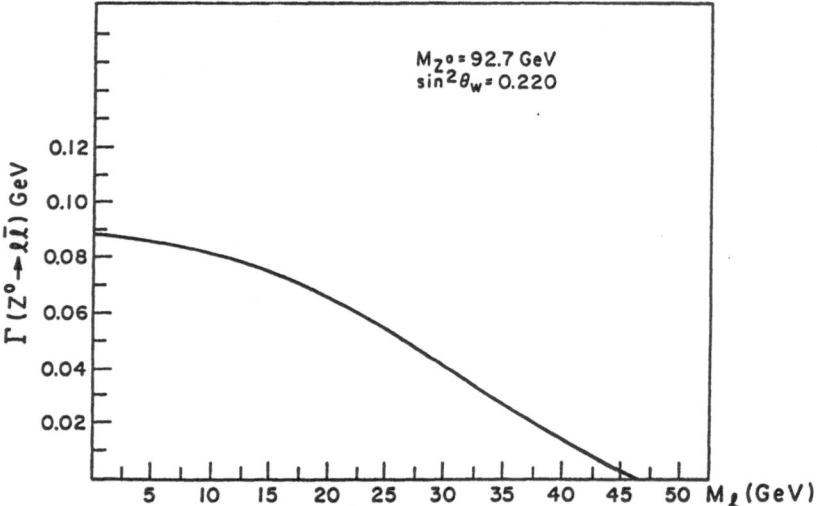

Fig.9: Decay rate of the Z° into quark pairs as a function of the quark mass

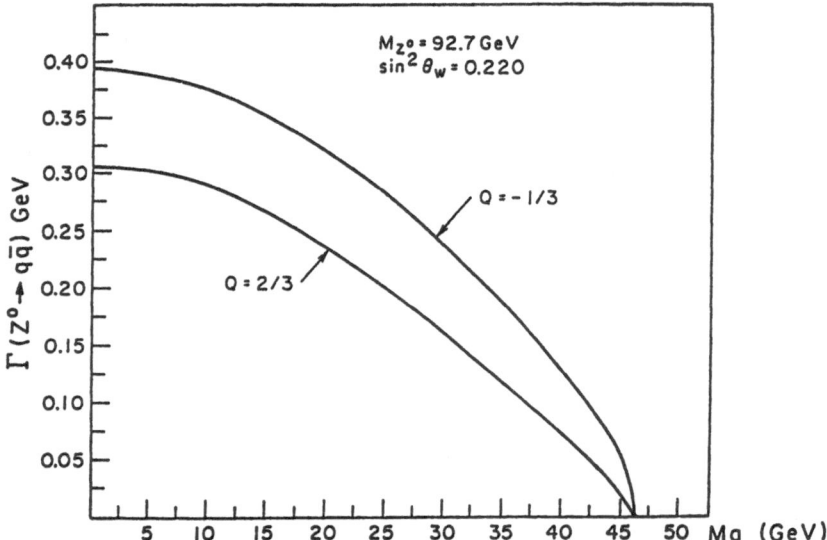

Fig.10: Decay rate of the Z° into charged lepton pairs as a function of the lepton mass

IX. The Z° and the Search for the Higgs

The observation of the Higgs meson is one of the major experimental questions that needs to be addressed in these experiments. Its existence is one of the cornerstones of the Weinberg Salam electroweak theory.

Because the coupling constant of the Higgs and gauge boson is proportional to the boson mass (14), the Z° decay into a Higgs and anything is a good candidate. It was pointed out[18] that the decay mode most suitable for experimental detections is

$$Z° \rightarrow H \ell\bar{\ell}$$

where the ℓ's are leptons. The Feynman diagram is shown in fig. 11. There have been various calculations of this decay rate[18,19] and the total rate versus Higgs mass is shown in fig. 12. We have also calculated the lepton Higgs angular correlation namely

$$\frac{1}{\Gamma(Z \rightarrow \ell\bar{\ell})} \frac{d\sigma(Z \rightarrow H\ell\bar{\ell})}{dx\,d\cos\theta} = \frac{\alpha}{4\pi \sin^2\theta_\omega \cos^2\theta_\omega} \left\{ \frac{1}{(x - M_H^2/M_z^2)^2 + \Gamma_z^2/M_z^2} \right\}$$

$$\times \frac{\{3E_\ell(M_z - E_H) - 2 E_\ell^2 + \vec{P}_\ell \cdot \vec{P}_H\}}{M_z^2} (x^2 - 4 M_H^2/M_z^2)^{1/2} \frac{E_\ell}{M_z + E_H - P_H \cos\theta} \quad (30)$$

where

$$E_\ell = \frac{M_z^2 - 2M_z E_H + M_H^2}{2(M_z - E_H + P_H \cos\theta)}$$

$$x = \frac{2E_H}{M_z} .$$

The results are shown in fig. 13. We notice that as the Higgs mass increases there is a stronger large angle correlation between the Higgs and the

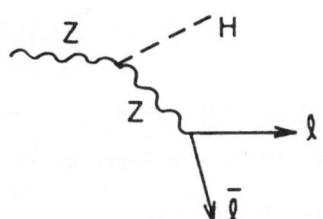

Fig.11: Feynman diagram for the decay $Z^o \rightarrow$ Higgs and lepton pairs

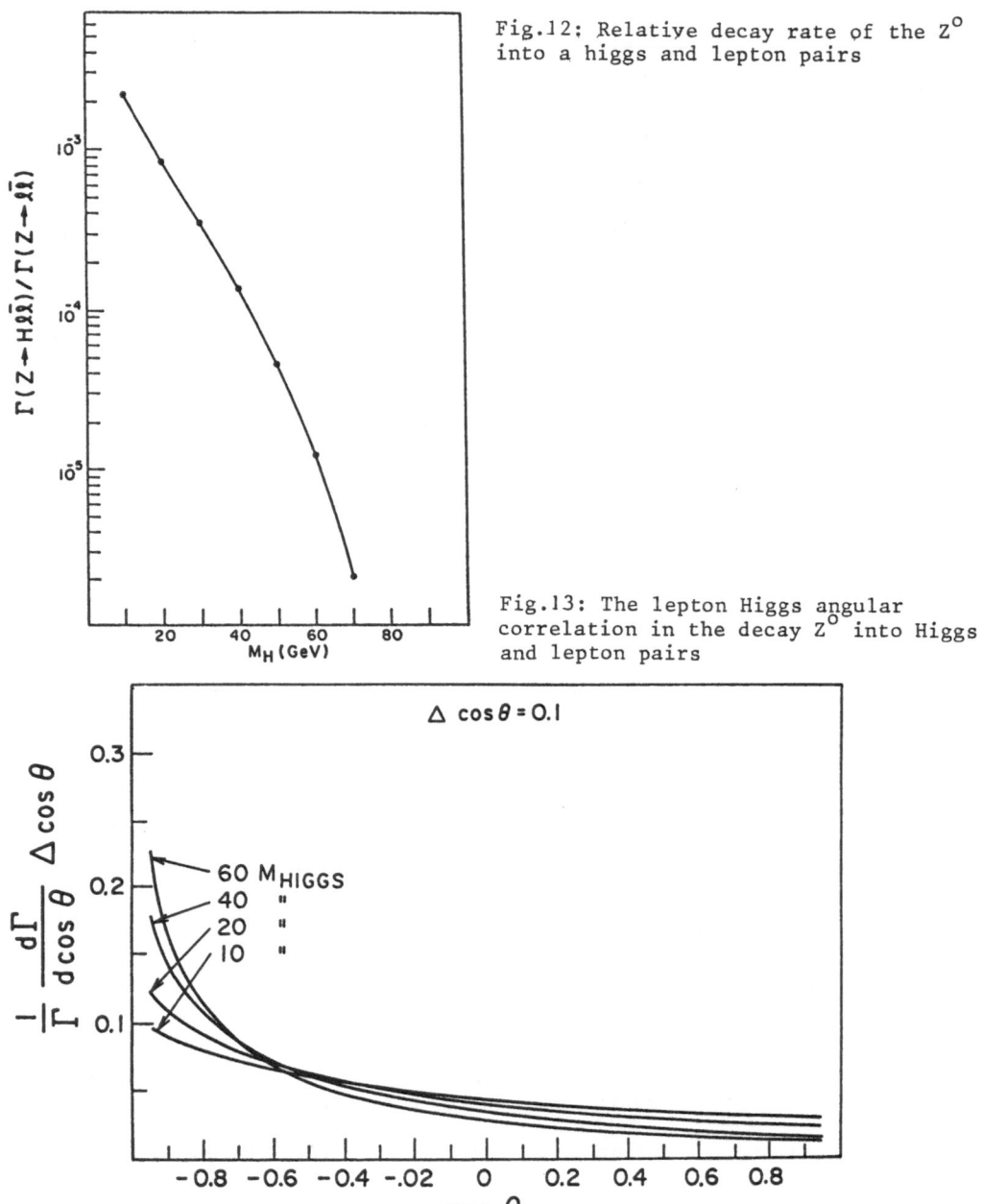

Fig.12: Relative decay rate of the Z^0 into a higgs and lepton pairs

Fig.13: The lepton Higgs angular correlation in the decay Z^0 into Higgs and lepton pairs

lepton. For example, for M_H = 60 GeV 40 % of the events have cos θ < - .7. This correlation in conjunction with others[19] should aid in the improvement of the signal over the background expected from the usual Z^0 decays.

The other possible signal where the Higgs appears is

$$Z^\circ \to H^\circ \gamma$$

but its rate is not favorable compared to the previous signal up to large values

of the Higgs mass.[20]   The decay $Z^\circ \to H^\circ H^\circ$ is possible from the Lagrangian terms

but is forbidden because of spin and statistics requirements.

X. Radiative Corrections

Because the Weinberg Salam electroweak theory will be tested to a high

degree of precision, as shown in the previous chapters, it is imperative to

determine the major corrections to the first order calculations.

In this theory there are several relations such as

$$G_F = \sqrt{2}\,\frac{g^2}{8M_w^2} \;;\qquad g' = g\,\tan\theta_\omega \quad,\qquad e = g\,\sin\theta_\omega$$

which could be altered when second order corrections are introduced. Great care

must be exercised in the correspondance between measurements and these para-

meters. For example, in the determination of G_F from a measurement of the μ

lifetime we already included in the lifetime formula the radiative corrections

to second order in α.[14]   This is necessary because the lifetime measurement

includes the muon decays where a photon is emitted. The Feynman diagrams for

this process are shown on fig. 14. There are a host of other diagrams which

lead to an effective renormalization of the SU(2) coupling constant $g \to \bar{g}$. This

leads to a redefinition of G_F

$$G_F = \sqrt{2}\,\frac{\bar{g}^2}{8M_w^2}$$

and it is this G_F that appears in the μ lifetime formula.[20]   Nevertheless, for

the energy regime below M_w, G_F is essentially a constant and is not changing

with energy (is not a running coupling constant). Some of the graphs used in

the renormalization of g are shown in fig. 15.

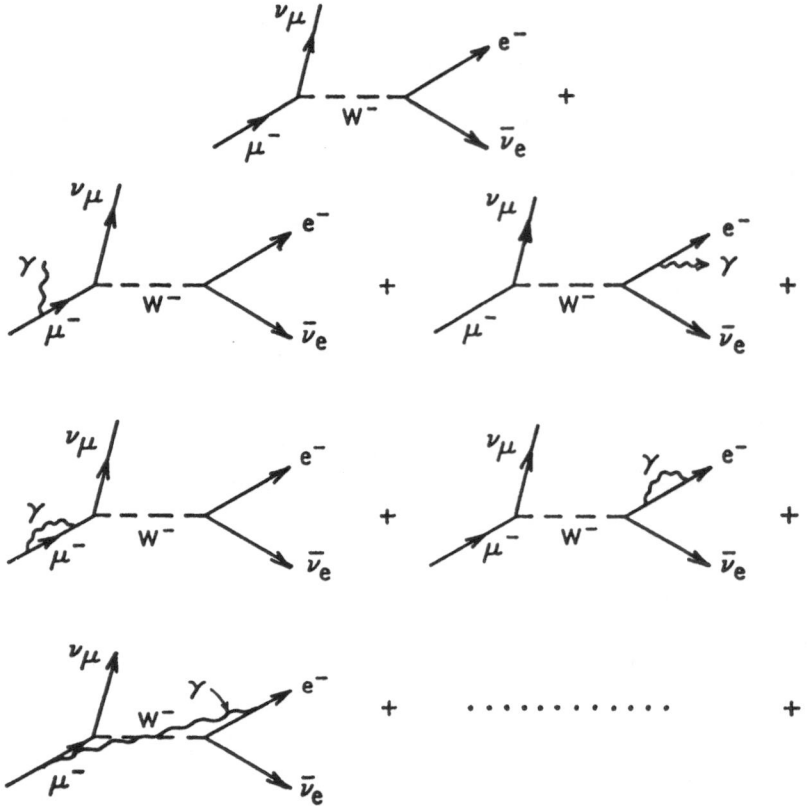

Fig.14: Some of the Feynman diagrams in μ^- decay that contribute
to the order α radiative corrections

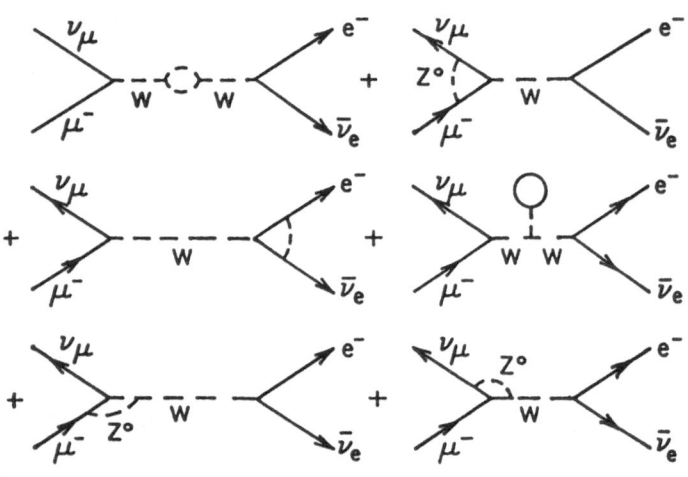

Fig.15: Feynman diagrams in μ decay that contribute to the
renormalization of the weak coupling constant

In the case of $e^+e^- \to q\bar{q}$ we have four quantities whose value we need to determine vis a vis radiative corrections. They are α, G_F, M_Z, and $\sin^2\theta_\omega$. In this review we have chosen to use the measured values of M_Z and $\sin^2\theta_\omega$ from the CERN UA1 and UA2 experiments,[10] and G_F as determined from μ decays. Under these conditions the value of α we should use is that at the mass of the Z,[14] namely $\alpha^{-1} \simeq 128$.

To carry out an accurate calculation of the process $e^+e^- \to f\bar{f}$ one needs to include γZ exchange mixing diagrams as shown in fig. 16. These terms are possible in terms of the electroweak gauge theory because both the photon and the Z are mixtures of the same SU(2) and U(1) fields.[14] These terms are outside the scope of these notes.

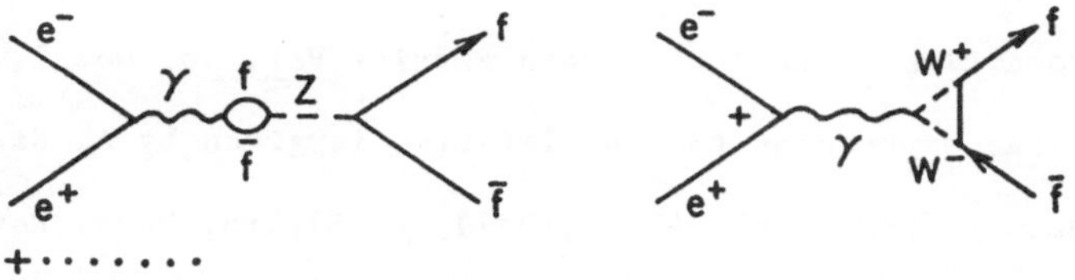

Fig.16: γ-Z mixing diagrams that contribute to the reaction $e^+e^- \to f\bar{f}$

XI. Acknowledgements

I would like to thank K. T. Mahanthappa, T. DeGrand, R. Marciano, A. Sirlin, M. Peskin, and W. Ford for many discussions and help in the preparation of this work.

References

1. S. Weinberg, Phys. Rev. Letters 19, 1264 (1967) and Phys. Rev. D5, 1412, (1972); A. Salam in Elementary Particle Theory, edited by N. Svartholm, Nobel Symposium No. 8 (Wiley, New York, 1969), p. 361; S. L. Glashow Nuclear Physics 22, 579, (1961) and S. L. Glashow et. al. Phys. Rev. D2, 1285, (1970).

2. E. Abers and B. Lee, Physics Reports 9, 16, 1973.

3. M. Kobayashi and T. Maskawa, Prog. Theor. Phys. 49, 652, (1973). See also

L. L. Chau Physics Reports 95, No. 1, 1983 for a comprehensive review of the experimental status.

4. N. Cabibbo, Phys. Rev. Letters 10, 531, (1963).

5. J. C. Pati and A. Salam, Phys. Rev. Letters 31, 661, (1973); R. N. Mohapatra and J. C. Pati, Phys. Rev. D11, 566, (1975); G. Senjanovic and R. N. Mohapatra, Phys. Rev. D12, 1502, (1975); H. Fritzsch and P. Minkowski, Nucl. Phys. B103, 61, (1976); M. A. Beg, R. V. Budny, R. N. Mohapatra, and A. Sirlin, Phys. Rev. Letters 38, 1252, (1977).

6. H. Harari and M. Leurer, Nuclear Phys. B233, 221, (1984); G. Beall, M. Bander, and A. Soni, Phys. Rev. Letters 48, 848, (1982); A. Gangopadhyaya, Phys. Rev. Letters 54, 2203, (1985). See also F. J. Gilman and M. H. Reno, Phys. Letters 127B, 426, (1983).

7. Review of Particle Properties — Rev. of Modern Physics Vol. 56, No. 2, Part II, April 1984. The expression for the lifetime is given by A. Sirlin and W. J. Marciano Nuclear Phys. B189, 442, (1981); A. Sirlin, Phys. Rev. D29, 89, (1984).

8. Here we only consider the classical decay channels. There are others like the Z° decay into Higgs which will be discussed latter. Z° decays into supersymmetric particles are not being considered in this work.

9. J. Ellis, M. K. Gaillard Annual Reviews of Nuclear and Particle Science Vol. 32, 443, (1982); D. Albert, W. Marciano, D. Wyler and Z. Parsa Nucl. Phys. B166, 460, (1980); W. Marciano, D. Wyler, and Z. Parsa Phys. Rev. Letters 43 , 22, (1979). See also Proceedings of the Cornell Z° Workshop CLNS 81-485 p. 127, 1981.

10. L. Di Lella Rapporteur Talk, 1985 Symposium on Lepton and Photon Interactions at High Energies, Kyoto, Japan.

11. In these relations, the coupling constants should be evaluated at the energy scale involved in the measurements. They differ at different energies because of vaccum polarization (radiative corrections) effect. I

will neglect them for the moment and come back to this question latter on.

12. An excellent discussion is given by Jonathan M. Dorfan SLAC-PUB-3407 August 1984.

13. The SLAC Z° Workshop SLAC-247.

14. W. J. Marciano, Phys. Rev. D20, 274 (1979); A. Sirlin, Phys. Rev. D22, 971, (1980); W. J. Marciano and A. Sirlin, Phys. Rev. D22, 2695, (1980); W. J. Marciano and A. Sirlin, Phys. Rev. Letters 46, 163, (1981) and Phys. Rev. D29, 945, (1984); B. W. Lynn, M. E. Peskin, and R. G. Stuart, SLAC-PUB-3725.

15. R. Steining. Private Communication. In addition 90 to 95 % of the beam will have a spread (sigma) of .2%.

16. I would like to thank M. Peskin for pointing out these possible problems.

17. G. Tarnopolsky, SLAC-PUB-2842, (1981).

18. J. D. Bjorken, SLAC-PUB-1866, (1977).

19. D. R. T. Jones and S. T. Petcov, Physics Letters 84B, 440, (1979); Ernest Ma and Jon Okhad, Phys. Rev. D20, 1052, (1979); J. Finjord, Physica Scripta 21, 143, (1980).

20. R. N. Cahn, M. S. Chanowitz, and N. Fleishon, Phys. Lett. 82B, 113, (1979).

Technical Limits on Colliders for High Energy Physics

W.K.H. Panofsky

Stanford Linear Accelerator Center, Stanford, CA 94305, USA

Jack Steinberger knows better than almost anyone that progress in high energy physics depends on a constructive interaction among those who built accelerators and colliders, those who design and build new experimental facilities, and those who make good measurements and interpret them. The only link in this chain to which Jack Steinberger has not personally made contributions is the design and construction of accelerators and colliders, and it is to the future of that topic to which I would like to address some remarks.

Jack Steinberger has used both electrons and hadron accelerators to greatest advantage. He is responsible for a detector to use what is perhaps the largest of high energy electron-positron colliders on storage rings -LEP- and his exploitation of the neutrino beams produced in proton colliders has provided definitive answers in that field. Stimulated by successful experiments such as his and the good initial performance of the Fermilab superconducting proton accelerator plans are advanced in the United States for a proton collider up to an energy of, perhaps, 20 TeV per beam, and the possibility of placing a hadron collider into the LEP tunnel is under investigation.

Accelerator builders face a fundamental dilemma. Protons are composite particles, while electrons down to dimensions of 10^{-16} cm are still point-like. As a result, the energy of collisions among protons is shared among the quarks and gluons. Therefore, depending on the reaction under study, the energy of protons has to be "derated" by about an order of magnitude relative to electrons in providing the same "reach" into the unknown. Thus electron accelerators of considerably lower energy are equivalent to the corresponding proton acceleerators. On the other hand, electron accelerators at ultra-high energies using known technologies are becoming excessively expensive. But this is not all there is to the electron-proton comparison. The total cross-section in proton-proton collisions is still on the increase, while the cross-section for producing events exhibiting new physical phenomena of interest is expected to decrease as the square of the relevant mass range to be investigated. Therefore, the signal-to-noise ratio for proton-proton colliders for <u>interesting</u> events is degenerating rapidly with energy, while the <u>total</u> event rate continues to increase. As a result the data analysis process applicable to hadron-hadron collider events is a frightening process which has to reject, hopefully successfully, a substantial number of supposedly unneeded features of the primary collision. While this can be accomplished with confidence in designing experiments with reasonably well-defined objectives, the lingering doubt remains that the increasing need for rejecting primary data might also reject unanticipated primary discoveries. Therefore, given the choice, most particle physicists would prefer an electron-positron collider of one-tenth of the particle energy to a corresponding proton-proton collider of roughly comparable luminosity. But does this choice exist? The answer is that in the TeV range it does not within present technologies.

The SSC is a bold step forward in extending the proton-proton collider technologies using superconducting magnets demonstrated so successfully at Fermilab into the multi-TeV energy range measured in terms of collision energy of the fundamental constituents. This proposal does not involve basically new technology but hopes to extend the energy range of the decreasing unit cost by economies of scale and simply good design. Corresponding matching opportunities for electrons do not exist today.

The reason for this conclusion is well known. Electrons in storage rings emit synchrotron radiation and the rate of energy loss per turn increases as the fourth power of the energy. The cost of storage rings in essence is composed of two terms: (1) proportional to the radius or physical size of the device, and (2) proportional to the fourth power of the energy divided by the orbit radius. If these two terms are matched a quadratic scaling law of cost vs. energy results. Thus it is likely that LEP with a circumference of 27 kilometers will be the last electron-positron collider built along the by-now established storage ring pattern.

We know that in principle the cost of proton colliders increases approximately linearly with energy, although the costs can be compensated to some extent by economies of scale and good design based on accummulated experience. How-

226

ever, even for protons synchrotron radiation will become dominant for energies exceeding that of the SSC by a substantial factor.

For the above reasons it is evident that if the exponential growth in energy for electron collisions and for proton collisions beyond the SSC is to be maintained new technology is needed.

If we are to extend electron-positron and eventually even proton-proton collisions to higher energies we must abandon storage rings, unless someone makes an invention to suppress synchrotron radiation. Thus the electron-positron collider of the future must be a linear collider in which two linear machines accelerating electrons and positrons are aimed at one another to produce the required rate of interaction.

Such linear colliders lose the primary advantage of storage rings which is the fact that the same electrons and positrons collide repeatedly before eventually being lost due to the finite lifetime of the stored beam. In consequence, as we shall see, useful designs for linear colliders require that both the total beam power required to achieve adequate reaction rates and the density of the particle beams during interaction become very large.

The reason for this conclusion derives from the unpleasant fact that the total energy of each beam will have to be thrown away after each collision between opposing bunches during which only an exceedingly small fraction of electrons and positrons produce events of interest. The only exception to this situation would arise if energy recovery of the "spent beam" would somehow be possible. This latter possibility has been explored by several physicists, in particular U. Amaldi, by using the spent beam to produce radiofrequency energy in a superconducting linear accelerator; that accelerator thus energized can then be used for accelerating the next bunch. Elementary arguments make it clear that there is hope for such energy recovery only if Q values of superconducting accelerators can be attained which are two orders of magnitude or so larger than those now deemed feasible. Current linear accelerators at room temperature are operating at Q values of the order of 10^4 and use duty cycles of the order of 10^{-5}. Thus if a superconducting accelerator running continuously is to utilize refrigeration power comparable to the rf power fed into conventional accelerators, then we require Q values of $10^4 \times 10^5 \times 10^3 = 10^{12}$, where the last factor of 10^3 is the reciprocal of the efficiency of refrigeration at temperatures usable for superconducting devices. The Q value of 10^{12} is about 10^2 to 10^3 larger than practical values attained to date. In addition, energy recovery in superconducting accelerators has numerous other practical problems which may or may not be more severe than those faced by other devices at room temperature which we will now discuss.

If energy recovery is proven infeasible, then we have to face the power consumption inherent in throwing the interacting bunches away after collisions. We can write an equation which directly relates the luminosity L of interaction as related to the frequency f of collision, and the cross section A of the interacting beams:

$$L = fN^2/A \tag{1}$$

where N is the number of particles per bunch. If we assume that the beam is circular in cross-section and has an invariant emittance ϵ_n, and if the opposing bunches are focused into collision with a focusing parameter β, then the average power P_B of each of the colliding beams is given by:

$$P_B = m_o fN\gamma \quad \text{and} \quad L = fN^2\gamma/4\pi\epsilon_n\beta \ . \tag{2}$$

The number of particles which we can handle per bunch is limited by the energy spread produced due to the beamstrahlung of the electrons or positrons during collision. The very factors, in particular the density of interaction, which enhance luminosity also increase the electromagnetic radiation experienced by an opposing particle. One might assume that this beamstrahlung can be calculated by the applicable classical formulas. However, using the ordinary expressions for synchrotron radiation, we find that the frequency spectrum of the emitted radiation extends up to a value scaling with the cube of the cyclotron frequency of the emitting particle in the electromagnetic field of the opposing bunch divided by the bending radius; in other words, the upper limit of the energy spectrum in general grows faster than the energy of the colliding particles themselves. Once the top energy of the radiated photons exceeds that of the primary beam the classical calculation is no longer valid; as a practical matter the photon spectrum is cut off at an energy equal to that of the primary beam. If we assume that the parameters of an electron-positron collider are of interest for the next generation

228

machines only if such a quantum mechanical cutoff in the synchrotron radiation spectrum is required, then we can write formulas relating the parameters of interest to the physicists to the those parameters which the machine builder would consider limiting in design.

The physicist has to specify three basic parameters;

    (a) the required beam energy

    (b) the luminosity

    (c) the tolerable energy spread δ_q produced in the interaction.

The latter is particularly important should there be resonances in the new energy range to be investigated. This importance is somewhat moderated by the fact that if there are resonances the cross sections would be larger at such a resonance and one could therefore afford to decrease the luminosity at the resonance in order to achieve a narrower energy width. With the exception of this unlikely situation reasonably broad energy spreads - say in the 50-30% range - could be tolerable. However it is unavoidable that within current quantum-mechanical wisdom the required luminosities would have to go up with the square of the energy in order to give reasonable rates for events of interest. A luminosity of at least $10^{33} cm^{-2} sec^{-1}$ in the 1 TeV range and $10^{34} cm^{-2} sec^{-1}$ in the 3 TeV range appear essential if event rates of interest of the order of 10^2 to 10^3 per year are to be anticipated.

In addition to these primary experimental parameters there are also secondary requirements of concern to physicists. Background due to beam gas interactions or radiation from upstream or downstream focusing devices must be held at tolerable limits. From the point of view of event reconstruction one would like to limit severely the expected number of events per bunch crossing. In addition, one would like to make it possible to place detectors very close to the interaction point in order to be able to detect short-lived decays. If possible the accelerating mechanism should preserve polarization so that events originating from states of specified angular momentum can be isolated. However, for these considerations let me assume that the primary factors, luminosity, energy and energy width are all that define the usefulness of a machine.

A problem is that at present it is not fully clear which machine parameters are controlling the economics of the overall machine. Average beam power combined with the finite efficiency of generating the beam is clearly a very important and probably the most important parameter. In other words, a linear collider at super high energy must operate heavily loaded.

Let me make a remark on the somewhat arcane "voodoo" economics in which the accelerator physicist defines the construction cost for a machine. In the past the government has been persuaded to supply appropriate adequate construction funds to build a new facility. With the exception of inflation all such monies are directly applied to the relevant construction cost. In contrast, when a public utility builds a power plant the money needed may be as large as three times the actual cost of construction due to the cost of raising the necessary capital and the accumulating interest rate during the construction period. Therefore, considering the large power probably required for an accelerator of the future, one might even entertain the notion that a power plant might be part of the projected construction cost. This, of course, is not a saving in real economic terms.

The attainable accelerating gradient and therefore the economically (or politically!) practical length of a machine may or may not be a constraining factor. Because of the high power costs the efficiency of extracting the stored energy in the accelerating devices by the beam is important. Since the total amount of energy per bunch is limited by the energy spread produced in the beam-beam interaction attainment of such extraction efficiency can become more difficult if the accelerating gradients are very large and therefore the energy stored in the accelerating structure is correspondingly increased. Therefore the shortest accelerator is not necessarily the most economical accelerator overall, although it is obviously much more attractive aesthetically and reduces the total impact of the installation.

Having said all this we can write a number of basic equations which define the various derived quantities for accelerator design in terms of the specified physical parameters. Since in the multi-TeV region the quantum mechanical cutoff of the synchrotron radiation spectrum will enter into the picture I am giving these equations only in that region.

$$N = \frac{4\pi}{(1.63)^3 \alpha^4} \left(\frac{\delta_q^3}{D} \right) \tag{3}$$

$$A = \frac{4\pi}{(1.63)^3 \alpha^4} \left(\frac{P_b \delta_q^3}{EDL} \right) \tag{4}$$

$$f = \frac{(1.63)^3 \alpha^4}{4\pi} \left(\frac{P_b D}{E \delta_q^3} \right) \tag{5}$$

$$\sigma_z = \frac{1}{4\pi} \left(\frac{P_b D}{r_e m_e c^2 L} \right) \quad . \tag{6}$$

In these equations D is the "disruption" parameter which is the ratio of the bunch length at interaction to the focal length produced by the electromagnetic focusing effect in the beam-beam interaction. This number cannot be very large because if it were each particle would undergo oscillations within the bunch of the opposing particle, and eventually these oscillations would become unstable. For moderate values of D this mutual focusing effect is beneficial in that it increases the average beam density and therefore the luminosity. For the SLC the disruption parameter leads to an enhanced luminosity of perhaps 2 or 3 once specified performance is attained. However, as we shall see below, the bunch length σ_z must be very short indeed for high multi-TeV devices and therefore the $D-$ factor will in practice have to be small.

The above equations are written for beams of circular cross sections. Additional advantages can be obtained using "flat beams" because then for a given beam density the total disruptive beam-beam interaction and the beamstrahlung effects are reduced. If for the time being we ignore this possibility and also ignore the enhancement produced by the beam-beam interaction (which at most can become a factor of 6), we can combine the above equations by eliminating the D factor and write a simple expression for the beam power as follows:

$$P_b = 5 \left(\frac{L}{10^{30}} \right) \left[\frac{\epsilon_n \beta \sigma_z}{\delta_q^3} \right]^{1/2} \quad . \tag{7}$$

Here the power per beam is measured in megawatts and the luminosity in $cm^{-2} sec^{-1}$. The quantity ϵ_n is the invariant emittance, that is $\gamma r r'$, and β is the usual focusing parameter and σ_z is the bunch length, all measured in centimeters. Note that the required luminosity would have to increase as the square of the energy.

Note that this equation is totally independent of the means which are employed to produce beams and bring them into collision. Such acceleration would be achieved by traditional rf structures or some new, for instance laser, devices.

It is interesting to plug numbers into this equation under the assumption that accelerator performance was no better than that employed in the design of the SLAC SLC, but assuming that we would be producing 5 TeV per beam, rather than 50 GeV per beam as is the case for the SLC. If we assume that the luminosity is to scale by the square of the energy from the design luminosity of $6 \times 10^{30} cm^{-2} sec^{-1}$ of the SLC, the invariant emittance ϵ_n is $3 \times 10^{-3} cm$, and the β value 1 cm, and if we assume a permissible energy spread δ_q of 30%, then we require a beam power well in excess of 10^4 Megawatts. This is clearly impractical and therefore one has to go considerably beyond SLC parameters if the linear collider idea without energy recovery is to become practical in the TeV range. Since the SLC is considered to be a "daring" machine by many, such a further extrapolation is clearly not an immediate prospect and thus the time scale at which the SSC (which is based on existing technology) and such a super linear collider can be achieved are many years apart.

In order to go beyond the SLC performance the following steps come to mind:

1. make the beam non-circular,
2. reduce the normalized (invariant) emittance,
3. reduce the bunch length.
4. reduce the β value in the interaction region.

A non-circular beam with aspect ratio R in principle introduces a factor $2/(1 + R)$ into the power equation. This is not a very steep dependence and there are limits to the practicality of focusing an asymmetric beam to the required small dimensions considering the extreme complexity of the final focus system when designed to handle beams of finite energy spread.

Reducing the normalized emittance ϵ_n is in practice a matter of designing a damping ring to "cool" the radial temperature of the electrons and positrons to an extent greater than that projected for the SLC. Increasing the brightness of the primary gun is not of major relevance to achieve reduction of the normalized emittance since the phase volume of the positrons is determined by the electromagnetic shower process in the positron converter and not by the driving beam.

The ultimate performance of damping rings has not been thoroughly studied but an improvement in ϵ_n by perhaps 2 orders of magnitude appears feasible. One should note, however, that if the particle energy is increased beyond the SLC and if the normalized emittance is decreased, then both of these factors decrease the actual beam diameters at the interaction point below the 1.2 micrometers now projected for the SLC. If we reduce the normalized emittance by 2 orders of magnitude and increase energy by 2 orders of magnitude, then the beam diameter is decreased by about 100 and the beam area by a factor of 10^4. We are talking about beams of radius of perhaps 10 angstrom units. Although beams of such dimensions have proven practical in scanning electron microscopes, the question how to design final focusing systems with sufficient freedom from aberrations, sufficient focusing strength, and mechanical stability has not even been considered.

The longitudinal dimension σ_z could also be shrunk from the 1 millimeter value of the SLC, let us say to 1 micron. All these factors combined would bring the average beam power into the few megawatt range. This might, perhaps, permit a practical design, provided the efficiency from the wall plug to the beam is not too small. Note, however, that the beam density would become of the order of $10^{27} electrons/cm^3$ (!) and a physical density of $10^7 g/cm^3$, very much larger than ordinary macroscopic densities.

A similar problem arises in respect to the accelerating structure. If we assume that a damping ring can produce normalized emittances well below those projected for the SLC we are facing an increased problem of growth of this emittance in the actual accelerating and final focusing structures.

Many different kinds of accelerating structures and methods have been proposed and are under intensive study. Time does not permit me to discuss these here. Some of these are plasma devices in which a laser beam or laser beams interfering with one another produce waves in a plasma which, in turn, accelerate the particles. Such devices appear attractive because of the potentially high gradient they might produce. However, I am personally pessimistic about their utility in this case because of two factors: (1) the overall efficiency of transfer of power from the source to the beam, and (2) the fundamental difficulty in controlling the exact micro-detail of the plasma to avoid growth in the emittance of the beam. Other accelerating methods use either the electric fields of lasers or microwave energy to produce acceleration.

The most predictable performance would involve "conventional" rf structures which, however, must be fed by power sources not as yet developed. All such sources are in effect tranformers from low voltage, high current devices to the high energy, low current beam to be produced for collision purposes. The "conventional" electron linac uses a multiplicity of such high current beams in the klystron tubes feeding the machine. The most challenging of such transformers are in essence two beam devices. In one design a hollow beam of high intensity is transmitted at the edge of a conventional rf structure where the magnetic field is large and the electric field is small. The beam to be accelerated passes at the axis where the electric field is high and the magnetic field is zero. In another version a many kilo-ampere beam is produced in an induction accelerator and rf energy is extracted periodically either by a free electron laser, that is by use of a wiggler, or with conventional rf cavities if the primary high current beam is bunched. The practicality of producing 10 kiloamp, multi-MeV beams has been demonstrated, in particular with the Advanced Test Accelerator in Livermore. The energy loss corresponding to that extracted is replenished by re-acceleration with induction cores. The microwave power is then fed from the extraction points to the high energy accelerator. Such two beam devices have an inherently high efficiency.

The problem with all such devices deals with wake field effects, in particular in respect to transverse deflecting modes. Any medium which supports the required longitudinal accelerating fields will also support transverse modes, and therefore the required alignment tolerances will become extraordinarily severe if, as indicated above, emittances much smaller than those projected for the SLC are to be employed. Since the optimization of parameters suitable for this purpose points to wave length much shorter than the customary S-band range, this situation would be aggravated.

Reducing the β value below 1 cm, or as far as that goes even to attain 1 cm for beam energies as high as 5 TeV, is an unsolved problem. An exciting possibility is to employ 2 bunches, one to produce the focusing field to contract the second bunch which is the one to yield the desired collision intensities. Such first order lenses produced by placing charge into the beam are potentially a great deal more powerful than existing external focusing lenses such as superconducting quadrupoles.

The situation can be summarized as follows: the construction of ultra-high energy electron-positron colliders does not involve foreseeable problems which cannot be solved without violating basic physical principles. However, in order to produce the high luminosity required for reasonable data rates at high energies, preferably all three quantities, that is invariant emittance, the bunch length, and the β value in the above equation (7) would each have to be reduced by roughly at least 2 orders of magnitude in order to reduce beam powers to what appear to be practical values. This, in turn, requires development of new power sources, the development of new final focus methods, unprecedented alignment techniques, possibly requiring continuous servo-loops, and unprecedented stability and freedom from noise of basic power supplies. All this will require not only development but also invention. However, I am confident that Steinberger's legendary productivity will endure long enough so he can do great experiments with that next generation of machines once it exists!

Experimental Observation of Some of the Properties of the Intermediate Vector Bosons

C. Rubbia

CERN, CH-1211 Genève 23, Switzerland
and
Harvard University, Cambridge, MA 02138, USA

1.-Introduction. A very large number of proton-antiproton collisions at very high centre of mass energies have been recorded at the CERN Super Proton Syncrotron (SPS) Collider during 1982 and 1983 [1—5]. These experiment have firmly established the existence of both the charged ($W^{\pm}$) and of the neutral (Z^0) Intermediate Vector Bosons. Masses as well some of the basic properties, namely decay branching ratios, the J=1 nature of the W's and the presence of characteristic parity violating asymmetry have been observed. The production of W's and Z's at the Collider proceeds by the Drell-Yan mechanism [6] in which a quark from the proton annihilates with an anti-quark from the antiproton. In the QCD description of such a production mechanism quarks are coloured, which gives a factor one third in the cross section and there are higher order corrections to the bare Drell-Yan process, including radiation of one or more gluons. This initial state bremsstrahlung is expected to give rise to i) a long tail in the transverse momentum distribution of the weak bosons, and ii) the occasional observation of one or more jets produced in association with such a high transverse momentum.

During the 1982 and 1983 the collider was operated at $\sqrt{s}$ =540 GeV. From an integrated luminosity of 136 nb^{-1} the UA1 experiment [7] has collected an essentially background free sample of 52 $W^{\pm} \Rightarrow e^{\pm} + \nu$ decays [1] , 14 $W^{\pm} \Rightarrow \mu^{\pm} + \nu$ decays [2], 4 $Z^0 \Rightarrow e^+ + e^-$ decays[3], 5 $Z^0 \Rightarrow \mu^+ + \mu^-$ decays [4] . Analysis of these samples showed that the production properties of the W's and Z's at the Collider are consistent with the QCD-improved Drell-Yan mechanism and that the decay properties are consistent with the expectations of SU(2) ⊗ U(1) Standard Model [8].

More recently, during the 1984 data taking period, the Collider has been operated at the increased centre of mass energy of $\sqrt{s}$ = 630 GeV and the UA1 Collaboration have recorded a further 263 nb^{-1} of data, tripling the size of the event samples, and enabling a more detailed examination of the production and decay properties of the W and Z to be made.

In this paper we start discussing the W production properties, obtained from an analysis of 172 events in which a W has been produced and subsequently decaying in the ($e^{\pm} \nu$) channel. With these improved statistics it is shown that the QCD-improved Drell-Yan mechanism gives a good description of the event rates, and of the W longitudinal and transverse momentum distributions. The observations of hadronic jets produced in association with the weak bosons is firmly established, and the properties of these jets are shown to be in agreement with the

237

expectations for jets arising from initial-state bremsstrahlung off the incoming annihilating partons. Next we shall give the Z^O production cross section, the $W^{\pm}$ and Z^O masses and widths, and the decay angular distributions extracted from the 172 W and 16 Z decays. Our values for the masses correspond to values of the Standard Model [8] parameters — ρ and $\sin^2\theta_w$ — which are in excellent agreement with current expectations [10,11] and previous measurements [1—5]. Finally an upper bond to the number of neutrino species can be set. Finally, there is no evidence in the current data sample for the existence of a massive object X which subsequently decays into a W and one or more hadronic jets.

2.-Detector. The UA1 detector is a general-purpose apparatus designed to study, as systematically as possible, the phenomenology of pp collisions at the collider. The apparatus and the data-taking conditions are essentially unchanged from those reported in our previous publications, and we refer the reader to Ref. [1] for more details. The main features of the UA1 detector and event reconstruction relevant to the present paper are the following:

i) *Momentum analysis and charge determination.* This is performed with accurate curvature measurement by the central detector, a large-volume drift chamber surrounding the crossing point and operated in a homogeneous 0.7 T magnetic field. The central detector is also used to determine the charged-particle topology of the event and, in particular, the isolation criteria for the electron track.

ii) *Electron detection.* Energetic electrons deposit almost all their energy in the UA1 lead-scintillator electromagnetic shower calorimeter, 27 radiation lengths deep, and deposit very little energy in the hadron calorimeter beyond this. The energy resolution of the electromagnetic calorimeter is approximately $\Delta E = 0.16\sqrt{E}$, where all energies are in GeV.

iii) *Neutrino detection.* The presence of neutrino emission is inferred from an apparent lack of momentum conservation in the two components transverse to the beam direction. The detection and measurement of neutrino emission by this method is made possible in the UA1 detector by the nearly complete solid-angle coverage of calorimeters down to 0.2^O from the direction of the beams. The accuracy in each component of missing transverse energy is $0.4\sqrt{E_t}$, where E_t is the scalar sum of the transverse energies deposited in all the calorimeter cells and all energies are in GeV.

iv) *Jet finding and reconstruction.* The UA1 jet-finding algorithm has been described elsewhere. Magnitudes are defined by the spatial position of the cell and energy deposition within the cell. The cells with transverse energy in excess of 2.5 GeV grouped in clusters and their associated momentum vector are added if the distance between them in [pseudorapidity-angle]-space $R \geq 1$ [$R \equiv (\Delta\eta^2 + \Delta\phi^2)^{1/2}$, where ϕ is in radians]. Cells with lower transverse energy are then included in the cluster if they are within $\Delta R \leq$

1 with respect to the cluster axis. Jets are retained if they have a transverse momentum in excess of 5 GeV/c, and an axis within the pseudorapidity interval $|\eta| \leq 2.5$. A detailed study of the relationship between the four-momentum of jets reconstructed by this method and the four-momentum of the underlying parton has been made using ISAJET [13] with the fragmentation function of Field and Feynman. The results of this study suggest that the jet energy is underestimated by about 20% and the jet momentum by about 15%. The implied scale shift, which is model-dependent, does not alter the essential results presented in this paper. For definiteness we do not correct the reconstructed jet four-momentum for this effect.

The precision with which we can measure the W and Z masses is determined by the precision with which we can measure electron energies in the range 40 GeV to 50 Gev, which, for a perfectly calibrated calorimeter, is approximately 2.5%. In practice we must add to this intrinsic resolution the additional smearing of the electron energy measurement coming from:

i) *cell-to-cell calibration differences:* the relative calibration of each e.m. calorimeter cell has been measured using minimum-bias data and cosmic-ray muons of known momenta . After this has been done, the response of any e.m. cell to a high -energy electron is known to $\pm 3\%$ relative to the average cell response;

ii) *pile-up from the underlying event*: the pile-up contribution coming from the energy deposited by additional particles entering the same calorimeter cells as the electron is subtracted from the electron cluster energy on an event-by-event basis. The energy subtracted is proportional to the sum of the energies E_π of all the additional charged tracks (momentum analysed in the CD and assumed to be pions) entering the same calorimeter cells as those entered by the electron. The correction assumes that there is an additional pile-up contribution from neutral pions of $0.5\, E_\pi$. The resulting pile-up correction is given by the expression $\Delta E = 1.1 E_\pi$, and is typically 1 GeV coming from fluctuations in the observed energy deposited by the charged pions, and in the charged-to-neutral pion ratio. An additional systematic error of 0.5 GeV is associated with the method.

Taking the above contributions into account, the overall resolution for measuring the highest-energy electrons from W and Z decays is 4%. However, the precision with which we measure the IVB masses is dominated by the systematic uncertainly in the absolute energy scale of the e.m. calorimeters. This uncertainty comes from two sources:

i)*Absolute calibration.* The absolute energy scale of e.m. calorimeter cells is obtained by illuminating them with an intense Co^{60} source at the beginning and end of each data-taking period. A test calorimeter cell has been calibrated directly with electron beams of known energy from the CERN SPS; the other e.m. cells are then corrected to give the same response of the Co^{60} source as that given by the calibrated test cell. This calibration method assumes that the test cell and all the e.m. cells in the detector are (mechanically) identical. This assumption introduces a systematics uncertainty of 3% in the absolute energy scale of the e.m. calorimeters.

ii) *Time dependence of the calorimeter response.* Owing to radiation damage, the light yield coming from the e.m. calorimeter cells decreased by about 11% during the 1984 data-taking period. We have monitored this effect by measuring the rate of isolated energetic neutral pions with transverse energies in excess of a fixed calorimeter threshold ($E_t > 20$ GeV) as a function of time. The neutral-pion rate indicates that the calorimeters have aged linearly with delivered integrated luminosity. We cannot, however, exclude the possibility that ageing has partially occurred at one, or several, discrete time during the running period. This uncertainty introduces an average systematic error of 1% on the measurement of the energy of the highest-energy electrons from W and Z decay.

The resulting overall systematic error on the electron energy scale, and therefore on the W mass determination, is 3.2%.

3.-Event Sample. Result are based on an integrated luminosity of 399 nb^{-1}, which is corrected for dead-time and other similar losses, and which includes the data samples from previous publications [1]. The trigger used throughout the data taking required the presence of an electromagnetic cluster (one or two adjacent electromagnetic calorimeter cells) with a transverse energy in excess of 10 GeV, and at an angle of more than 5 $^{\circ}$ with respect to the beam direction. After complete off-line reconstruction, about 5×10^5 events had at least one electromagnetic (e.m.) cluster with transverse energy $E_t \geq 15$ GeV. To extract an essentially background-free sample of W decays, the following cuts have been applied to the data:

i) *Cluster validation.* To ensure that we have a reliable reconstruction for the e.m. cluster, we require that, after removing the central detector track with the highest transverse momentum (the electron candidate), the sum of the transverse momenta of all other tracks entering the two calorimeter cells associated with the cluster is less than 3 GeV/c. We further require that the centroids of the energy depositions in the four longitudinal samplings of the e.m. calorimeter are consistent with a single e.m. cluster.

ii) *Associated track requirement.* If the e.m. cluster is at an angle of more than 15 with respect to the horizontal, where the central detector has good efficiency, we require the presence of a charged track in the central detector associated with the cluster, and with transverse momentum consistent with being in excess of 15 GeV/c.

iii) *Loose isolation.* To exclude two-jet events in which one jet fakes the electron signature, we demand that the electron is isolated in a cone in (pseudorapidity, azimuthal angle)-space with radius $\Delta R = 0.4$, where $\Delta R \equiv (\Delta \eta^2 + \Delta \phi^2)^{1/2}$, and ϕ is in radians. The isolation conditions within this cone are that: a) after removing the track with

240

the highest transverse momentum, the sum of the transverse momenta of all other tracks in the cone is less than 10% of the e.m. cluster transverse energy; and b) the cluster transverse energy is at least 90% of the total transverse energy (e.m. calorimeter plus hadronic calorimeter) inside the cone.

iv) Tight isolation. To obtain further rejection against two-jet events we require additional isolation inside a large cone with radius $\Delta R = 0.7$. The isolation conditions within this cone are a) after removing the track with the highest transverse momentum, the sum of the transverse momenta of all other tracks in the cone is less than 3 GeV/c; and b) the sum of the transverse energy depositions in the calorimeter, after excluding the e.m. cluster, is less than 3.2 GeV.

v)Electromagnetic shape. To reject hadronic showers faking an electron signature we require that a) the energy E_{had} deposited in the hadron calorimeter cell immediately behind the e.m. cluster, is either less than 2% of the e.m. cluster energy or less than 0.6 GeV (1.5 GeV) for clusters with pseudorapidity $|\eta| \leq 1.5$ ($|\eta| \geq 1.5$). We further require that b) the quality parameter χ_R is less than 60, where χ_R measures the goodness of fit to the electron hypothesis; χ_R has been developed from a study of electron test-beam data, making full use of the longitudinal profile of the shower measured in the four samplings of the electromagnetic calorimeter, and taking into account the electron energy and angle of incidence. The expected χ_R distribution for isolated electrons coming from $W^{\pm} \rightarrow e^{\pm} + \nu$ decays is shown in Fig. 1c.

vi) Neutrino emission. Finally, we require that the missing transverse energy in the event is in excess of 15 GeV after validation during which obvious instrumental problems are removed. If the electron cluster is within 15 of the vertical direction there must be no reconstructed jet in the event (with transverse energy in excess of 7.5 GeV in the calorimeters, or transverse momentum in excess of 5 GeV in the central detector) which is also going in the vertical direction (within 15°).

After these requirements have been imposed on the data, we are left with a sample of 173 events. One event has a very low electron-neutrino transverse mass, and a topology consistent with the hypothesis W $\rightarrow$ tb [12] (or possibly p$\bar{p} \rightarrow$ tt + X) with a subsequent semileptonic decay of the t-quark. We exclude this event from our sample by requiring that the transverse mass of the electron-neutrino system is in excess of 20 GeV/c .

We are left with a sample of 172 events for further analysis, of which 59 events come from collisions at $\sqrt{s}$= 546 GeV and 113 events come from collisions at $\sqrt{s}$= 630 GeV.

Next we estimate the background in the W sample. After our selection the distributions showing the electron quality and isolation for the sample of 172 W candidates are consistent with expectations for an essentially background-free sample of isolated electrons (Figs. 1a-d). We

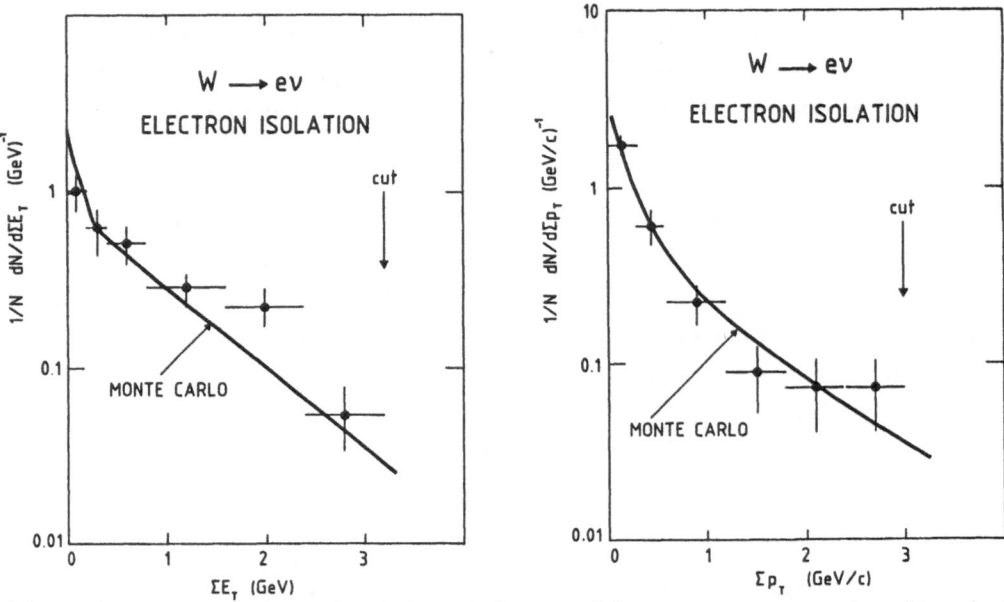

Figure 1. Electron quality and isolation: a) the sum of the transverse energy depositions in the calorimeter in a cone of $\Delta R = 0.7$ around the electron, after excluding the electron cluster. b) The sum of the transverse momenta of charged tracks in the central detector in a cone of $\Delta R = 0.7$ around the electron, after excluding the electron track

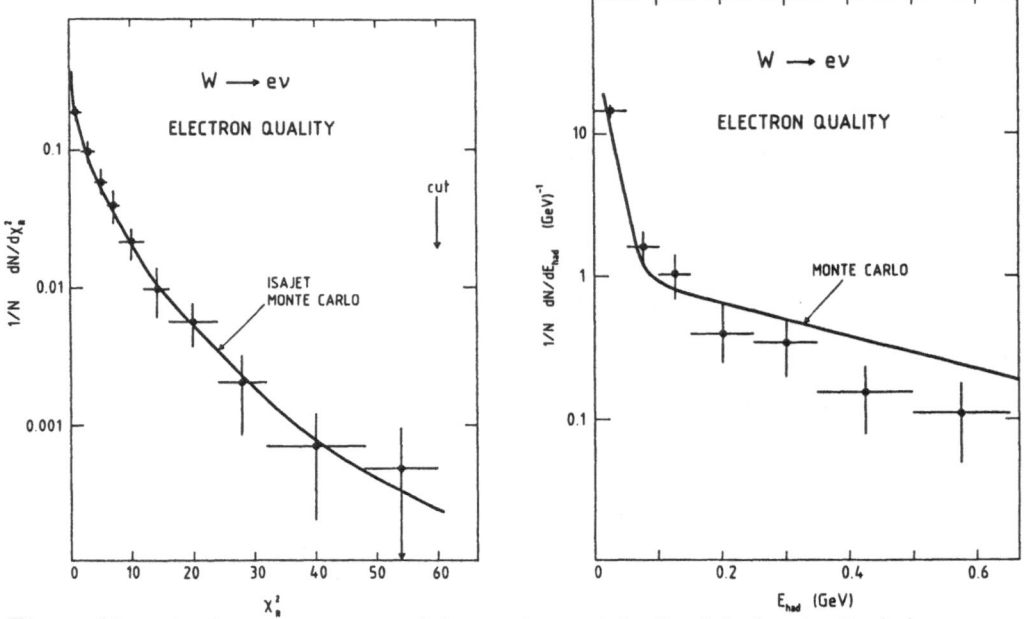

Figure 1(cont.): c) χ_R , a measure of the goodness of the fit of the longitudinal shower profile to the electron hypothesis; d) the energy deposition in the hadron calorimeter cells immediately behind the electron cluster

have estimated, in detail, the possible sources of background in our sample of $W^{\pm} \Rightarrow e^{\pm} + \nu$ decays.

   i) hadronic interactions. Hadronic jets can fake an isolated electron signature a) if they fragment in such a way that one energetic charged pion overlaps with one or more neutral pions, or b) if they contain a genuine energetic electron arising from heavy-flavour decays which overlaps with one or more neutrals. In both these cases we would expect a two-jet topology with little or no missing transverse energy in the event. To estimate the background from these sources, we drop the last requirement imposed in the selection, namely the presence of a missing transverse energy $\Delta E_M \geq 15$ GeV, and examine the missing transverse energy for the enlarged sample. For those events with $\Delta E_M \leq 15$ GeV the distribution is well described by the experimental resolution in the measurement of this quantity (Fig.2), and furthermore almost all these events have a two-jet topology.

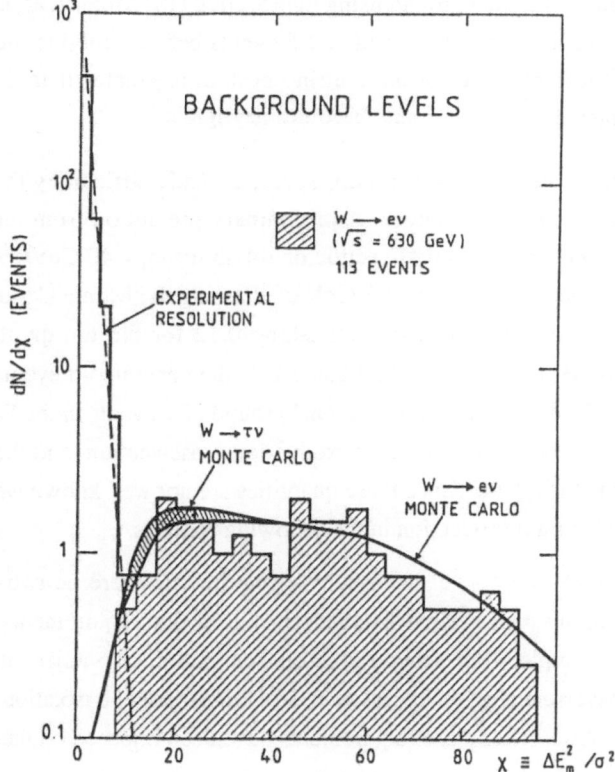

Figure 2. *Background from two jet-fluctuations. The missing energy squared $\Delta E_M{}^2$, divided by the experimental resolution is shown for the sample of 1287 events which come from the W selection procedure, without the requirement $\Delta E_M < 15$ GeV. The contribution of the 113 W candidates (cross-hatched) agrees very well with the predictions of the ISAJET Monte Carlo. On the other hand the events for which $\Delta E_M < 15$ GeV, is well described by the expected resolution for events with no true missing energy*

Extrapolating the resolution curve in the region $\Delta E_M \geq 15 \text{GeV}$, we estimate a background in our $W^\pm \Rightarrow e^\pm + \nu$ sample of 3.4 $\pm$ 1.8 events at $\sqrt{s}$ = 540GeV and 1.9 $\pm$ 0.6 events at $\sqrt{s}$ = 630 GeV. As a final check, each of the 172 W events has been visually inspected on an interactive graphics display. Only 7 events have a hadronic jet coplanar with the electron and the beams. We retain these events in the data sample since they are consistent with events in which the jet arises from initial-state bremsstrahlung off the incoming partons and is only accidentally back-to-back with the electron in the transverse plane.

ii) W $\Rightarrow \tau\nu$. We have evaluated this background using the ISAJET [13] Monte Carlo together with a full simulation of the UA1 detector. We estimate a background of (11.8$\pm$ 0.6) events from this source, of which 2.7 events are associated with the decay $\tau \Rightarrow \pi^\pm\pi^0\nu$ and 9.1 events are associated with the decay $\tau \Rightarrow e^\pm\nu\nu$.

iii)W $\Rightarrow$ tb. We have evaluated this background using the ISAJET [13] Monte Carlo, taking m_t = 40 GeV/c^2 [12]. We estimate a background of 1.5 events before applying the requirement that the transverse mass of the electron-neutrino system is greater than 20 GeV/c^2. After this cut has been applied, this background becomes negligible.

iv)tt production. Evaluation of background from this source is made difficult by the lack of knowledge of the tt production cross section. A preliminary prediction from the EUROJET Monte Carlo[12] gives a production cross-section of 1.4 nb for m_t = 40 GeV/c^2 , and a transverse momentum cut on the t-quarks of $p_t \geq 3$ GeV/c. Passing the Monte Carlo generated events through the W selection procedure and taking 0.12 for the top quark semileptonic decay branching ratio B($t \Rightarrow be\nu$) , we find that 1% of the generated tt events survive the W selection cuts. This leads us to an estimated background of 7 events in the W sample. However, this estimate is sensitive to m_t, the tt production cross-section, and the semileptonic branching ratio for the top quark. Since these quantities are not well known we make no correction for background from tt production in the following analysis.

We consider next the sample of $Z^0 \Rightarrow e^+ + e^-$ decays. We require that there be two isolated e.m. clusters, at least one of which satisfies the isolated electron selection requirements used in defining the $W^\pm \Rightarrow e^\pm + \nu$ sample. The second e.m. cluster (one or two adjacent e.m. calorimeter cells with cluster transverse energy in excess of 8 GeV) must pass the isolation and quality cuts listed above for the W case. After these requirements have been imposed on the data, we are left with a sample of 39 events having two isolated e.m. clusters, at least one of which is a good electron candidate with transverse energy in excess of 15 GeV. The two-cluster invariant mass distribution for this sample of events is shown in Fig. 3.

There are 21 events with a two-cluster mass m < 50 GeV/c^2,no events in the interval 50 GeV/c^2 < m < 80 GeV/c^2 , and a cluster of 18 events with m > 80 GeV/c^2 , peaking at a mass of ~ 93 GeV/c^2 . The shape of the low-mass part of the distribution is consistent with our expectation for

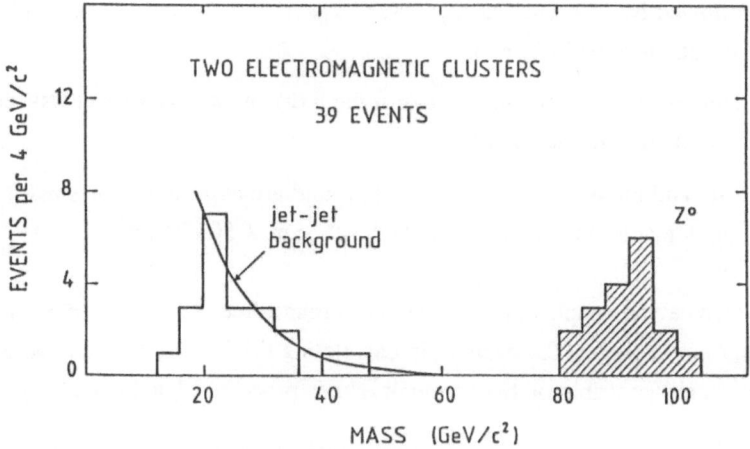

Figure 3. Invariant mass distribution of the pairs of electromagnetic clusters passing the cuts described in the text. The curve shows the expected shape for events arising from jet-jet fluctuations. The high mass peak is clearly separated from the background

background coming predominantly from two-jet events. This background shape has been evaluated from a sample of isolated candidates which have no track associated with the isolated e.m. cluster, but otherwise satisfy the isolated electron selection criteria. We then require the presence of a second e.m. cluster in the event satisfying the second-cluster isolation and quality cuts used in the Z selection. The resulting shape of the two-cluster mass distribution gives a good description of the low-mass part of Fig. 3, although we do not exclude a contribution from processes producing genuine electrons (e.g. $Z \rightarrow \tau\tau$, $Z \rightarrow$ tt, bb production, Drell-Yan production, etc). To exclude the low-mass background, we require $m > 50$ GeV/c^2.

We are left with 18 events, which have been examined by physicists on a high-resolution graphics display. In 16 events the CD information is consistent with the presence of a high transverse momentum (> 7 GeV/c) positively charged track associated with one electromagnetic cluster and a high transverse momentum negatively charged track associated with the other electromagnetic cluster. In the remaining two events the CD information is inconsistent with the presence of a track with transverse momentum in excess of 7 GeV/c associated with one of the two clusters. Noting that the expected background from $Z \rightarrow \tau\tau$ decays producing two electromagnetic clusters with $m > 80$ GeV/c^2 is 0.30 ± 0.14 events (evaluated using the ISAJET [13] Monte Carlo together with a full simulation of the UA1 detector), we exclude these two events. This leaves 16 $Z^0 \rightarrow e^+ + e^-$ candidates for further analysis, four of which are from the 1983 data-taking period and have already been published [3].

We have estimated, in detail, the possible sources of background contributing to our sample of $Z^0 \rightarrow e^+ + e^-$ decays. Given the large gap in the mass distribution between the 16 Z candidates and the low-mass events, only W and Z related backgrounds seem to be relevant. Using the ISAJET [13] Monte Carlo together with a full simulation of the UA1 detector, we have evaluated the following background processes:

i) $Z^0 \Rightarrow \tau\tau$, followed by $\tau \Rightarrow e\nu\nu$ decays. We expect ~ 0.1 events in the region populated by our Z events ($m > 80$ GeV/c^2).

ii) $Z^0 \Rightarrow$ bb, with subsequent semileptonic decays of the b (b) quark. We expect less than 8×10^{-2} events (90% c.l.) with $m > 80$ GeV/c^2 .

iii) $Z^0 \Rightarrow$ tt , W $\Rightarrow$ tb, and p$\bar{p} \Rightarrow$ tt + X. These background are expected to be small ($\leq$ 0.2 events) for top-quark masses in the interval 30 GeV/c^2 < m < 50 GeV/c^2 .

We conclude that we have a sample of essentially background-free $Z^0 \Rightarrow e^+ + e^-$ decays, the total estimated background being ≤ 0.4 events. In calculating the Z cross-section, we will neglect the potential (but small) contribution from t-quark related processes, leaving a negligible background (≤ 0.2 events).

4.-Production properties of W particles. The production properties of the Intermediate Vector Bosons can best be studied using W events, since they are statistically ten times more abundant than the Z events. The integrated luminosity of the experiment is 136 nb^{-1} at $\sqrt{s} = 546$ GeV and 263 nb^{-1} at $\sqrt{s} = 630$ GeV, known to about $\pm 15\%$ uncertainty. We have used two techniques to estimate the efficiency of our selection requirements : i) the ISAJET [13] Monte Carlo together with a full simulation of the UA1 detector, and ii) a randomized W-decay technique in which the 172 real W events are used, but the electron is replaced by a generated electron coming from a random (V-A) decay of the produced W. The two methods give consistent results, which are shown in Table 1.

We find that the efficiency of our trigger and selection requirements is 0.69 ± 0.03 for the $\sqrt{s} = 546$ GeV data sample , 0.62 ± 0.03 for the $\sqrt{s} = 630$ GeV sample. The cross-section is then :

$$(\sigma . B)_W = 0.55 \pm 0.08 \ (\pm 0.09) \text{nb}$$

at $\sqrt{s} = 546$ GeV, where the last error takes into account the systematic errors. This cross-section is consistent with the previously published value of 0.53 nb [1] The small change is due to an improved and extended selection, and to a more thorough evaluation of the selection efficiencies. Our result is in good agreement with the corresponding result from the UA2 experiment [5] $0.53 \pm 0.10 \ (\pm 0.10)$ nb, and is not far from the prediction of Altarelli et al. of $0.38 \ _{-0.05}^{+0.12}$ nb, where we take the branching ratio B = 0.089.The corresponding experimental result for the 1984 data at $\sqrt{s} = 630$ GeV is

$$(\sigma . B)_W = 0.63 \pm 0.05 \ (\pm 0.09) \text{nb}.$$

This is in agreement with the theoretical expectation [14] of $0.47 \ _{-0.08}^{+0.14}$ nb. We note that the 15% systematic uncertainty on these results disappears in the ratio of the two cross-sections. We obtain the result

246

Table 1

Selection efficiency. The efficiency for each cut in the selection leading to the sample of 172 $W^{\pm} \rightarrow e^{\pm} \nu_e$ events is shown, evaluated by the two methods described in the text. The efficiency given for each requirement is for those events passing all preceding requirements.

Cut	Efficiency			
	$\sqrt{s}$ = 546 GeV		$\sqrt{s}$ = 630 GeV	
	Events with random decay	ISAJET	Events with random decay	ISAJET
Electron E_T > 15 GeV	0.86	0.87	0.85	0.84
i) Cluster validation	0.97	0.96	0.96	0.95
ii) Associated track requirement	0.97	0.98	0.95	0.99
iii) Loose isolation (in cone ΔR = 0.4)	0.98	0.99	0.98	0.98
iv) Tight isolation (in cone ΔR = 0.7)	0.98	0.98	0.93	0.98
v) Electromagnetic shape:				
a) hadronic energy	0.95	0.95	0.95	0.95
b) χ^2_R	0.99	0.98	0.99	0.99
vi) Neutrino emission	0.98	0.97	0.97	0.96
Trigger efficiency	0.96	0.96	0.96	0.96
Overall efficiency	0.69 ± 0.03	0.68 ± 0.03	0.62 ± 0.03	0.66 ± 0.03

$$\sigma_W(\sqrt{s} = 630 \text{ GeV}) / \sigma_W (\sqrt{s} = 546 \text{ GeV}) = 1.15 \pm 0.17,$$

in agreement with the theoretical expectation of 1.24.

The longitudinal momentum distribution of the W boson is expected to reflect the structure functions of the incoming annihilation partons, predominantly u-quarks and d-quarks. Unfortunately we do not measure the longitudinal momentum of the W directly since we do not measure the longitudinal component of the neutrino momentum. We can overcome this difficulty by imposing the mass of the W on the electron-neutrino system. This will yield two solutions for the longitudinal component of the neutrino momentum, one corresponding to the neutrino being emitted backwards. Hence we have two solutions for x_W, the Feynman x for the W. In practice, in one third of the events one of these two solutions for x_W is trivially unphysical (x_W > 1), and in another one-third the ambiguity is resolved after consideration of energy and momentum conservation in the whole event. In the cases where the ambiguity in x_W is resolved, it tends to be nearly always the lowest of the two x_W solutions which is chosen. In the following analysis the lowest of the two solutions will be used for all the events.

Taking 84 GeV/c^2 for the mass of the W, the resulting x_W distribution is shown in Fig. 4a for the $\sqrt{s}$ = 546 GeV and $\sqrt{s}$ = 630 GeV data samples separately. There is some indication of the expected softening of the W longitudinal momentum distribution with increasing $\sqrt{s}$. The data are in reasonable agreement with the expectation resulting from the structure functions of Eichten et al. [15] with $\wedge$ = 0.2 GeV. Using the well-known relations

$$x_p x_{\bar{p}} = m^2_W/s \qquad \text{(energy conservation)}$$

and

$$x_W = |x_p - x_{\bar{p}}| \qquad \text{(momentum conservation)},$$

we can determine the parton distributions in the proton and antiproton sampled in W production. The results are shown in Fig. 4b. The proton- and antiproton-parton distributions are consistent with each other, and are well described by the structure functions of Eichten et al. For those events in which the charge of the electron, and hence the W, is well determined, we can identify the proton (antiproton) parton with a u-(d) quark for a W$^+$, or a d- (u) quark for a W$^-$. There are 118 events in which the charge of the electron track is determined to better than 2 standard deviations (from infinite momentum). The resulting u- and d-quark distributions for these events are shown in Figs. 4c and 4d, respectively. Once again there is agreement with the predictions of Eichten et al.

The W transverse momentum $p_T^{(w)}$ is obtained by adding the electron and neutrino transverse momentum vectors. The resulting distributions for $p_T^{(w)}$ is shown in Fig. 5. The distribution peaks at a value $p_T^{(w)} \approx 4$ GeV/c (primarily reflecting the experimental resolution of the measurement of the missing transverse energy in the event), and has a long tail extending to $p_T^{(w)} \approx 40$ GeV/c . The electron isolation criteria that we have used in defining the W$^\pm \rightarrow$ e$^\pm \nu_e$ event sample introduce a bias again the more active [highest $p_T^{(w)}$] events. We have studied this bias using the randomized W decay technique described in Section 3. The average loss of events due to the isolation criteria rises from 4% for the lowest $p_T^{(w)}$ events to about 14% for the higest $p_T^{(w)}$ events. The effect is therefore small, and we can safely neglect it. The shape of the measured W transverse momentum distribution is in reasonable agreement with the expectations of the QCD-improved Drell-Yan model. As an example of the QCD predictions in Fig. 4, we show the curve from Altarelli et al. [14] using the structure functions of GHR [16] with $\wedge$ = 0.4 GeV. This gives an excellent description of the data.

5.- Associated W and jet(s) Production. The QCD-improved Drell-Yan mechanism predicts that the highest transverse momentum W's are produced in association with one ore more gluons radiated off the incoming annihilating quarks. If these gluons are sufficiently energetic and

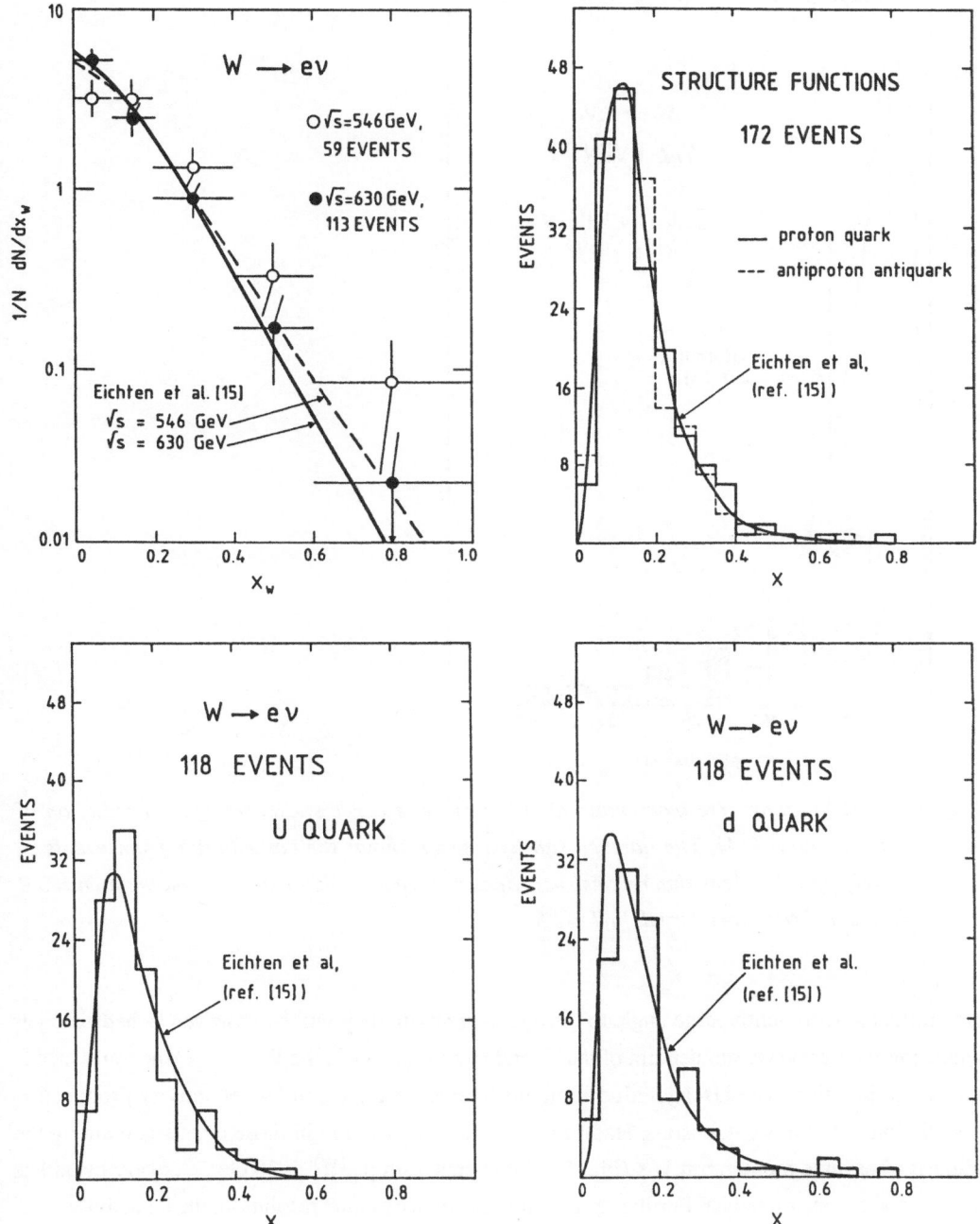

Figure 4. *Longitudinal momentum distributions (Feynman-x) for: a) W ; b) proton and antiproton partons ; c) u-quarks sampled by W production ; d) d-quarks, sampled by W production. Data are compared with the predictions of the structure functions from Eichten et al.,based on neutrino experiment, corrected for q^2 effects, experimental biases and detection efficiencies*

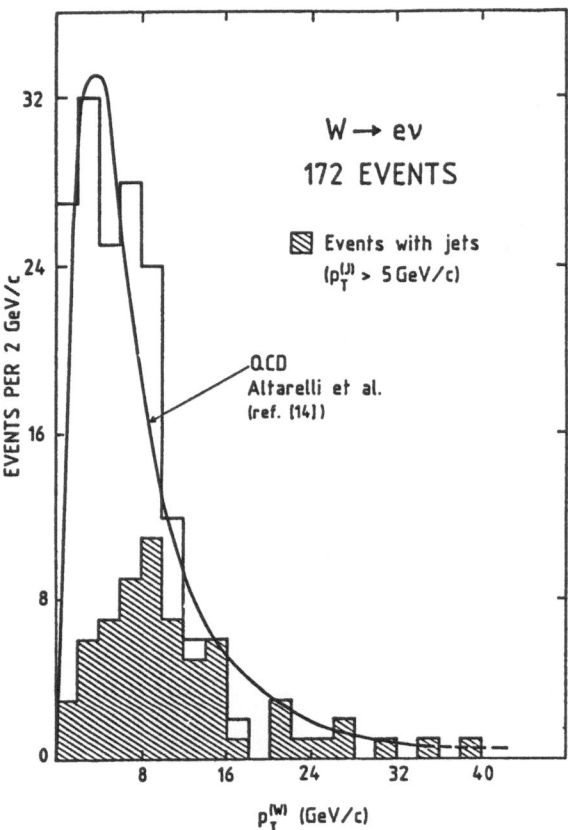

*Figure 5.The W transverse momentum distribution. The curve shows the QCD prediction of
Altarelli et al [14]. The hatched sub-hystogram shows the contribution from events in
which the UA1 algorithm reconstructs one or more jets with transverse momentum > 5.0
GeV/c and rapidity interval | η | < 2.5*

are emitted at sufficiently large angle to the beam direction, they will be observed as hadronic jets
balancing the transverse momentum of the W and recoiling against the W in the plane transverse to
the beam direction. The UA1 jet algorithm has been used to search for jet activity produced in
association with the weak bosons. Hadronic jets are indeed found in those events containing the
highest transverse momentum W's (Fig. 5). For events with $p_T^{(w)}$ > 20 GeV/c, every event has
at least one reconstructed jet. Furthermore, within the experimental resolution, the jet activity
observed in W events does indeed balance $p_T^{(w)}$ (Fig. 6). The resolution curve shown in Fig. 6
is rougly a Gaussian with a width of ≈ 0.25 $p_T^{(w)}$; it has been obtained using the ISAJET [13]
Monte Carlo together with a full simulation of the UA1 detector. A detailed study of the properties
of the jets reconstructed in W events shows that they are fully consistent with the expected
properties of jets arising from initial-state gluon bremsstrahlung.

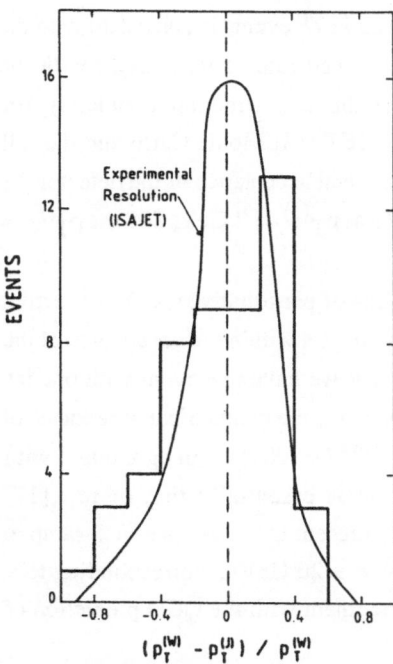

Figure 6. The W-jet momentum balance in the transverse plane. The unbalance between the W and the jet is shown as a fraction of the W transverse momentum for events in which $p_T{}^{(w)} > 5GeV/c$. The continuous curve is the prediction of ISAJET for the expected experimental resolution on this quantity

In 38% of the W sample, one or more hadronic jets with transverse momentum $p_T{}^{(J)}$ in excess of 5 GeV/c are observed. The number of observed events (Table 2) decreases with increasing jet multiplicity by a factor of roughly 0.3 per additional jet. Defining $R^{(w)}$[17]

$$R^{(w)} = \sigma[W + jet(s)] / \sigma[w],$$

where $\sigma[w]$ is the cross-section for producing a W with no observed jets, we find $R_w = 0.61 \pm 0.10$. A similar preliminary analysis for Z^0 events gives $R_z = 0.7 \pm 0.3$ and the ratio of these ratios

Table 2

Rate of occurrence of jets in W events

$\sqrt{s}$ (GeV)	Number of events						Total number of jets
	Total	≥ 1 jet	1 jet	2 jet	3 jet	4 jet	
546	59	17	15	1	1	0	20
630	113	48	35	7	4	2	69
TOTAL	172	65	50	8	5	2	89

$R = R_Z / R_W = 1.1 \pm 0.5$. Clearly the rate of jet activity observed in Z^0 events is consistent with the rate observed in W events. Before we can compare the observed rate of jet activity with the expectations from perturbative QCD, we need to know the experimental efficiency for reconstructing low transverse momentum jets. Using the ISAJET [13] Monte Carlo and the full simulation of the UA1 detector, we find that within the geometrical acceptance of the detector the reconstruction efficiency rises from zero at $p_T^{(J)} = 0$ to 20% at $p_T^{(J)} = 7$ GeV, 80% at $p_T^{(J)} = 13$ GeV/c and is 100% for $p_T^{(J)} \geq 20$ GeV/c.

To make a quantitative comparison with the predictions of perturbative QCD, we restrict ourselves to the region in which the reconstruction efficiency for jets within the acceptance of the detector is 100%, namely $p_T^{(J)} > 20$ GeV/c. At $\sqrt{s} = 546$ GeV we have one event with one jet passing this cut. After correcting the jet reconstruction efficiency for the geometrical acceptance of the detector (0.9) the resulting cross-section is $0.012\,_{-0.007}^{+0.014}$ (± 0.002)nb, in agreement with the QCD prediction of 0.010 ± 0.001 nb. This prediction is essentially that of ref. [17] recalculated to take into account the cuts used in defining the present UA1 $W^\pm \to e^\pm \nu_e$ sample. At $\sqrt{s} = 630$ GeV we have six events with one jet with $p_T^{(J)} > 20$ GeV/c, corresponding to a cross-section of 0.41 ± 0.015 (± 0.0006)nb, once again in agreement with the QCD prediction of 0.016 ± 0.002nb.

The jet transverse momentum distribution is shown in Fig. 7. This distribution essentially reflects the $p_T^{(W)}$ distribution and is well described by the expectation from QCD perturbation

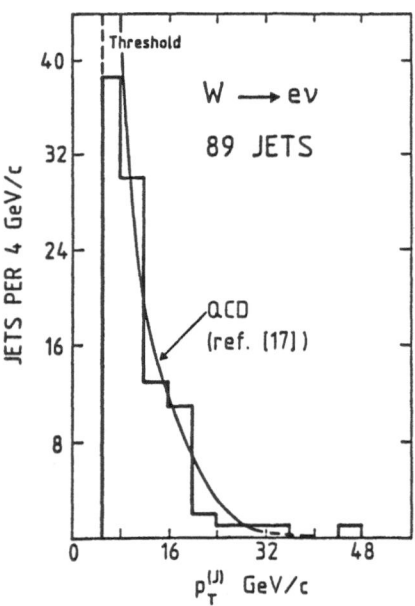

Figure 7. Jet transverse momentum distribution for all jets observed in association with the W. the curve is the QCD prediction , normalized to the tail of the distribution, namely to the region in which we expect good reconstruction efficiency for jets

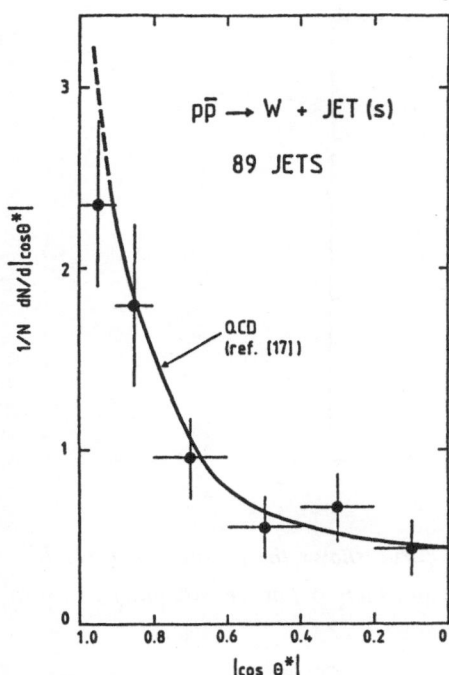

Figure 8. The angular distribution for jets reconstructed from W events. The distribution in cosθ* is shown, θ* being the angle between the jet and the average beam direction in the rest frame of the W and the jet(s). The curve shows the QCD prediction

theory for jets arising from initial-state bremsstrahlung. Furthermore, as expected, the angular distribution of these jets is strongly peaked in the beam directions. This can be seen in Fig. 8 where the distribution of cos θ* is shown, θ* being the angle between the jet and the average beam direction in the rest frame of the W and the jet(s). In the region in which the experimental acceptance is reasonably constant (|cos θ*|< 0.95) the shape of the angular distribution is well described by the QCD expectation from bremsstrahlung jets, which is basically $(1- |\cos \theta^*|)^{-1}$. Finally, we examine the invariant mass of the (W + 1jet)-system for those events in which one jet has been reconstructed by the UA1 jet algorithm (Fig. 9). The shape of this mass distribution is a little broader than the expectation from ISAJET [13]. It is, however, well described by a simple Monte Carlo in which the observed W four-vectors are randomly associated with the four vectors from our sample of jets, suggesting that the shape of the (W + 1jet) mass plot is controlled more by the proton structure functions than by the QCD matrix element.

We conclude from this mass distribution that there is no evidence for the production of a massive-state X which subsequently decays into a W plus one jet. For a mass $m_x > 180$ GeV/c^2, with X subsequently decaying into a W and a single hadronic jet, we place an upper limit on the production cross-section for X of $\sigma \cdot B_{(x \to W+jet)} / \sigma_W < 0.01$ at 90% confidence level.

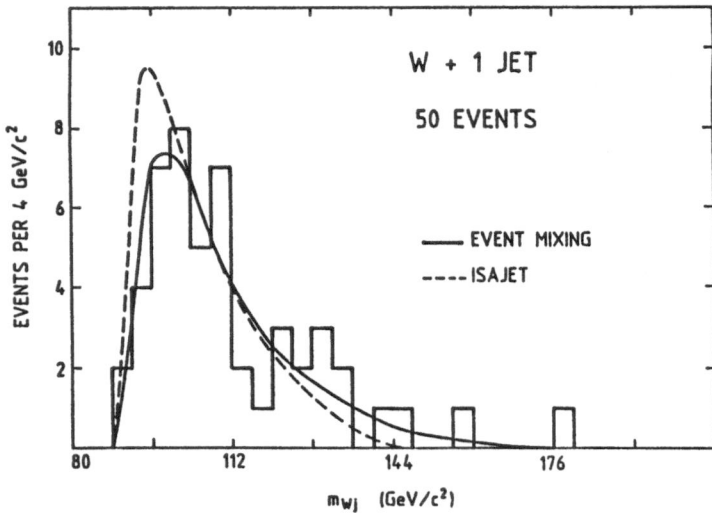

Figure 9. The (W + jet)-mass distribution . The curve shows the prediction of ISAJET (dashed curve) and of event mixing (solid curve) ,in which W four vectors and jet vectors from __different__ events are randomly associated

6.-Comparing W and Z productions: number of neutrino species. The production cross section for W production has been already given above . To obtain the corresponding results for Z production, we have evaluated the efficiency of our selection requirements by a Monte Carlo technique in which the underlying events from the sample of 172 $W^{\pm} \Rightarrow e^{\pm} + \nu$ decays have been used. The W decays in these events have been replaced by randomized $Z^{0} \Rightarrow e^{+} + e^{-}$ decays, with the Z retaining the same momentum vector as that of the replaced W. We find that the efficiency of our trigger and selection requirement is 0.69 ± 0.03 for the $\sqrt{s}$ = 546 GeV data sample, and 0.61 ± 0.03 for the $\sqrt{s}$= 630 GeV data sample. The cross-section is then

$$(\sigma \cdot B)_{Z} = 42^{+25}_{-18}(\pm 6) \ pb$$

at $\sqrt{s}$ = 546 GeV. This cross-section is consistent with the corresponding result from the UA2 experiment [5], 110 ± 40 (± 20)pb, and with our published value for the decay $Z \Rightarrow \mu^{+}\mu^{-}$ [4], 100 ± 50 (± 15) pb. Taking B ($Z \Rightarrow e^{+}e^{-}$) = 0.032, the measured cross-section is also in excellent agreement with the prediction of Altarelli et al. [15] of 41^{+13}_{-7} pb.

The corresponding experimental result for the 1984 data at $\sqrt{s}$ = 630 GeV is

$$(\sigma \cdot B)_{Z} = 74^{+23}_{-20}(\pm 11) \ pb \ .$$

This is in agreement with the theoretical expectation of 51^{+16}_{-8} pb.

To compare the W and Z production cross-section, we define

$$R \equiv (\sigma \cdot B)_{W} / (\sigma \cdot B)_{Z}$$

Combining results from the two energies,

$$R = 9.6^{+3.0}_{-2.1} \cdot$$

This result is in excellent agreement with the Standard Model QCD expectation [18] for R of $9.2 \pm$ 0.6, which is insensitive to $\sqrt{s}$ for the centre-of-mass energy range considered here. The production properties for the W [19] have been shown to be in good agreement with the QCD predictions. It is therefore reasonable to use the QCD prediction[14] for the ratio of the production cross-section $\sigma_W/\sigma_Z = 3.3 \pm 0.2$,to extract a value for the ratio of branching ratios

$$R_B \equiv B(W^{\pm} \Rightarrow e^{\pm} + \nu)/B(Z^0 \Rightarrow e^+ + e^-) = 2.9^{+0.9}_{-0.6}(\pm 0.1)$$

where the first error reflects the statistical error on the number of W and Z events, and the second reflects the theoretical uncertainty associated with uncalculated higher-order diagrams and the choice of Λ_{QCD} . In the Standard Model [8], with a top-quark mass $m_t = 40$ GeV/c^2, and three generations, the expected value for R_B [18] is 2.72, in excellent agreement with the measured value. In a modified Standard Model [8] in which there are further (> 3) generations containing additional light neutrino types, the expected values for R and R_B would be different. If there were no associated new charged leptons or quarks lighter than the W, we expect R to increase with the number of generations owing to additional $Z^0 \Rightarrow \nu\nu$ decays.

The relationship between R and the number of light neutrino types N_ν has been calculated by Deshpande et al. [18]. Using their result, and the experimental limit $R \leq 13.5$ at 90% c.l.and we obtain a limit on the number of light neutrino types, $N \leq 10$ at 90% c.l.

7.- Determination of the W mass. To measure the W mass we restrict ourselves to the subsample of $W^{\pm} \Rightarrow e^{\pm} + \nu$ decays in which both the electron and the neutrino are more than 15^0 away from the vertical direction in the transverse plane. This ensures the best accuracy in the electron and neutrino transverse energy determinations, and the subsequent mass measurement. We are left with a sample of 148 W decays for which we estimate a background of 15.6 ± 3.6 events. The electron and neutrino transverse-energy distributions are shown in Figs. 10a and 10b, respectively, for our sample of well-measured $W^{\pm} \Rightarrow e^{\pm} + \nu$ decays. The shaded parts of these distributions show the expected contributions from the various background sources. We note that i) the background events tend to have low electron and neutrino transverse energies, and ii) the background-subtracted distributions have the expected Jacobian shape. The exact shapes of the electron and neutrino transverse energy distributions depend on the underlying W mass and decay width, the experimental selection biases and transverse energy resolutions, and the underlying W transverse-momentum distribution. To determine the W mass from our sample of 148 decays, we prefer to fit a distribution which is not sensitive to the underlying W transverse-momentum distribution. We use the electron-neutrino transverse-mass distribution,

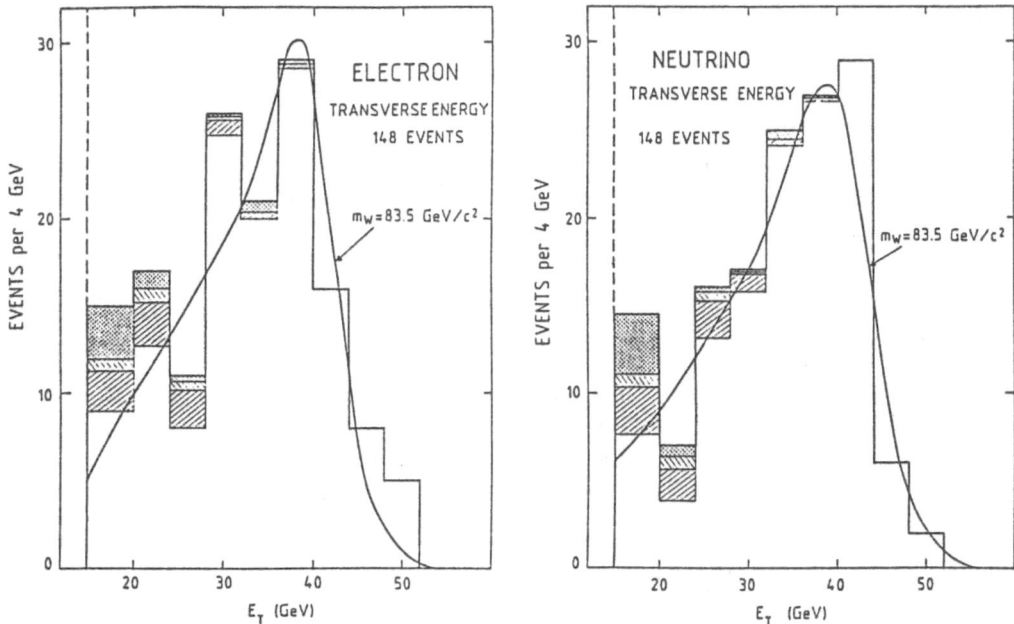

Figure 10. *Lepton (electrons and missing energy, i.e. neutrino) transverse energy distribution for well measured* $W^{\pm} \Rightarrow e^{\pm} + \nu$ *events. The background contributions are: jet-jet fluctuations—cross hatched ;* $W^{\pm} \Rightarrow \tau^{\pm} + \nu$ *followed by* $\tau \Rightarrow$ *hadrons —top left to bottom right hatching ; and* $W^{\pm} \Rightarrow \tau^{\pm} + \nu$ *followed by* $\tau \Rightarrow e\nu\nu$ *— top right to bottom left hatching. The curve shows predictions for background subtracted distributions and mass of W equals to 83.5 GeV/c^2*

$$m_T \equiv [\ 2\ E^e_T\ E^\nu_T\ (\ 1 - \cos\phi)\ ]^{1/2}$$

where E^e_T and E^ν_T are the lepton transverse energies, and ϕ is the angle between the electron and neutrino momentum vectors in the plane transverse to the beam directions. The electron-neutrino transverse-mass distribution is shown in Fig. 11a for the sample of 148 $W^{\pm} \Rightarrow e^{\pm} + \nu$ decays. The expected number of (shaded) background events decreases with increasing m_T and is negligible for $m_T \leq 60$ GeV/c^2. We define the enhanced transverse-mass distribution (Fig. 11b) as the m_T distribution for those events in which both the electron and the neutrino have energies in excess of 30 GeV. We are left with a background -free ($\leq$ 1event) sample of 86 $W^{\pm} \Rightarrow e^{\pm} + \nu$ decays.

Extensive Monte Carlo studies have shown that a good first approximation to the underlying W mass can be obtained by taking the point at which the upper edge of the enhanced transverse-mass distribution has fallen to half of the peak value. Thus, on the basis of Fig. 11b, we see that our data correspond to a W mass of ~83 GeV/c^2 . A maximum-likelihood fit to the enhanced transverse-mass distribution has been performed to find the most probable value for the mass m_W and decay width Γ_W of the W.

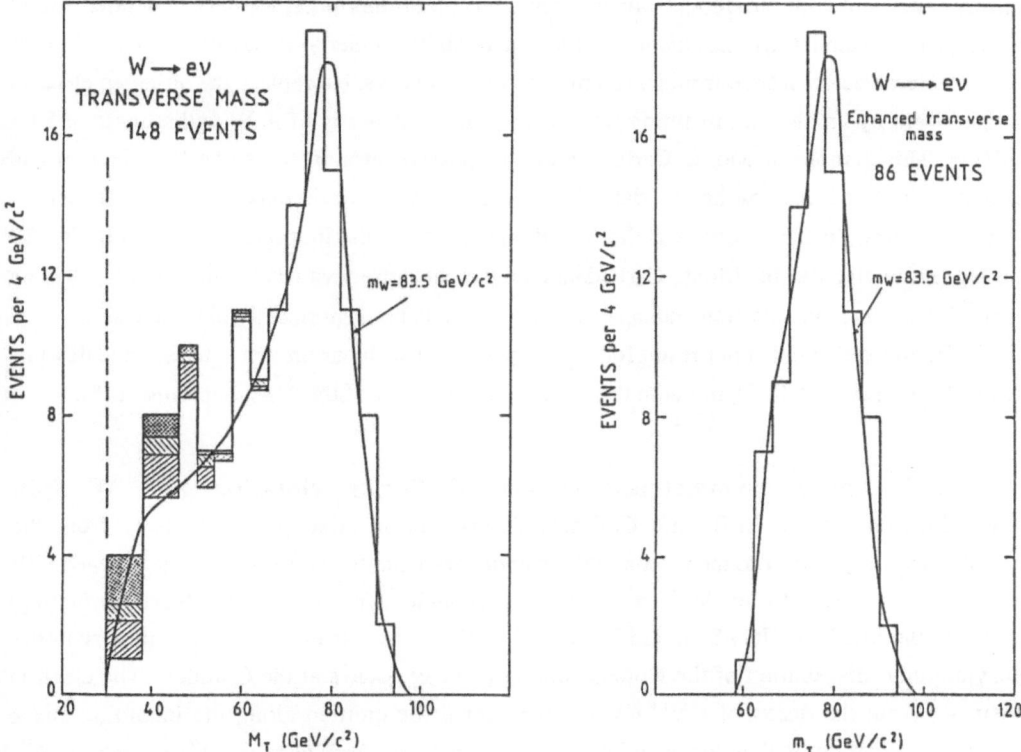

Figure 11. The electron-neutrino transverse mass distribution for the sample of 148 well measured $W^\pm \Rightarrow e^\pm + \nu$ decays and the subsample of these decays in which both electron and neutrino have transverse energy in excess of 30 GeV. This is the so-called enhanced transverse mass distribution in which events decaying at large angles are selected. Then the transverse mass approaches the true invariant mass

The result is

$$m_W = 83.5 \pm 1.05 \,(\pm 2.7\,) \text{ GeV/c}^2 \quad \Gamma_W = 2.7^{+1.4}_{-1.5} \text{ GeV/c}^2$$

where the second error on the measurement of m_W comes from the systematic uncertainty in the energy scale of the experiment. The error quoted in Γ_W is the statistical error on the result. If we now include the systematic error (coming from the precision with which we know the experimental resolution function) we are only able to quote an upper limit , $\Gamma_W \leq 6.5 \text{ GeV/c}^2$, again at 90% c.l. .

We note that our result for m_W is rather independent of Γ_W . As a check of our fitting procedure we have used a second independent program which, for each event, makes a full kinematical fit to the electron four-vector, the transverse components of the neutrino momentum, and the energy flow in the underlying event. After making reasonable assumptions for the longitudinal-and transverse-momentum distributions of the W's, we obtain results which are

257

entirely consistent with our quoted values for m and Γ_W. Furthermore, we have generated 15000 W decays by a Monte Carlo technique in which we retain the underlying events and reconstructed W momentum vectors from our initial sample of 172 W decays, but replace the observed electron in each event by an electron resulting from a random V - A decay of a W with a mass of 83.5 GeV/c^2. This sample of Monte Carlo events has passed through the same W selection and reconstruction procedure as the real data. The resulting Monte Carlo expectations for the lepton transverse-energy distributions, and the m and enhanced-m distributions, are shown in Fig. 10 and Fig. 11. In all cases the Monte Carlo data recovers the correct values for m_W and Γ_W, giving us confidence that our selection and analysis procedures do not significantly bias the results.

Finally we note that our result for m_W is consistent with our previous measurements [m_W = (80.9±1.5) GeV/c^2] [1,2] and with the corresponding result of the UA2 experiment [5].

8.-W Decay asymmetries: effects of Parity violation and W Spin determination At the CERN pp Collider, W production arises predominantly from the annihilation of a quark from the proton with an antiquark from the antiproton . The charged IVB is expected to couple to left-handed, but not right-handed, fermions. This (parity-violating) feature of the SU(2) ⊗ U(1) Standard Model [8] therefore results in a charge asymmetry in the decay angular distribution of the leptons coming from W decays at the Collider. The electron (positron) from the decay of a W^+ (W^-) will prefer to be emitted along the incoming quark (antiquark) direction, with a decay angular distribution of the form $(1 + \cos \theta^*)^2$, where θ^* is the emission angle of the electron (positron) with respect to the proton (antiproton) direction in the W centre-of-mass frame. To study this decay asymmetry, we restrict ourselves to the subsample of well-measured $W^\pm \rightarrow e^\pm + \nu$ decays for which the sign of the charge of the electron (positron) is well determined by the measurement of the curvature of the CD track in the UA1 magnetic field. Requiring that the track sagitta is greater than 2 st. dev. from zero, we are left with 103 $W^\pm \rightarrow e^\pm + \nu$ decays, of which 44 are W^+ decays and 59 W^- decays. Comparing the number of observed W^+ and W^- events we find

$$\sigma \cdot B_{W^+} / \sigma \cdot B_{W^-} = 0.75 \pm .15$$

consistent (within 2st. dev.) with equal W^+ and W^- $\sigma \cdot B$'s In 73% of our sample of W^+ and W^- events the W centre-of-mass frame is well defined, i.e. there is no kinematical ambiguity [19] in reconstructing the longitudinal momentum of the neutrino once the W mass has been imposed on the (eν) system. The decay angular distribution for this subsample of 75 $W^\pm \rightarrow e^\pm + \nu$ decays is shown in Fig. 12.

The data have been corrected for the experimental resolution, geometrical acceptance, selection efficiencies, reconstruction biases, and for the expected background contribution from jet-jet fluctuations and W decays (7.4±1.2 events). The resulting angular distribution shows a striking asymmetry which is consistent with the Standard Model [8] (V - A) expectation of $(1 + \cos \theta^*)^2$.

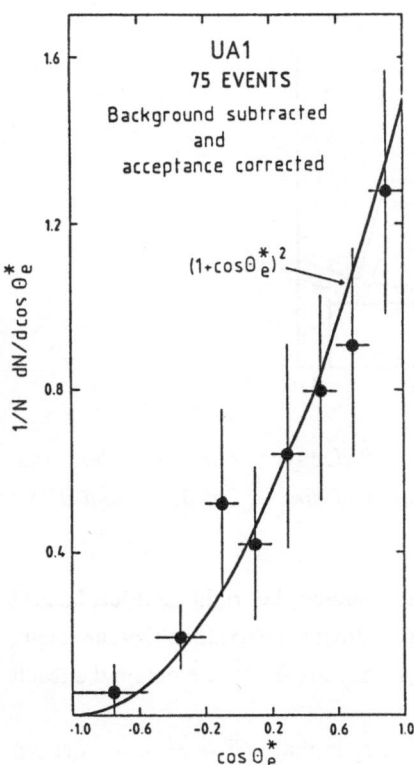

Figure12. *The angular distribution of the emission angle θ^* of the electron (positron) respect to the proton (antiproton) direction in the rest frame of the W. Only those events for which the lepton charge and the decay kinematics are well determined have been used. The continuous curve is the (V - A) expectation of $(1 + \cos \theta^*)^2$*

Defining the asymmetry parameter:

$$A_W \equiv (N^+ - N^-)/ (N^+ + N^-)$$

where N^+ (N^-) are the number of events in the positive (negative) half of the $\cos \theta^*$ plot, we obtain $A_W = 0.77 \pm 0.04$, in agreement with the pure (V - A) value of 0.75. It has been shown [20] that for a particle of arbitrary spin J, one expects

$$< \cos \theta^* > = < \lambda > < \mu >/ J(J + 1)$$

where $< \lambda >$ and $< \mu >$ are, respectively, the global helicity of the production system (ud) and of the decay system (eν). For (V -A) one then has $< \lambda > = < \mu > = -1$ leading to a maximal value $< \cos \theta^* > = 0.5$. For J = 0 one obviously expects $< \cos \theta^* > = 0$, and for any other spin value $J \geq 2$, $< \cos \theta^* > \leq 1/6$. Experimentally, we find $< \cos \theta^* > = 0.43 \pm 0.07$, which supports both the J = 1 assignment and maximal helicity states at production and decay.

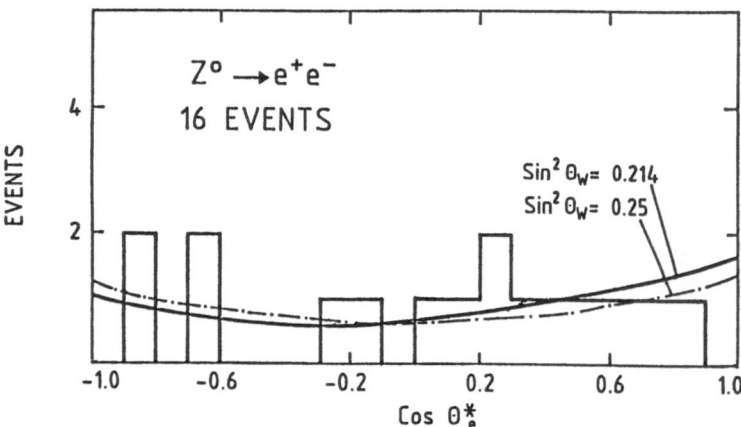

Figure 13. The same as for Figure 12 except for the Z^0 decays. The lines show the expectations of the Standard Model [8] for values of $\sin^2 \theta_w$ of 0.214 and 0.250 respectively

Note that the choice of sign $\langle \lambda \rangle = \langle \mu \rangle = \pm 1$ cannot be separated, i.e. right- and left-handed currents, both at production and at decay, cannot be resolved without a polarization measurement.

Finally, fitting the angular distribution to the form $(1 + \alpha_w \cos \theta^*)^2$, we obtain the result $\alpha_w = 0.79^{+0.15}_{-0.17}$.

Clearly we have observed a very significant asymmetry in the $W^\pm \rightarrow e^\pm + \nu$ decays produced at the Collider. In contrast with this, we expect (and observe) only a relatively small asymmetry in the corresponding angular distribution for $Z^0 \rightarrow e^+ + e^-$ decays (Fig. 13). The asymmetry parameter for the Z is $A_z = 0.25 \pm 0.24$ which, with the current limited statistics, is consitent with zero.

9.-Mass of the Z^0 and parameters of the Standard Model. To measure the Z^0 mass precisely, we restrict ourselves to the subsample of Z decays which pass the following cuts:

i) To ensure the best accuracy in the electron energy determinations, we require that both electrons are more than 15% away from the vertical direction in the transverse plane. This removes one event.

ii) To ensure that both electron cluster energies are well measured, we further require that, for both electrons, there be less than four additional charged tracks entering the same calorimeter cells as the electron energy uncertain.

We are left with a sample of 14 $Z^0 \rightarrow e^+ + e^-$ decays for the Z^0 mass determination, with an expected background contamination of ≤ 0.1 events. The mass distribution for the sample of 14 well-measured high-mass $e^+ + e^-$ pairs is shown in Fig. 14. The distribution has a mean value of $m(e^+e^-) = 92.9$ GeV/c^2, and r.m.s. width of 4.8 GeV/c^2. To determine the Z mass

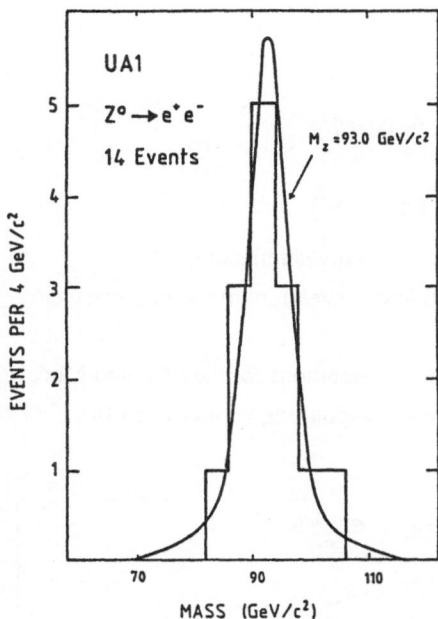

Figure 14. *Invariant mass distribution for the well measured electromagnetic clusters with an invariant mass > 50 GeV/c². The curve shows the expectation for M_z = 93.0 GeV/c²*

and decay width, a maximum likehood fit of a Breit-Wigner distribution smeared by a Gaussian resolution function has been made to the e e mass-distribution.

The result of this fit are:

$$m_z = 93.0 \pm 1.4 \ (\pm 3.0) \ \text{GeV/c}^2 \qquad \Gamma_z = 3.9^{+2.3}_{-1.5} \ \text{GeV/c}^2$$

where the second error on the measurement of m_z comes from the systematic uncertainty in the energy scale of the experiment. Our result for m_z is consistent with our previous measurements (m_z = 93.9 ±2.9 GeV/c² [3,4]) and with the corresponding measurement from the UA2 experiment [5]. The error quoted on Γ_z is the statistical error on the result. If we now include the systematic error we can, once again, only quote an upper limit on the width , $\Gamma_z \leq 8.3$ GeV/c², at 90% c.l. Using the fitted values of m_z and Γ_z the predicted shape of the Z mass distribution (Fig. 14) gives a good description of the data. As a final check of the fitting procedure, we have generated Monte Carlo event samples with various values for the Z mass and width. Generating 14 Z decays in each event sample, we find that our fitting program recovers the correct mass. The decay width, however, is systematically underestimated by 10 % to 15 % in the region of interest. Our quoted value of has been corrected for this systematic effect by using the Monte Carlo to calibrate the fitting procedure for the Z width.

We can now compare our measurement of m_w and m_z with the prediction of the Standard Model [8]. Our result for the mass difference is

$$m_z - m_w = 9.5^{+1.8}_{-1.7} \ (\pm 0.5) \ \text{GeV}/c^2$$

Defining the Standard Model [8] parameters, we obtain:

$$\sin^2\Theta_w \equiv (38.65/m_w)^2 \qquad = 0.214 \pm 0.006 \ (\pm 0.015)$$
$$\sin^2\Theta_w \equiv 1 - (m_w/m_z)^2 \qquad = 0.194 \pm 0.032$$
$$\rho \equiv m_w^2/(m_z \cos\Theta_w) = 1.026 \pm 0.037 \ (\pm 0.019) \ .$$

In Fig. 15 we show our measurements together with the error curves reflecting the uncertainty in the energy scale (at the 68% and 90% confidence levels) in a) the m_w versus m_z plane, and b) the versus $\sin^2\Theta_w$ plane.

These results are in excellent agreement with current expectations for the Standard Model [10,11], and also with our previous results [1,3] and the corresponding results from the UA2 collaboration [5].

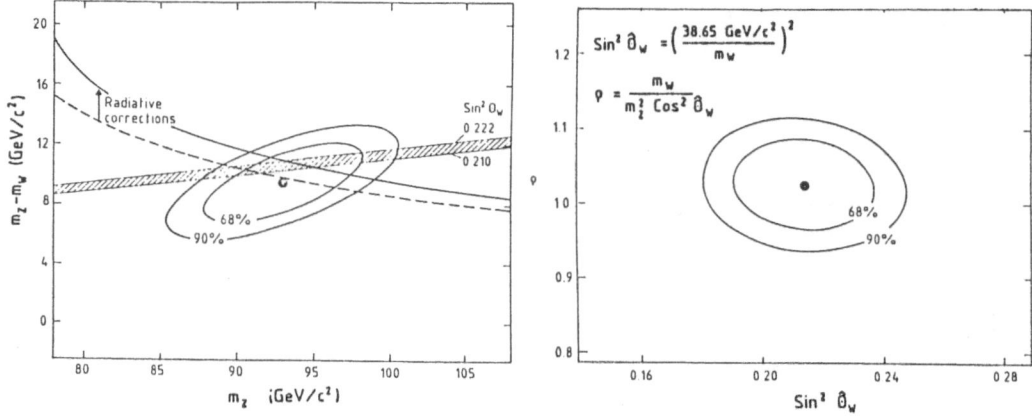

Figure 15. *Fit results for (a) —the m_w vs. m_z plane. The broken line shows the results of the Standard Model for $\rho = 1$. The shaded line is the expectation from neutrino experiments (b)— ρ vs. $\sin^2\Theta_w$ as determined from the mass measurements.*
In both graphs the contours are for 68% and 90% confidence levels

10.-Limits for higher mass W's and for supersymmetric decays of W particles. No $W^{\pm} \Rightarrow e^{\pm}$ candidates have been observed with electron-neutrino transverse mass in excess of the expected distribution for W decays (Fig. 11). Similarly, no $Z \Rightarrow e^+ e^-$ candidates have been observed with an electron-positron invariant mass m in excess of the expected distribution for Z decays (Fig. 9). These results can be used to set limits on the production and decay of very massive W-like (W') and Z like (Z ') particles decaying into electron-neutrino and electron-positron pairs, respectively. Using standard couplings and quark distributions [14] to evaluate the cross-sections at $\sqrt{s} = 630$ GeV we find $(\sigma . B) < 10$ pb, and $(\sigma . B) < 13$ pb at 90% confidence level, corresponding to $m_{w'} > 210$ GeV/c^2, and $m_{z'} > 160$ GeV/c^2, respectively.

We have also looked for unusual W decays in the electron plus missing transverse energy channel inclusively. The observed density of W decays in the (cos θ, m_t)plane, where θ is the angle between the electron and the beam direction, is well described by our expectation for $W^{\pm} \Rightarrow e^{\pm} + \nu$ decays together with a small contribution from known backgrounds. We have used this result to obtain an upper limit on the supersymmetric decay of the W in the ($e_s \nu_s$) channel [21], with the subsequent decay e $\Rightarrow$ e γ_s. Assuming a massless γ_s for a given e_s and ν_s mass the branching ratio for the decay W $\Rightarrow e_s + \nu_s$ is known, enabling us to use the upper limit on the decay rate to exclude the region shown in Fig. 16.

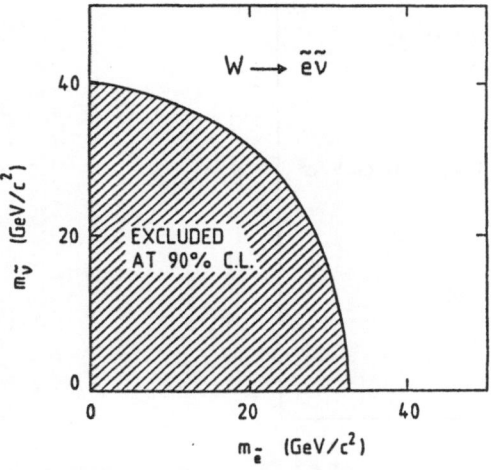

Figure 16. Limits on supersymmetric decays of the W in the supersymmetric electron and neutrino, as a function of the masses

11.- Concluding remarks. The experiment UA1 has extracted essentially background-free samples of 172 $W^{\pm} \Rightarrow e^{\pm} + \nu$ decays and 16 $Z^{0} \Rightarrow e^{+} + e^{-}$ decays from data, corresponding to an integrated luminosity of 399 nb^{-1} , taken at the CERN SPS $p\bar{p}$ Collider. A summary of the quantitative results coming from the analysis of these data samples is given in Table 3. There is excellent agreement between the predictions of the SU(2) ⊗ U(1) Standard Model [10,11] and the experimental data. Data at high energies confirm beautifully the results observed at low energies, and in particular the neutrino experiments.

Most of the errors in the parameters listed in the Table 3 are due to instrumental limitations of the technologies presently used. In particular they are not determined by statistical errors only. During a run presently completed another 600 nb^{-1} have been collected, a three fold increase of the statistical sample reported in the present paper. In addition a tenfold increase of the avaliable luminosity is foreseen with the advent of an additional element in the accumulation of antiprotons, the ACOL ring. Therefore about 10,000 nb^{-1} look like a reasonable goal for the foreseable future. In order to make use of this over all increase in the statistics of a factor of order thirty times, a

Table 3

A summary of the quantitative results coming from the analysis of the
$W^{\pm} \rightarrow e^{\pm} \nu_e$ and $Z^0 \rightarrow e^+e^-$ event samples. The first errors are
statistical and the second errors systematic. An explanation of the measured
and derived quantities can be found in the text.

a) Measured quantities

$W^{\pm} \rightarrow e^{\pm} \nu_e$	$Z^0 \rightarrow e^+e^-$
$\sigma \cdot B_W = 0.55 \pm 0.08 \, (\pm 0.09)$ nb ($\sqrt{s} = 546$ GeV)	$\sigma \cdot B_Z = 42 ^{+25}_{-18} \, (\pm 6)$ pb ($\sqrt{s} = 546$ GeV)
$\sigma \cdot B_W = 0.63 \pm 0.05 \, (\pm 0.09)$ nb ($\sqrt{s} = 630$ GeV)	$\sigma \cdot B_Z = 74 ^{+23}_{-20} \, (\pm 11)$ pb ($\sqrt{s} = 630$ GeV)
$\sigma \cdot B_{W^+}/\sigma \cdot B_{W^-} = 0.75 \pm 0.15$	
$m_W = 83.5 ^{+1.1}_{-1.0} \, (\pm 2.7)$ GeV/c^2	$m_Z = 93.0 \pm 1.4 \, (\pm 3.0)$ GeV/c^2
$\Gamma_W \leq 6.5$ GeV/c^2 (at 90% c.l.)	$\Gamma_Z \leq 8.3$ GeV/c^2 (at 90% c.l.)
$A_W = 0.77 \pm 0.04$	$A_Z = 0.25 \pm 0.24$
$\langle \cos \theta^* \rangle = 0.43 \pm 0.07$	
$\alpha_W = 0.79 ^{+0.15}_{-0.17}$	

b) Derived quantities

Mass-derived	$\sigma \cdot B$-derived
$(m_Z - m_W) = 9.5 ^{+1.8}_{-1.7} \, (\pm 0.5)$ GeV/c^2	$R = 9.6 ^{+3.0}_{-2.1}$
$\sin^2 \hat{\theta}_W = 0.214 \pm 0.006 \, (\pm 0.015)$	$R_B = 2.9 ^{+0.9}_{-0.6} \, (\pm 0.1)$
$\sin^2 \theta_W = 0.194 \pm 0.032$	$N_\nu \leq 10$ (at 90% c.l.)
$\varrho = 1.026 \pm 0.037 \, (\pm 0.019)$	

novel type of calorimeters is needed, much more stable and which can be calibrated to the accuracy of few tenths of a percent. To that effect we have developed a new method of detection based on the collection of free electrons produced in TMP (tetra-methil-pentane) by the passage of ionizing radiations. First results are promising.

The new calorimetry and the increased luminosity of the collider should permit an ultimate determination of the SU(2) $\otimes$ U(1) Standard Model parameters with an accuracy which is about ten times better than the values given in Table 3. In particular for the first time one shall be able to detect the effects of the radiative corrections.

12.-References.

[1] G. Arnison et al. (UA1 Collaboration), Phys. Lett. 122B (1983) 103 ; 129 B (1983) 273.

[2] G. Arnison et al. (UA1 Collaboration), Phys. Lett. 134B (1984) 469.

[3] G. Arnison et al. (UA1 Collaboration), Phys. Lett. 126 B (1983) 398.

[4] G. Arnison et al. (UA1 Collaboration), Phys. Lett. 147B (1984) 241.

[5] M. Banner et al. (UA2 Collaboration), Phys. Lett. 122B (1983) 476.

   P. Bagnaia et al. (UA2 Collaboration), Phys. Lett. 129B (1983) 130: Z. Phys. C 24 (1984)1.

   J.A. Appel et al. (UA2 Collaboration), CERN-EP/85-166, submitted to Z. Phys. C.

[6] S. Drell and T.M. Yan, Phys. Rev 25 (1970) 316 , and Ann Phys. 66 (1971) 578.

[7] UA1 Proposal, A 4 solid-angle detector for the SPS used as a proton-antiproton collider at a centre-of-mass energy of 540 GeV, CERN/SPS 78-06 (1978).

[8] S. Weinberg, Phys. Rev Lett. 19 (1967) 1264.

   A. Salam, Proc. 8th Nobel Symp., Aspenasgarden, 1968 (Almqvist and Wiksell, Stockholm, 1968), p. 367.

[10] F. Bergsma et al. (CHARM Collaboration), A precision measurement of the ratio of neutrino-induced neutral-current and charged-current total cross-sections. Contributed paper to the Int. Symp. on Lepton and Photon Interactions at High Energies, Kyoto, Japan, 1984: preprint CERN-EP/85-113 (1985).

   A. Blondel (CDHS Collaboration), Talk presented at the EPS International Europhysics Conference on High-Energy Physics, Bari, Italy, 18-24 July 1985.

[11] W. J. Marciano and A. Sirlin, Phys. Rev. D29 (1984) 945.

[12] The EUROJET Monte Carlo program contains all first order and second order QCD processes.[see B. van Eijk, Note UA1/Tn/84-93(1984) and A.Ali, E. Pietrarinnen and B. van Eijk, to be published].The fragmentation functions are the one given by C. Peterson et al. Phys. Rev. D27,(1983),105

[13] F.E. Paige and D.Protopopescu, ISAJET program, BNL 29777 (1981).

[14] G. Altarelli, R. K. Ellis, M. Greco and G. Martinelli, Nucl. Phys. B246 (1984) 12.

   G. Altarelli, R.K. Ellis and G. Martinelli, Z. Phys. C 27 (1985) 617.

   W.J. Stirling, Proc. of the Workshop on Drell-Yan Processes, Batavia, 1982 (FNAL), Batavia, III., 1983), p. 131.

[15] E. Eichten et al. Rev. Mod. Phys. 56(1984) 579

[16]M. Glück, E.Hoffman and E. Reya, Z. Phys. C13(1982) 119

[17] S. Geer and W.J. Stirling, Phys. Lett. 152B(1985)373

[18] N.G. Deshpande et al. Phys. Rev. Lett. 54 (1985) 1757.

   N. Cabibbo, Proc. Third Topical Workshop on pp Collider Physics, Rome, 12-14 January 1983, page 567 (Geneva 1983).

   F. Halzen and K. Mursula, Phys. Rev. Lett. 51 (1983) 857.

   D. Cline and J. Rohlf, UA1 technical note TN 83-40 (1983), unpublished.

   K. Hikasa, Phys. Rev. D29 (1984) 1939.

[19] G. Arnison et al. (UA1 Collaboration), Nuovo Cimento Lett 44 (1985) 1.

[20] M. Jacob, Nuovo Cimento 9 (1958) 826.

[21] R. Barbieri, N. Cabibbo, L. Maiani and S. Petrarca, Phys. Lett. 127B (1983) 458.

R. M. Barnett, K.S. Lackner and H. E. Haber, Phys. Rev. Lett. 51 (1983) 176.

H. Baer, J. Ellis, D. V. Nanopoulos and X. Tata, Phys. Lett. 153B (1985) 265.

Reminiscences of Columbia

N.P. Samios

Brookhaven National Laboratory, Upton, NY 11973, USA

In these brief remarks on the occasion of Jack Steinberger's Festschrift, it is my intent to reminisce and comment on the grand Columbia years of the late 50's. I do this instead of contributing a scientific article since such documents ultimately find themselves in the literature while unique occasions such as this afford the opportunity to discuss impressions, ambiance, and activities that are usually omitted from such discourses. The beginning, for purposes of this article, occured in 1954. It was at the time that three young graduate students, Mel Schwartz, Jack Leitner and myself agreed to do our thesis under Jack Steinberger - in effect a package deal. There are two major occurrences that took place during this same time, namely, the innovation of the bubble chamber by Glaser and the introduction of the strangeness quantum numner by Gell-Mann and Nishijima. This then set the stage for the physics which we pursued over the next five years.

A series of bubble chambers were built and used starting with a test propane chamber, a cylinder 4 in. in diameter and 4 in. in depth. This was superseded by 1) propane chamber, 6 in. in diameter, 4 in. in depth, no magnetic field, 2) propane chamber, 12 in. in diameter, 8 in. in depth placed in a 13.4 kG magnetic field and 3) hydrogen bubble chamber, 12 in. in diameter and 6 in. in depth, again with a 13.4 kG magnetic field. All these experimental instruments were essentially designed and built by a small group of experimentalists in a relatively short span of time. By 1958 a lengthy paper was published involving analysis of exposures of all chambers in π^- beams at several energies yielding hundreds of strange particles. Although strange particles had been discovered in cosmic rays, their number was limited and production mechanisms unknown. The Cosmotron, a 3 GeV synchrotron became operational in 1952. The pioneering experiments were performed by Shutt and associates in a cloud chamber at this accelerator, again with limited data but sufficient to demonstrate associated production of strange particles. With the availability of hundreds of events, due to the increased target density affected by a liquid in a bubble chamber versus a gas in a cloud chamber, detailed studies of the production and decay of strange particles were carried out. This contributed immensely to determining the mass, spin, lifetime and branching ratios of these new important particles. (There was some evidence (meager) that attributed a very high spin to the Λ^0 and some speculation that the long lifetime was due to barrier effects of large spins - these were laid to rest.) We discovered the Σ^0 hyperon and demonstrated that the rate for Σ^- beta decay was much lower than that expected from ordinary (non-strange) beta decay. Although we searched for parity non-conservation in hyperon decay, we didn't see it until after the famous Co^{60} parity experiment of Wu and collaborators. It should be noted some of the systematic measurements, such as K_s branching ratio have significance to today vis-a-vis $\Delta I = 1/2$ rule whose origin is still not understood.

It was also at this time that the Cosmotron developed magnet difficulties and required modification involving a downtime of approximately one year. As such we brought the hydrogen bubble chamber to Nevis and performed a series of experiments with stopping π^+ and π^- mesons. This resulted in a determination of the π^0 parity, a study of Dalitz pair distributions, a precise measurement of the electron spectrum from μ^+ decay (the ρ value) and the observation of pion beta decay. In this latter instance, we confirmed the CERN result. It should be noted that the hydrogen chamber was cycled 90 times per minute, yielding one million pictures (triads) in the π^- exposure. Quite a feat in 1958.

This work was done during an exciting time. The study of weak decays was the center of attention. Parity non-conservation and V-A coupling were uncovered. Strange particles Λ's, Σ's, K's were investigated in depth. This set the stage for the physics for the next ten years. Here we parted ways. Jack Steinberger and Mel Schwarz continued their investigation of the weak interactions, their most significant result being the demonstration of the existence of two types of neutrinos in 1962 followed by a program of studies involving K^0 decays investigating the CP question. I left Columbia at this time to come to BNL and embarked on studies of particle spectroscopy. This resulted in the discovery of a host of resonances, the most notable being the ϕ, Ξ^- (1530) and the Ω^-, all being important in establishing SU(3) as the symmetry of hadrons. Indeed a productive and exciting time.

As one looks back at this era one cannot help to be impressed by the rapid progress of physics during the time and the ability of a small group of people to influence the paths of progress both experimentally and theoretically. In fact the people who were present at Columbia during this era are quite impressive: A. Pais, M. Gell-Mann, T. D. Lee, S. Weinberg, R. Saber, L. Lederman, C. Rubbia, C. Townes, I. I. Rabi and others. A truly remarkable cast and era.

Jack Steinberger – The Best Teacher I Have Known

M. Schwartz

Stanford Linear Accelerator Center, Stanford, CA 94305, USA

I first met Jack Steinberger one morning in late September of 1951 as I began my third undergraduate year at Columbia. Having decided to enroll in a course normally taken by fourth year undergraduates entitled "Atomic Physics and Introductory Quantum Mechanics", I was looking forward with some trepidation to my first real view into the twentieth century. In fact this course was to become the great turning point in my career. It would introduce me to the person who throughout many years to come would be my role model, my mentor and father figure, and last but not least the best experimental physicist that I have ever been associated with.

In prior years at Columbia I had been taught by a succession of outstanding distinguished professors, most of whose active years were behind them. Courses were typically well-polished lecture series given by gentlemen clad in suits and ties (as was required in those days). Never was it apparent that Columbia was the place at which some of the greatest discoveries of the postwar period were being made. (Indeed a total of five independent Nobel Prizes in Physics would subsequently be awarded for work carried out at Columbia in the fifties.)

On that first day of class Jack walked in wearing an open-necked workshirt with red eyes and a touch of exhaustion. He had just spent the night running an experiment at Nevis (the Nevis Cyclotron Laboratory, twenty miles north of Columbia, on the Hudson) and was in no shape to give a lecture. In addition, he hadn't bothered to prepare and was hoping to wing it. This was Jack's first formal lecture to undergraduates; he had just arrived from Berkeley having achieved fame as the discoverer of the π^0 meson. Jack was barely thirty and already one of the outstanding particle physicists in the country.

Needless to say, that first lecture was, in a formal sense, a disaster. But for me it was an eye-opener. As I left the room that day I knew beyond a shadow of a doubt that Jack was the most exciting person that I had ever met and I wanted nothing more than the opportunity of working with him.

Jack Steinberger's teaching style was unique. Because he so rarely prepared his lectures, they became more like a "think along" than a normal class. For the average student it was just a confusing mess. But for me it was one of the great experiences of my life. To be able to see a step ahead and help Jack think his way through a problem gave me a sense of participation that I had never had before. From those days on I knew that I wanted nothing more than to be a particle physicist.

I finally completed all of my formal course work toward the Ph.D. in the spring of 1954 and went to see Jack with the intent of becoming his graduate student. He suggested that I ought to consider theory rather than experimental physics. Somehow he had an inordinate admiration of theoretical physics, perhaps because he had tried his hand at it for a short period of time before coming to Columbia. To Jack who had been a student of Fermi, the theorist always seemed to have the upper hand. Fortunately for both of us, this was not to be the case during the halcyon days of the Fifties and Sixties. Those would be the days in which experiments would lead the way and the theory would follow. In any case, I had my heart set on working with Jack and would not be diverted.

I should point out that by this time Jack had increased his stature immensely by having determined the parity of the pion (through the reaction $\pi^- + d \rightarrow 2n$) and having helped determine the spin of the pion through measuring the reaction $\pi^+ + d \rightarrow 2p$ (the inverse of which had been determined at Berkeley).

So there, in the summer of 1954, I and two other aspiring young particle physicists (Jack Leitner and Nick Samios) came to work for Jack at the Nevis Cyclotron Laboratories. At that time Jack had just completed an exceedingly important "negative" experiment in which he had shown that π^+ does <u>not</u> decay into e^+ and γ (to a limit of 10^{-5}). He was enormously intrigued by this result and indeed, if it hadn't been for the invention of the bubble chamber by Don Glaser, he would probably have spent a few more years pursuing this limit. As it was, the newly invented bubble chamber proved to be too exciting a diversion.

The invention of the bubble chamber by Donald Glaser was without doubt one of the most important milestones in the development of particle physics. Its importance was particularly significant in the study of the then recently discovered strange particles, most of which only traveled a few centimeters before undergoing decay. The only techniques available for studying these shortlived particles prior to the bubble chamber were cloud chambers and nuclear emulsions. Both of these techniques had major shortcomings. The cloud chambers had too low density to be useful for studying the production of large numbers of events under controlled conditions in hydrogen. The emulsions were useless for the study of neutral particle production and decay.

The first bubble chambers put together by Glaser were made entirely of glass and because of their fragile nature, were not terribly useful as experimental instruments. We (Jack along with Leitner, Samios and myself) tried to build one of these, but before we finished we heard that Glaser was going to be out at the Brookhaven Cosmotron with his new "dirty" chamber having an aluminum body and thick plate glass windows. Filled with pentane, it reputedly was producing excellent quality tracks. We all drove out together to meet Glaser (Leon Lederman joined us on this trip) and immediately ran into trouble because Jack did not have proper clearance to visit Brookhaven. After some fancy footwork, Jack was allowed in for the day and we got to attend Glaser's seminar and have a look at the chamber.

At that point Jack decided to move full speed ahead toward the production of a series of bubble chambers. Obviously the ideal chamber from the point of view of Jack's physics interests was a liquid hydrogen chamber, but that would clearly take a long time. So we compromised and decided to construct a small chamber making use of propane as the filling. No one had ever tried propane in a bubble chamber, but Jack felt (correctly as it turned out) that propane had as much likelihood of working as pentane. We were on our way.

By December of 1954 we had completed our first four-inch diameter propane bubble chamber and were then faced with the chore of making it work. As an expansion mechanism Jack had designed an O-ring sealed piston driven by air pressure which was released by a solenoid valve to trigger the expansion. We tried to guess a possible operating point with respect to pressure and temperature, but try as we would, we got no bubbles. The entire month of January 1955 was spent in frustration as the chamber refused to work. Only at the high end of the operating range did we see any bubbles at all and those were along the walls of the chamber. We were almost ready to give up on propane and go over to pentane (a la Glaser) when the American Physical Society annual meeting rolled around and we went down to hear Glaser's invited talk.

Jack and I went down to NYU on Saturday morning where the talk was being held and listened in rapt attention as Glaser described the details of his propane chamber operation. At one point he noted that the temperature region of sensitivity <u>began</u> a bit above where bubbles were made at the walls. Suddenly we sensed what was wrong and raced back to Nevis that Saturday evening. We raised the temperature some ten degrees above where we had been and <u>lo and behold</u> we got tracks (caused by a gamma-ray source). That night we went to Jack's home in Hastings and celebrated over drinks. For Jack who had been through many successes before, this was probably old-hat. For me it was the most exciting time of my life.

Now the work really began in earnest. The four-inch chamber was really too small to do any physics with and we launched into the construction of a six-inch diameter, four-inch deep propane chamber to use in a real experiment. Completion of the chamber took several months and then we had to decide on a physics program.

The decision as to what to do with the chamber was almost obvious. At Brookhaven, Ralph Shutt had just completed an outstanding program making use of his hydrogen diffusion cloud chamber to produce hyperons and show that Λ^0's were always accompanied by K mesons. This confirmed the theory of "associated production", a theory which explained the fact that Λ's and K's decayed slowly by means of weak interactions but were produced by means of strong interactions. Associated production required that there be a new quantum number, strangeness, attached to hyperons and kaons. The quantum number was conserved in the strong interactions and violated in the weak interactions. However, try as he might, Shutt could never get more than a handful of Λ^0's because of the low density of the hydrogen gas. On the other hand the propane bubble chamber would provide an excellent vehicle for expanding his work and for producing some of the other strange hyperons.

So we packed our chamber and all of its associated paraphernalia onto an old Navy truck and brought it out to the Cosmotron. The first problem to overcome was the matter of Jack's clearance. Overcoming this barrier took many months and in the intervening period Jack's presence at the Cosmotron could not be advertised too widely. (In the end Jack's AEC clearance was granted after he was given Navy clearance to work at the Columbia Hudson Laboratories.) Nevertheless we set up a π^- beam at about 1.9 GeV and began taking pictures.

Unfortunately, there was a basic design flaw in the chamber. The O-rings on the piston began to destroy themselves in a matter of a few hours and we really obtained very little data before we had to shut down the chamber. It took a number of months before we could redesign and rebuild the chamber so as to make it a reliable instrument.

By late 1955 we had obtained enough film to allow us to complete an experiment. Although we saw no more than about 100 strange particles of various sorts, we had multiplied by a factor of about 20 the total available information on the production of Λ^0 and Σ^- .

Analysis of the data began in earnest in early 1956. Those were very exciting times in particle physics and in a short time we were able to establish lifetimes for various strange particles as well as differential cross sections for their production.

As a side note, it will be remembered that those were the days in which the decay of the K^+ into both $\pi^+\pi^-$ and $\pi^+\pi^-\pi^0$ channels gave rise to the major problem of the times—the apparent violation of parity in strange particle decay. Many alternative explanations were offered, but I remember very clearly the day in the spring of 1956 when Jack came back to Nevis after a talk with T. D. Lee. He said that Lee had suggested a method of investigating parity conservation in Λ^0 decay by studying the "up-down" asymmetry in the Λ^0 decay process. Excitedly we examined the data, but unfortunately there were far too few events to make any statement. Within six months of course the issue was resolved by the Wu, Ambler experiment. By the middle of 1957 the expected asymmetry was also seen in Λ^0 decay, first by the Alvarez group and very shortly thereafter by ourselves.

By the summer of 1956 all of our data had been analyzed and published and I left Columbia for a two-year postdoc position at the Cosmotron. During this period I collaborated closely with Jack in the continuing experimental program making use of a new twelve-inch propane chamber. During the same period Jack built a twelve-inch diameter liquid hydrogen chamber which in the years ahead would provide large amounts of outstanding data in the study of strange particle production at various pion energies. But in a sense, our most exciting years in the bubble chamber business were over. We had carried out the very first real experiment with this type of chamber and had made some major strides in the understanding of strange particle production and decay.

One unfortunate circumstance came in our way in 1957 and prevented us from making the maximum use of our new hydrogen chamber. The Cosmotron broke down and was to remain down for a period of eighteen months, during which time there was very little that could be done. Jack submitted a request to Berkeley to permit us to move the chamber out West for a period of time, but that request was unceremoniously denied. The Berkeley Lab was very much of a closed institution in those days and Jack was not considered to be a welcome competitor. This attitude was to hurt them within the community of particle physics many years later when very few people felt inclined to put a new machine under Berkeley control. Instead it went to Fermilab.

In 1958 I came back to Columbia as a colleague of Jack's and we continued our close collaboration for a period of time. The high point of this next period was of course the neutrino experiment which began construction in 1960 and ran successfully in 1962. The neutrino experiment was a three-way collaboration between Jack, Leon Lederman and myself and succeeded in proving that there were two neutrinos in nature. In a sense this was a fitting culmination to Jack's unsuccessful search for the $\mu \rightarrow e + \gamma$ decay because it provided a complete explanation for the absence of that decay.

Looking back over that period of more than a decade, it appears clear to me that Jack was without doubt the most important influence on my life and career. Jack is not only a great physicist and teacher. He is a man of consummate honesty and innate modesty. His ability to see through sham and posturing has not made him many friends, but it has given me a sense of respect for him that exceeds that which I have for anyone else in the world of particle physics. Jack is unique. There is no one like him.

J. Steinberger and the Early Days of Pion Physics

G.C. Wick

Emeritus of Columbia University, New York
and
Scuola Normale Superiore, Pisa

I have known Jack Steinberger and followed his work since a quite early time in his career. Physicists from more recent generations than I will probably connect Jack's name with things that he did later. I derive a certain pleasure from recalling some episodes which show Jack still young and inexperienced groping to find his way in the maze of modern physics, and finding it in fact rather quickly, displaying in the process some of the qualities, that we have all learned to admire and that have made his strength in the daily struggle to "wring significant facts from an inflexible nature".

I met Jack when he was one of that exceptional collection of students, who had flocked around Fermi at the University of Chicago soon after the end of the second world war. I was then teaching at Notre Dame University, close to Chicago; this made it possible for me to attend quite often the seminar which Fermi ran together with Teller. I thus became acquainted with many of those students, with Jack in particular when I discovered that we had a common interest in a problem, in slow-neutron physics, arising out of some puzzling results obtained by Don Hughes and his team on the polarization of a neutron beam by magnetized iron. The method seemed to be quite a bit more effective than was predicted theoretically. My own interest in slow neutrons went back to the days, when I was assistant to Fermi in Rome. I seem to remember that I had some idea where the discrepancy found by Hughes came from, and I had made some progress on some aspects of the problem; Jack, on the other hand, had worked on another aspect of the problem by doing some useful calculations on the form-factor of the d-electrons in the iron atom. We decided to join forces and soon were able to publish a reasonable explanation of the results of Hughes; as I recall, the paper was well received.

This collaboration had left me with the impression that Jack was headed towards becoming a successful theorist. I was surprised, therefore, when his next paper turned out to be about an experiment. I had lost sight of him, having left the Middle-West to fill a physics chair in Berkeley; meanwhile Jack's interests had turned to the study of cosmic ray mesons, a subject where, entirely by chance, our interests again overlapped: I had worked on mesons with Bernardini while I was still in Italy. The subject was still clouded in a thick fog of misunderstandings, partly due to the assumption (quite natural at first) that the mesotrons discovered by Anderson (today's μ-mesons, of course) were the same thing as the particle hypo-

thesized by Yukawa. In the paper I have mentioned, Steinberger described how, following a suggestion by Fermi, he had designed an experiment to determine the energy
spectrum of the electrons emitted in the decay of a μ-meson, by measuring their
range in a hydrocarbon absorber. This was not a very accurate way to measure the
energy, and in fact not much later Leighton et al. published cloud-chamber results
that were definitely better. Jack's method, however, was quick and simple and good
enough to allow him to establish the basic fact that the decay-mode of the μ-meson
was not a two-body decay; this was a crucial point on the process of clarification
of the meson puzzle. According to my memory, Jack's data were amongst the earliest
evidence on this score.

After doing this work, Jack was invited by Oppenheimer to spend a year at the
Institute in Princeton, where he again showed his versatility by switching back to
theoretical work of a more sophisticated kind than anything he had done before and
in a field where he was complete novice. The problem was to calculate the rate of
decay of a neutral pion into two gamma rays, and I should mention that it was also
tackled at about the same time (and with results similar to Jack's) by Fukuda,
Miyamoto and other students of Tomogana and Yukawa. I shall ask for the reader's
patience if I discuss Jack's paper at some length. It is a remarkable paper both
for what it tells us about Jack's way of doing physics, and for its relevance to
the later development of the theory.

To begin with, the episode is a good example of Jack's instinct to attack important problems, undaunted - as I have indicated - by their possible difficulties. I
should recall that at that moment the neutral pion itself was a bit of a mistery.
The belief in its existence was mainly tied to a brilliant conjecture of Kemmer
about the symmetry of the pion field required by the so-called charge independence
of nuclear forces. Recently Oppenheimer had speculated on some interesting consequences of the existence of π_0's on the behaviour of cosmic radiation in the high
atmosphere; he had pointed out that a π_0 could decay into a pair of gamma-rays via
a virtual transition to a nucleon-antinucleon pair, and his student Finkelstein
had made a crude estimate of the probability of this process. Altogether it seemed
plausible that the process could be an important source of energetic showers in
the high atmosphere. The problem now was to replace the crude order-of-magnitude
estimate with a "genuine" calculation, and this is just what Jack set out to do.
It is entirely to his credit, that within a very short time he was able to master
all the new ideas and technical tricks that were needed to do an adequate job,
according to the standards of the times. Which, as our readers will recall,
required the evaluation of a triangle graph, in which one of the vertices, representing the pion-nucleon interaction could be of two different types: γ^5 (i.e.
pseudoscalar) or pseudovector; in the latter case the evaluation was especially
delicate, since a rather strong divergence had to be eliminated by a somewhat deba-
table "regularization" procedure. The result for this case was a bit surprising,

since it contradicted an alleged "equivalence theorem" between the two forms of the interaction.

When this work was published, there was no direct confirmation of the existence of the process; a reliable measurement of the decay-rate was even further away in the future. It turned out later that Jack's formula for the case of the pseudoscalar vertex is pretty good, but for rather subtle reasons which were not understood at the time. It does not in any way detract from Jack's merits, that the full significance of his results was only claryfied by later theoretical ideas, such as Nambu's "soft pion theorems" and Adler's discovery of the axial-vector-current anomaly. It simply would not have been possible, in 1949, to foresee these later developments; but it is only fair to say that the calculations of Steinberger and of Fukuda et al. paved the way for them.

I now wish to tell you briefly about the next occasion I had to exert a minor influence on Jack's carreer;it happened more or less by chance, and I have no intention to claim any particular merit or foresight in the affair. It so happened that my chair in Berkeley was endowed with some extra money to pay a salary to an assistant, and when that became available I thought of offering the job to Steinberger at the end of his year in Princeton.

So he came to Berkeley and although things did not quite turn out the way I expected, this move was, I think, a good thing both for him and for physics in Berkeley, even though, for reasons too long to explain here, he did not remain there very long. What happened was that, not long after arriving in Berkeley to take up his job as my assistant, Jack showed up one day and with some embarassment told me that while he, of course, understood that I probably expected him to work on some theoretical project I was interested in, he had become aware recently of an experiment he could do at McMillan's synchrotron and was very interested in this possibility. I do not remember the gist of the conservation in great detail, but I think I told him that I did not feel I had any proprietary right on his work and that as long as he did good physics, he had my blessing, regardless of the type of work he did. Not much later I met McMillan and I still recall with some satisfaction what he told me. "It was a real pleasure - he said - to see Jack at work with his machine; Jack was a born experimenter". I knew then, that I had made the right decision in giving Jack a free hand. To be quite truthful, I should admit that I do not know whether I could have stopped him anyway; Jack has a strong will, and I have never been very good at throwing my weight around.

But let me conclude briefly with the rest of the story. The Radiation Laboratory had just then become the main center of research on artificially produced mesons; it had the right kind of machines to do that. First pions produced by the proton beam of the big cyclotron had been detected by Lattes with the emulsion technique he had learned from Powell and Occhialini. When Jack arrived, Luis Alvarez and his team had just successfully completed an experiment in which pions were detected using an electronic detection system, which made essential use of the very short length of the cyclotron pulse. Jack thought that it would be interesting to observe photoproduction of charged pions by the synchrotron beam. The energy of the machine was obviously sufficient, but the greater length of the synchrotron pulse did not allow one to employ the technique invented by Alvarez. Jack, however, had a somewhat different idea, which - he thought - could surmount the difficulty; his paper with Bishop was in fact his first important experimental contribution to pion physics.

I was even more excited by the work he next did with Panofsky on the photoproduction of neutral pions. This was in my opinion, and I do not think it is only my opinion, the first unmistakable evidence on the existence of the neutral pion. I have mentioned before that there was some indirect evidence from the study of cosmic rays. And I should also add that there had been an interesting result of York and Mover on gamma rays emitted by an internal target struck by the proton beam of the Berkeley cyclotron. There was a strong inclination to interpret these gamma-rays as the result of the decay of neutral pions, but alternative interpretations, e.g. direct production of gammas in the collision of protons with the target, could not be excluded. Clearly what was needed was a coincidence experiment, detecting the simultaneous emission of a gamma-ray pair by a neutral pion. Steinberger and Panofsky realized that the photoproduction process established by Steinberger and Bishop was probably a very convenient source of π_0's for this experiment. It was to be expected, from the results on charged pions, the the π_0's would be fairly energetic and peaked forward, resulting in a rather characteristic energy spectrum amd angular correlation of the photon pairs. The subsequent detailed verification of these expectations in the experiment of Steinberger and Panofsky remains up to this day one of the pieces of work that I have read about with the greatest pleasure; this work laid to rest all doubts about an issue of fundamental significance for the future development of the theory of nuclear forces.

And this is perhaps a good point where to end my story. Not much later than the events I have recounted, both Jack and I left Berkeley and moved to different places. The world of physics is small: we have met many times again and we have remained good friends, but our orbits have never again been as closely intertwined as in the period I have recalled.

Is KNO Scaling Valid for Two Jet Events in e^+e^- Collisions?

C.N. Yang

Institute for Theoretical Physics, State University of New York at
Stony Brook, Stony Brook, NY 11794, USA

KNO scaling, a theoretical hypothesis proposed[1] in 1972, was approximately confirmed experimentally[2] for $p\bar{p}$ collisions in 1981. This hypothesis (or approximate scaling[3]) implies that the mean fluctuation of the charged-particle multiplicity, $[\overline{(\Delta n)^2}]^{1/2}$, is proportional to $\bar{n}$. Such a behavior is highly nonstochastic. Yet it seems a priori clear that for a collision with the emission of dozens of particles, some aspect of the process must be stochastic. In other words, both stochastic and nonstochastic elements seem to be present in the CERN $p\bar{p}$ collision experiments. The question then becomes: which aspect was stochastic, which nonstochastic.

This question was resolved recently[4]. Furthermore extrapolating from this development T. T. Chou and I were able to[5] arrive at a simple description of the single particle spectrum in terms of a concept called partition temperature. In our view these developments give a satisfactory phenomenological description of inclusion single particle emission in hadron-hadron collisions, and supply the underlying physical reason for KNO scaling in such collisions.

Can the same ideas be applied to 2 jet events in e^+e^- collision?

Chou and I answered this question in the affirmative[6]. A most important prediction of this line of reasoning is that KNO scaling, exact or approximate[3], does not obtain in e^+e^- collision. Instead we predict that for these events the multiplicity distribution is Poisson, (therefore stochastic). These predictions are in opposition to current concepts[7] on the subject, but we believe they will be verified in future TRISTAN, SLC and LEP experiments. I shall now summarize these developments and give details of those arguments which have not yet appeared in the literature.

In our view for each angular momentum of the incoming system, the
multiplicity distribution follows a Poisson law. In $p\bar{p}$ collision, the angular
momentum extends over a wide range (from 0 to ~ 2000 ℏ for $\sqrt{s}$ = 540 GeV,) each
of which yields a Poisson distribution. The superposition of this multitude of
Poissons is, in our view, what gives rise to the wide fluctuation of
multiplicities that characterizes approximate[3] KNO scaling. In e^+e^-
collision, on the other hand, the reaction diagramwise goes through an
intermediate state of one virtual photon. The angular momentum is thus 0 or 1ℏ
and does not vary over a wide range. The multiplicity being the superposition
of two almost identical Poissons is then an essentially Poisson distribution.

Is this in agreement with experimental data? The best data available was
given by the TASSO group[8] for charged multiplicity distribution in e^+e^- 2 jet
events at total energy W = 14, 22 and 34 GeV. (Ref. 8, Fig. 2, reproduced in
Fig. 1). These are approximate Poisson distributions but have been interpreted
as being consistent with KNO scaling.

It is clear that these data are not massive enough, and the range of
values of $\bar{n}$ not large enough, for a decisive conclusion to be reached about
whether KNO scaling obtains for e^+e^- collision or not. But we want to point
out some striking characteristics of these curves which in our view favor
Poisson distributions: Each curve exhibits a region of n_{ch}, to the left of the
peak, i.e. on the low n_{ch} side of the peak, where the curve has a positive
curvature. The presence of this region makes the peak almost left-right
symmetrical. Furthermore as $\bar{n}$ increases (when W increases from 14 to 34 GeV),
this region widens toward the right. These features are absent in $p\bar{p}$ curves
and are characteristic of Poisson distributions (see Fig. 1).

Furthermore, if one takes an ingenuous attitude, one would find that the
multiplicity distributions for e^+e^- and $p\bar{p}$ collisions plotted in Fig. 8 of
Ref. 8, reproduced here as Fig. 2, exhibit a striking qualitative difference:
The former is simple and the latter complex. We believe this difference
reflects the fact that the former is a pure Poisson and therefore simple, while
the latter is a complex superposition of many Poisson distributions.

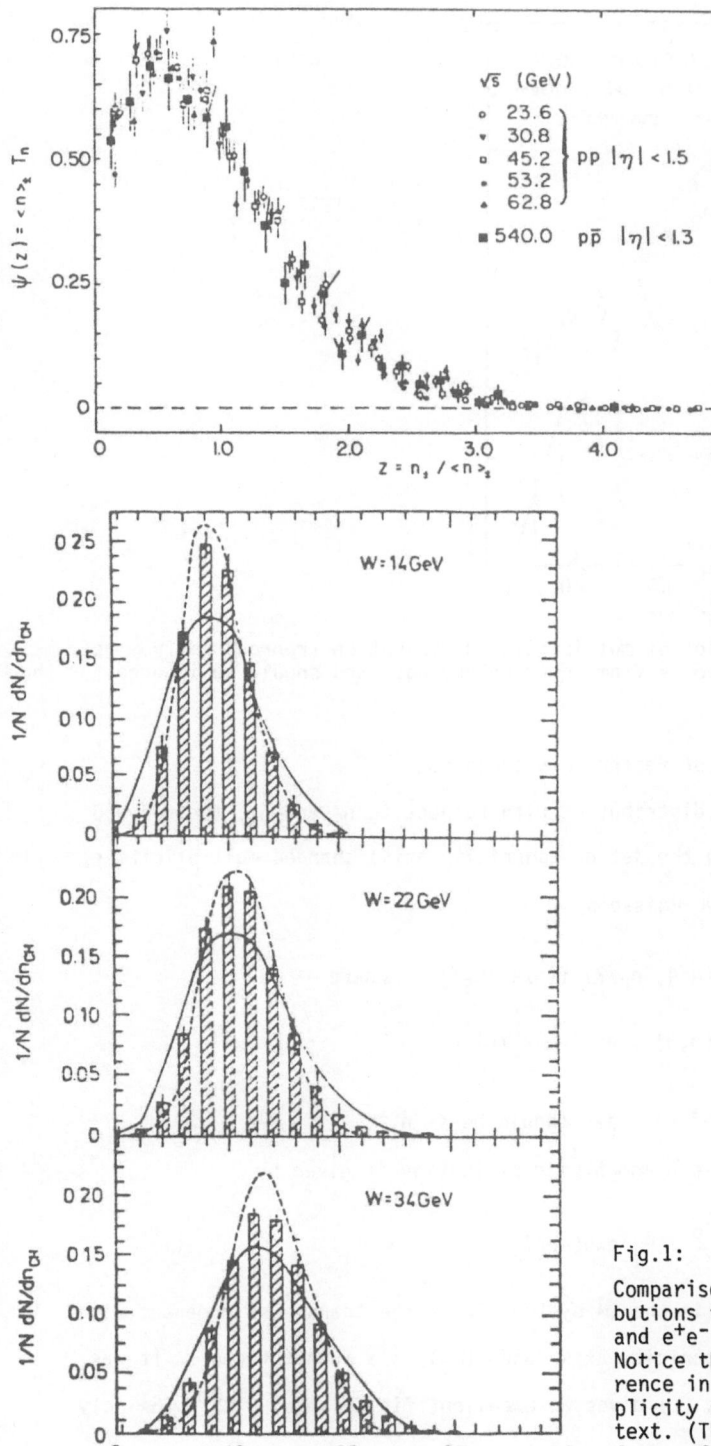

Fig.1:

Comparison of multiplicity distributions from p̄p collisions (above) and e⁺e⁻ collisions (below). Notice the characteristic difference in shape on the low multiplicity side of the peak. See text. (The p̄p data is reproduced from ref. 2 and the e⁺e⁻ data from ref.8)

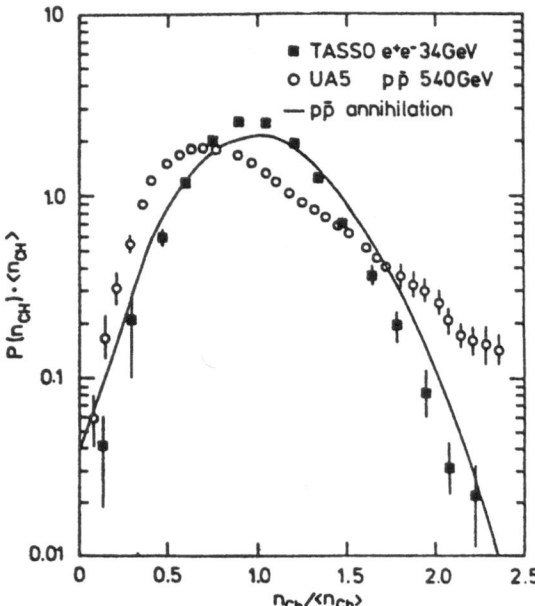

Fig.2: Semilogarithmic plot of multiplicity distribution (reproduced from ref. 8).
See text. The curve is from lower energy data and should be ignored for the
present purpose

Further predictions of reference 6 include:

(A) The probability distribution with respect to n_F and n_B, the forward
and backward (relative to the jet or sphericity axis) charged multiplicities,
should be a product of two Poissons

$$f(\bar{n}/4,\ n_F/2)\ f(\bar{n}/4,\ n_B/2) \qquad \text{where}$$

$$f(a,n) \equiv \exp(-a)a^n/n! \ .$$

(B) The net charge of each jet should be $\ll \bar{n}$.

(C) The single particle momentum distribution is given by

$$\frac{d^3p}{E}\ G(P_T)\exp(-E/T_p) \qquad\qquad (1)$$

where all quantities are in the CM system, P_T is the transverse momentum with
respect to the jet or sphericity axis, and $G(P_T)$ is a cutoff factor. It was
found in reference 6 that (1) gives an excellent fit to TASSO data[8] The only
adjustahble parameter is T_p, the partition temperature, which was found to have

the values 1.6 GeV and 3.3 GeV respectively for total CM energy W = 14 GeV and 34 GeV.

REFERENCES

1. Z. Koba, H. B. Nielsen and P. Olesen, Nucl. Phys. <u>B40</u>, 317 (1972).

2. UA5 collaboration, K. Alpgard et al., Phys. Lett. <u>107B</u>, 310 (1981);
    UA1 collaboration, G. Arnison et al., Phys. Lett. <u>107B</u>, 320 (1981).

3. We interpret the spirit of KNO scaling as the hypothesis that $\overline{(\Delta n)^2}$ is proportional to $\bar{n}^2$, (which we define as approximate KNO scaling), while for pure stochastic processes $\overline{(\Delta n)^2}$ is proportional to $\bar{n}$. (In our view, exact KNO scaling is unlikely to be a physically sound hypothesis.)

4. T. T. Chou and Chen Ning Yang, Phys. Lett. <u>135B</u>, 175 (1984). Crucial experimental data came from the UA5 collaboration: K. Alpgard et al., Phys. Lett. <u>123B</u>, 361 (1983).

5. T. T. Chou and Chen Ning Yang, Phys. Rev. Lett. <u>54</u>, 510 (1985).

6. T. T. Chou and Chen Ning Yang, Phys. Rev. Lett. <u>55</u>, 1359 (1985).

7. See e.g. S. Barshay and Y. Yamaguchi, Phys. Letters <u>51B</u>, 376 (1974); S. Barshay, Phys. Letters <u>116B</u>, 193 (1982); Review article by R. Felst in <u>Proceeding of the 1981 International Symposium on Lepton and Photon Interactions at High Energies</u>, Bonn, edited by W. Pfeil.

8. M. Althoff et al., Z. Phys. <u>C22</u>, 307 (1984).

Bibliography of Research Published in Journals

On the range of electrons in meson decay
J.Steinberger, Phys. Rev. 75, 1136 (1949)

On the use of subtraction fields and the lifetimes of some types of mesons
J.Steinberger, Phys. Rev. 76, 1180 (1949)

The detection of artificially produced mesons with counters
J.Steinberger and A.S.Bishop, Phys. Rev. 78, 493 (1950)

Evidence for the production of neutral mesons by photons
J.Steinberger, W.K.H.Panofsky, J.Steller, Phys. Rev. 78, 802 (1950)

Measurement of $\pi-\mu$ decay lifetime
O.Chamberlain, R.F.Mozley, J.Steinberger, C.Wiegand, Phys. Rev. 78, 394 (1950)

Total cross-sections of π mesons on protons and several other nuclei
Chedester, Isaacs, Sachs and J.Steinberger, Phys. Rev. 82, 958 (1951)

The spin of the pion
Durbin, Loar, J.Steinberger, Phys. Rev. 83, 646 (1951)

The production of positive mesons by photons
J.Steinberger and A.S.Bishop, Phys. Rev. 86, 171 (1952)

Total cross sections of 60 MeV mesons in H_2 and D_2
Isaacs, Sachs and J.Steinberger, Phys. Rev. 85, 803 (1952)

Internal pair production of γ rays of mesonic origin
Lindenfeld, Sachs, J.Steinberger, Phys. Rev. 89, 531 (1953)

Differential cross-section for the scattering of 58 MeV π^+ mesons in hydrogen
Bodanksy, Sachs, J.Steinberger, Phys. Rev. 90, 996 (1953); Phys. Rev. 93, 1367 (1954)

The mass difference of neutral and charged π mesons
Chinowsky and J.Steinberger, Phys. Rev. 93, 586 (1954)

Absorption of negative pions in D_2: parity of the pion
Chinowsky and J.Steinberger, Phys. Rev. 95, 1561 (1954)

Search for β decay of the pion
S.Lokanathan and J.Steinberger, Nuovo Cimento Supplemento Vol.II, Series X (1955) 151

Electrons from muon capture
J.Steinberger and H.B.Wolfe, Phys. Rev. 100, 1490 (1955)

Reaction $\pi^- + d \rightarrow 2n + \pi^0$: parity of the π^0
W.Chinowsky and J.Steinberger, Phys. Rev. 100, 1476 (1955)

Properties of heavy unstable particles produced by 1.3 BeV π^- mesons
R.Budde, M.Chretien, J.Leitner, N.P.Samios, M.Schwartz and J.Steinberger, Phys.
Rev. 103, 1827 (1956)

π^--p elastic scattering at 1.44 GeV
M.Chretien et al., Phys. Rev. 108, 383 (1957)

Demonstration of parity nonconservation in hyperon decay
Phys. Rev. 108, 1353 (1957)

β decay of the pion
G.Impeduglia, R.Plano, A.Prodell, N.Samios, M.Schwartz, J.Steinberger, Phys. Rev.
Letters 1, 249 (1958)

Demonstration of the existence of the Σ^0 hyperon and a measurement of its mass
R.Plano, N.Samios, M.Schwartz, J.Steinberger, Nuovo Cimento 5, 216 (1957)

Experimental situation on parity doubling
F.Eisler et al., Phys. Rev. 107, 324 (1957)

Systematics of Δ^0 and θ^0 decay
F.Eisler et al., Nuovo Cimento 5, 1700 (1957)

π-p elastic scattering at 1.44 GeV
M.Chretien et al., Phys. Rev. 108, 393 (1957)

Experimental detection of Δ^0 and Σ^0 spins
F.Eisler et al., Nuovo Cimento 7, 222 (1958)

Mass of the Σ^0
Phys. Rev. 110, 226 (1958)

Parity of the neutral pion
R.Plano, A.Prodell, N.Samios, M.Schwartz and J.Steinberger, Phys. Rev. Letters 3,
525 (1959)

Leptonic decay of a Σ hyperon
P.Franzini and J.Steinberger, Phys. Rev. Letters 6, 281 (1961)

Observation of high energy neutrino interactions and the existence of two kinds of
neutrinos
C.Danby et al., Phys. Rev. Letters 9, 36 (1962)

Production of pion resonances in $\pi^\pm$p interactions
C.Alff et al., Phys. Rev. Letters 9, 322 (1962)

Decays of the ω and η mesons
C.Alff et al., Phys. Rev. Letters 9, 325 (1962)

A 170 l. liquid H_2 bubble chamber
A.Prodell and J.Steinberger, Rev. Sci. Instr. 33, 1327 (1962)

Lifetime of the ω meson
N.Gelfand et al., Phys. Rev. Letters 11 436 (1963)

Width of the ϕ meson
N.Gelfand et al., Phys. Rev. Letters 11, 438 (1963)

Compilation of results on the two pion decay of the ω
G.Lutjens and J.Steinberger, Phys. Rev. Letters 12, 517 (1964)

β decay of the Σ^+ and Σ^- hyperons and validity of the $\Delta S = \Delta Q$ law
U.Nauenberg et al., Phys. Rev. Letters 12, 679 (1964)

Test of the validity of the $\Delta S = \Delta Q$ law in K^0 decay
L.Kirsch et al., Phys. Rev. Letters 13, 35 (1964)

$\Sigma^0-\Lambda^0$ relative parity
C.Alff et al., Proc. Sienna Conf., Vol.1, Soc. Ital. Fisica 1963

Parity of the neutral pion
N.Samios et al., Phys. Rev. 126, 1844 (1962)

Annihilations of antiprotons in hydrogen at rest into two mesons
C.Baltay et al., Phys. Rev. Letters 15, 532 (1965)

Test of charge conjugation invariance in $\bar{p}p$ annihilation at rest
C.Baltay et al., Phys. Rev. Letters 15 (1965) 591

K_S-K_L interference in the $\pi^+\pi^-$ decay mode
C.Alff-Steinberger et al., Phys. Lett. 20, 207 (1966)

Further results from the interference of K_S and K_L in the $\pi^+\pi^-$ decay mode
Phys. Lett. 21, 595 (1966)

Measurement of the charge asymmetry in the decay $K_l^0 \rightarrow e^\pm + \pi^\pm + \nu$
S.Bennett et al., Phys. Rev. Letters 19, 993 (1967)

Nonorthogonality of the long and short lived neutral kaon states ...
S.Bennett et al., Phys. Rev. Letters 19, 997 (1967)

Measurement of the K_S-K_L regeneration phase in copper
S.Bennett et al., Phys. Lett. 27B, 239 (1968)

η_{+-} phase, Reϵ and the superweak model
S.Bennett et al., Phys. Lett. 27B, 248 (1968)

$\Delta S-\Delta Q$ rule in the decay $K^0 \rightarrow e^\pm + \nu + \pi^\mp$
S.Bennett et al., Phys. Lett. 27B, 244 (1968)

Charge asymmetry in the decay $K_L \rightarrow \pi^+\pi^-\pi^0$
A.Scribano et al., Phys. Lett. 32B, 224 (1970)

Charge asymmetry in the K_{e_3} decay and Reϵ
J.Marx et al., Phys. Lett. 32B, 219 (1970)

Construction and performance of large multiwire proportional chambers
P.Schilly et al., Nucl. Instr. Meth. 91, 221 (1971)

A measurement of the total cross-sections for lambda hyperon interactions on
protons and neutrons in the momentum range from 6 GeV/c to 21 GeV/c
S.Gjesdal et al., Phys. Lett. 40B, 152 (1972)

Search for the decay $K_S \rightarrow 2\mu$
S.Gjesdal et al., Phys. Lett. 44B, 217 (1973)

Observation of the decay $K_L^0 \rightarrow \mu^+\mu^-$
W.C.Carithers et al., Phys. Rev. Letters 30, 1336 (1973)

A new determination of the $K^0 \rightarrow \pi^+\pi^-$ parameters
C.Geweniger et al., Phys. Lett. 48B, 487 (1974)

Measurement of the charge asymmetry in the decays $K_L^0 \rightarrow \pi^\pm e^\mp \nu$ and $K_L^0 \rightarrow \pi^\pm \pi^\mp \nu$
C.Geweniger et al., Phys. Lett. 48B, 483 (1974)

Measurement of the kaon mass difference M_L-M_S by the two regenerator method
C.Geweniger et al., Phys. Lett. 52B, 108 (1974)

A measurement of the K_L-K_S mass difference from the charge asymmetry in semileptonic decays
S.Gjesdal et al., Phys. Lett. 52B, 113 (1974)

The phase $\phi^{\pm}$ of CP violation in the $K^0 \rightarrow \pi^+\pi^-$ decay
S.Gjesdal et al., Phys. Lett. 52B, 119 (1974)

Search for $\Delta S = 2$ decays of neutral Ξ hyperons
C.Geweniger et al., Phys. Lett. 57B, 193 (1975)

Measurement of the $\bar{\Lambda}$-p total cross-section in the momentum range from 4 to 14 GeV/c
F.Eisele et al., Phys. Lett. 60B, 297 (1976)

Measurement of the vector form factor in the decay $K_L \rightarrow \pi e \nu$
S.Gjesdal et al., Nucl Phys. B109, 118 (1976)

Study of the photon spectrum in the decay $K_S^0 \rightarrow \pi^+\pi^-\gamma$
H.Taureg et al., Phys. Lett. 65B, 92 (1976)

Measurement of the Σ^0 lifetime
F.Dydak et al., Nucl. Phys. B118, 1 (1977)

Opposite sign dimuon events produced in narrow band beam neutrino and antineutrino beams
M.Holder et al., Phys. Lett. 69B, 377 (1977)

Like sign dimuon events produced in narrow-band neutrino and antineutrino beams
M.Holder et al., Phys. Lett. 70B, 396 (1977)

Observation of trimuon events produced in neutrino and antineutrino interactions
M.Holder et al., Phys. Lett. 70B, 393 (1977)

Is there a high-y anomaly in antineutrino interactions?
M.Holder et al., Phys. Rev. Letters 39, 433 (1977)

Measurement of the neutral to charged current cross section ratio in neutrino and antineutrino interactions
M.Holder et al., Phys. Lett. 71B, 222 (1977)

Study of inclusive neutral current interactions of neutrinos and antineutrinos
M.Holder et al., Phys. Lett. 72B, 254 (1977)

A detector for high energy neutrino interactions
M.Holder et al., Nucl. Instr. Meth. 148, 235 (1978)

Results of a beam dump experiment at the CERN SPS neutrino facility
T.Hansl et al., Phys. Lett. 74B, 139 (1978)

Performance of a magnetized total absorption calorimeter between 15 GeV and 140 GeV
M.Holder et al., Nucl. Instr. Meth. 151, 69 (1978)

Observation of a neutrino event with four energetic muons
M.Holder et al., Phys. Lett. 73B, 105 (1978)

Search for single positive muon production in neutrino interactions
M.Holder et al., Phys. Lett. 74B, 277 (1978)

Origin of trimuon events in high energy neutrino interactions
T.Hansl et al., Phys.Lett. 77B, 114 (1978)

Characteristics of trimuon events observed in high energy neutrino interactions
T.Hansl et al., Nuclear Physics B142, 381 (1978)

Inclusive interactions of high energy neutrinos and antineutrinos in iron
H.deGroot et al., Zeitschrift für Physik C1, 143 (1979)

Comparison of moments of the valence structure functions with QCD prediction
H.deGroot et al., Phys. Lett. 82B, 292 (1979)

QCD analysis of charged current structure functions
H.deGroot et al., Phys. Lett. 82B, 456 (1979)

Trimuon events observed in high energy antineutrino interactions
H.deGroot et al., Phys. Lett. 85B, 131 (1979)

Investigation of like-sign dimuon production in neutrino and antineutrino interactions
H.deGroot et al., Phys. Lett. 86B, 103 (1979)

The response and resolution of a iron-scintillator calorimeter for hadronic and
electromagnetic showers between 10 GeV and 140 GeV
H.Abramowicz et al., Nucl. Instr. Meth. 180, 429 (1981)

A measurement of the ratio of longitudinal and transverse structure functions in
neutrino interactions between 30 and 200 GeV
H.Abramowicz et al., Phys. Lett. 107B, 141 (1981)

Limit on right handed weak coupling parameters from inelastic neutrino interactions
H.Abramowcz et al., Zeitschrift für Physik C12, 225 (1982)

Determination of the gluon distribution in the nucleon from deep inelastic neutrino
scattering
H.Abramowicz et al., Zeitschrift für Physik C12, 289 (1982)

Evidence for ψ production by neutrinos via neutral currents
H.Abramowicz et al., Phys. Lett. 109B, 115 (1982)

Prompt neutrino production in a proton beam dump experiment
H.Abramowicz et al., Zeitschrift für Physik C13, 179 (1982)

Tests of QCD and non-asymptotically free theories of the strong interaction by an
analysis of the nucleon structure functions xF3, F2 and $\bar{q}$
H.Abramowicz et al., Zeitschrift für Physik C13, 199 (1982)

Experimental study of opposite-sign dimuons produced in neutrino and antineutrino
interactions
H.Abramowicz et al., Zeitschrift für Physik C15, 19 (1982)

Neutrino and antineutrino charged-current inclusive scattering in iron in the
energy range $20 < E_\nu < 300$ GeV
H.Abramowicz et al., Zeitschrift für Physik C17, 283 (1983)

A search for ν_μ oscillations in the Δm^2 range from 0.3 to 90 eV^2
F.Dydak et al., Phys. Lett. 134B, 281 (1984)

Measurement of neutrino and antineutrino structure functions in hydrogen and iron
H.Abramowicz et al., Zeitschrift für Physik C25, 29 (1984)

Measurement of the neutral to charged current cross section ratios in neutrino and
antineutrino nucleon interactions and determination of the Weinberg angle
H.Abramowicz et al., Zeitschrift für Physik C28, 51 (1985)

Index of Contributors

Springer Tracts in Modern Physics

Editor: G. Höhler
Associate Editor: E. A. Niekisch

Springer-Verlag
Berlin Heidelberg
New York Tokyo

Topics in Current Physics

Founded by H.K.V.Lotsch

Springer-Verlag
Berlin Heidelberg
New York Tokyo